Biomathematics

Volume 12

R. Gittins

Canonical Analysis

A Review with Applications in Ecology

With 16 Figures

Springer-Verlag
Berlin Heidelberg NewYork Tokyo

Robert Gittins

Department of Plant Sciences
University of Western Ontario
London, Ontario, Canada N6A 5B7

AMS-MOS Classification (1980): 62-02, 62H20, 62H17, 62J05, 92-02, 92A17

ISBN 3-540-13617-7 Springer-Verlag Berlin Heidelberg New York Tokyo
ISBN 0-387-13617-7 Springer-Verlag New York Heidelberg Berlin Tokyo

Printed in Germany

Typesetting: H. Stürtz AG, Würzburg. Offsetprinting: Saladruck, Berlin.
Bookbinding: Lüderitz & Bauer, Berlin

2141/3020-543210

For my wife Ong
and daughter Geneviève

Preface

Relationships between sets of variables of different kinds are of interest in many branches of science. The question of the *analysis* of relationships of this sort has nevertheless rather surprisingly received less attention from statisticians and others than it would seem to deserve. Of the available methods, that addressing the question most directly is canonical correlation analysis, here referred to for convenience as canonical analysis. Yet canonical analysis is often coolly received despite a lack of suitable alternatives. The purpose of this book is to clarify just what may and what may not be accomplished by means of canonical analysis in one field of scientific endeavor.

Canonical analysis is concerned with reducing the correlation structure between two (or more) sets of variables to its simplest possible form. After a review of the nature and properties of canonical analysis, an assessment of the method as an exploratory tool of use in ecological investigations is made. Applications of canonical analysis to several sets of ecological data are described and discussed with this objective in mind. The examples are drawn largely from plant ecology. The position is adopted that canonical analysis exists primarily to be *used;* the examples are accordingly worked through in some detail with the aim of showing how canonical analysis can contribute towards the attainment of ecological goals, as well as to indicate the kind and extent of the insight afforded. The applicability of canonical analysis is of course not confined to plant ecology. Indeed, a surprisingly wide class of problems that arise in such allied disciplines as animal ecology, biogeography, climatology, epidemiology, palaeoclimatology, geology, psychology, taxonomy, agriculture, forestry and fisheries can be formulated and analyzed in terms of relationships between two or more sets of variables. Potential applications in these and other fields no doubt will occur to readers with interests in particular areas.

The book is intended primarily for ecologists and others interested in relationships between variables of different kinds. More generally, the aim is to raise issues encountered in analyzing systems of this sort so as to encourage interaction between statisticians and ecologists. The work may appeal to statisticians by exposing areas where further theoretical work is needed and by providing a source of worked numerical examples. The book may also serve a useful purpose by drawing the attention of ecologists to opportunities afforded by canonical analysis not only as a means of addressing a varied assortment of ecological problems, but also as a unifying structure in terms of which many statistical methods can be compared and related.

The book has been aimed at the graduate or research level. Suitable prerequisites would be a course in linear algebra and one in multivariate analysis. Nahikian's *A Modern Algebra for Biologists* or G. Strang's *Linear Algebra and its Applications* would provide much of the necessary mathematical background. With respect to statistical requirements, P.E. Green's *Analyzing Multivariate Data* covers the relevant material at a level appropriate to understanding the book. The same author's *Mathematical Tools for Applied Multivariate Analysis* provides a useful exposition of the same ground from a geometric and more intuitively appealing viewpoint. The text is in three parts. Part I reviews the theoretical foundations of the subject, Part II consists of accounts of applications of canonical analysis to real data, and Part III is devoted to assessing the applications and to a consideration of the prospects for canonical analysis in ecology and more generally. Chapter 1 provides a general introduction to the study of relationships between two sets of variables. Chapter 2 deals with the formulation and derivation of canonical analysis before proceeding to review properties of the solution and the computation of canonical correlations and variates. In Chap. 3 several numerical indices which are helpful in interpreting the results of analysis are introduced and a number of extensions and generalizations of canonical analysis described. The chapter closes with an account of hypothesis testing in canonical analysis. Two special cases of canonical analysis important in applications form the subject matter of Chaps. 4 and 5, namely *canonical variate analysis* and *dual scaling,* respectively. The second of these chapters concludes the first part of the book.

The applications of Part II cover field surveys and designed experiments though the balance between the two is heavily in favor of the former. Chapter 6 describes a canonical analysis of spatial variation in the representation of three plant species. Soil/vegetation relationships are the focus of attention in Chaps. 7 and 8. In Chap. 9 canonical analysis is applied in a study of the temporal stability of vegetation. Chapter 10 deals with the analysis of vegetation structure. Chapter 11 reports a comparative study of the multivariate responses of several grass species to an imposed treatment regime in a controlled experiment. In Chap. 12, the final chapter of Part II, canonical analysis is used to explore relationships between the occurrence of several species of large herbivores and environmental conditions in an East African rangeland. Part III consists of Chaps. 13 and 14. In Chap. 13 the results of the seven applications are reviewed and the worth of the analyses in ecological terms assessed. Chapter 14 is to some extent speculative. In recent years statisticians have devoted a great deal of attention to two topics which have a strong bearing on canonical analysis – statistical data analysis and linear regression methodology. The spirit of this work is well-conveyed by Gnanadesikan's text *Methods for Statistical Data Analysis of Multivariate Observations*. In Chap. 14 developments in statistical data analysis and modern regression methodology are reviewed and an attempt is made to embed canonical analysis in the context provided by them. In effect, the chapter seeks to anticipate and delineate changes in the way in which canonical analysis will come to be used during the course of the next decade. The text closes with Appendices devoted to multivariate regression and to tabulations of the data and other information pertaining to the worked examples.

Part of the book was written while I was visiting the Department of Plant Sciences at the University of Western Ontario. My debt to Professor L. Orlóci for making my visit to Canada possible and for his support and encouragement during this period is considerable. I am also much indebted to Dr. N.A. Campbell of the Division of Mathematics and Statistics, CSIRO, Australia who read the theoretical chapters in manuscript and made valuable suggestions for their improvement. Portions of the manuscript were also read by Professor S.W. Nash of the Department of Mathematics, University of British Columbia and by Mr. J.C. Gower of the Statistics Department, Rothamsted Experimental Station. Both suggested improvements for which I am most grateful. Naturally, responsibility for all errors and obscurities which remain is mine. Dr. J. Ogden kindly placed the rain forest data used in the analyses of Chaps. 8 and 9 at my disposal and helped in interpreting the results obtained. Finally, it is a pleasure to thank Mrs. Eleanor Lowther for her skill and patience in so expertly preparing the typescript for publication.

October 1984 Robert Gittins

Contents

1. Introduction

1.1 The study of relationships

Ecology deals with relationships between plants and animals and between them and the places where they live. Consequently, many questions of interest to ecologists call for the investigation of relationships between variables of two distinct but associated kinds. Such relationships may involve those, for example, between the plant and animal constituents of a biotic community. They might also involve, as in plant ecology, connections between plant communities and their component species, on the one hand, and characteristics of their physical environment on the other. As another example, comparative relationships among a number of affiliated species or populations with respect to a particular treatment regime in a designed experiment might be studied. In more general terms, the question which arises calls for the exploration of relationships between *any* two or more sets of variables of ecological interest.

Despite the importance and generality of the question, comparatively few formal models are available for its investigation. The most obvious procedures are multivariate regression analysis and canonical correlation analysis. Surprisingly, substantive applications of these methods in ecology are few. In practice, all too often other less suitable forms of analysis are pressed into service for the purpose. Among these multiple regression analysis and principal component analysis are frequently encountered, although procedures such as clustering and the minimum spanning tree are also sometimes employed. Where multiple regression analysis is used, the p variables of one or other set are first dismembered into single variables. A series of p multiple regression analyses is then performed between the separated variables and the q variables of the second set. This procedure unfortunately results in the discarding of information on relationships within the response set, the consequences of which cannot be dismissed lightly. Not only is an unwieldy set of analyses substituted for a single, comprehensive analysis, but, more importantly, the opportunity to consider *linear combinations* of the response variables is lost and with it possibilities for dimensionality reduction and graphic display. Furthermore, in statistical applications, more realistic probability statements in hypothesis testing and interval estimation are likely to result where relations within the response variables are taken into account. Estimating p multiple regressions in relation to a dismembered p-variate response carries with it the assumption that the correlation matrix of the responses is an identity matrix, an assumption which is rarely well-founded in practice.

Principal component analysis, or related methods such as principal coordinate analysis and factor analysis, are also commonly used to investigate relationships between two sets of variables. Various analytical strategies exist. Separate components analyses may for example be performed on each set of variables and an attempt then made to relate the resulting components or factors across domains. Alternatively, a low-dimensional projection of the sample into a subspace specified by the principal components of one or other set may first be obtained which is then used as framework onto which the variables of the 'second' set are disposed. In yet other cases the distinction between the two sets of variables is disregarded, both sets being combined and analysed in a single analysis. All such uses of principal component analysis involve projecting the sample into a subspace which is specified solely on the basis of maximizing what in this context is an arbitrary mathematical criterion. Yet, in any study of two sets of variables the opportunity exists to map the sample into subspaces whose specification is guided at least in part by ecological considerations pertinent to the problem at hand.

Appeal to the algebraic models or structures underlying multiple regression and principal component analyses shows that in fact neither is appropriate for the analysis of joint observations on two sets of $p \geq 2$ and $q \geq 2$ variables (e.g., see Timm, 1975; Morrison, 1976; Gnanadesikan, 1977; Mardia, Kent & Bibby, 1979).

Relationships between two measurement domains may be regarded as either *symmetric* or *nonsymmetric*. In the symmetric case the variables of each domain are on an equal footing. Interest centers on the extent of the variance common to both sets, or, alternatively, on the proportion of the total variance of each set that is predictable from the other. In the nonsymmetric case the relationship is directed in the sense that the variables of one set are unequivocally recognized as antecedents or predictors and those of the other as responses. Interest focuses on the proportion of the total variance of the response measures which is attributable to the predictors. Where there is $p=1$ response variable and $q \geq 2$ predictors, it is not difficult to see that multiple correlation and multiple regression provide suitable methods for the analysis of symmetric and nonsymmetric relationships, respectively. These methods each generalize readily, extending analysis to situations in which both measurement domains are multivariate ($p, q \geq 2$). The corresponding procedures are *canonical correlation analysis* and *multivariate regression analysis*. These procedures, therefore, would appear at least at first sight to be appropriate for the exploratory analysis of a large class of ecological problems. Part of the appeal of canonical correlation and multivariate regression is that they allow the *simultaneous* entry of both sets of measures into an analysis. It is therefore disconcerting to find that neither procedure is well known to ecologists. Apart from a study by Buzas (1971), few serious attempts to apply multivariate regression have been made. Canonical analysis in one form or another has been used somewhat more extensively. Yet, the results reported have frequently been disappointing. In reality the distinction between canonical correlation and multivariate regression is not quite as sharp as indicated above. Canonical correlation analysis can in fact be used for exploring nonsymmetric relationships, and, indeed, was initially proposed with this purpose in mind.

There is a further procedure for investigating relationships between two measurement domains. This is the *inter-battery factor analysis* of Tucker (1958). The rationale of inter-battery factor analysis is to identify constructs or 'factors' which can be considered to underlie *both* domains. In other words, the objective is to extract a single set of factors which account for the observed interrelations among the two domains. Browne (1979) has recently shown that, while the underlying rationales of inter-battery factor analysis and canonical analysis are entirely different, numerically the procedures are substantially equivalent.

1.2 Objectives

It seems desirable to attempt to establish precisely what canonical correlation and multivariate regression might contribute towards the solution of ecological problems. We shall confine our attention here to canonical correlation analysis. Such a limitation on coverage will serve to keep the presentation within reasonable bounds. A strong *prima facie* case nevertheless exists for similarly examining multivariate regression. With the aim of furthering this goal a summary of multivariate regression is given in an appendix (A.1).

We have remarked that the appeal of canonical analysis arises in part from facilitating the joint analysis of two sets of variables. Nevertheless, in ecological applications the method has often proved disappointing (Austin, 1968; Cassie & Michael, 1968; Barkham & Norris, 1970; Gauch & Wentworth, 1976; W.T. Williams, 1976, pp. 66–7; Gauch & Stone, 1979; Sabo, 1980; van Hecke, Impens, Goossens & Hebrant, 1980). Consequently, canonical analysis is often viewed with indifference or even suspicion by ecologists. The linear, orthogonal nature of the solution, in particular, is often held to be unrealistic in ecological contexts and largely responsible for many of these failures (Goodall, 1970; Austin, 1972; Dale, 1975; Gauch & Wentworth, 1976; Johnson, 1980). W.T. Williams (1976, p. 66), however, considered the high unique variances characteristic of many biological systems the likely cause. Canonical analysis has also met with disfavor on grounds of its supposed cumbersomeness (Burt, Reid & Williams, 1976).

Having regard for the intrinsic appeal of certain features of canonical analysis on the one hand, and of the widespread disquiet expressed by ecologists concerning the method together with the paucity of suitable alternatives on the other, effort directed towards better defining what may in fact be accomplished by means of canonical analysis would be worthwhile. The present monograph is a contribution towards this end. The principal aims of the work are to review the nature and properties of canonical analysis and to provide a basis for an assessment of the method as an analytical tool for use in connection with exploratory studies in ecology. A subsidiary objective is to show that canonical analysis is also useful as a means of unifying several seemingly unrelated analytical methods widely used in ecological studies. With these objectives in mind, after first describing the theoretical foundations of the method, seven examples of canonical analysis are reported. The first two examples are in the nature of experiments designed to throw light on the ability of the method to recover *known* relationships from among sets of ecological variables. These are followed by five further analyses which illustrate something of the flexibility and power of the procedure which arises in part from various specializations on the nature of the variables employed.

Before turning to a formal treatment of canonical analysis we first sketch in general terms something of its nature and essential features.

1.3 Canonical analysis: overview

The notion that field observations in ecology are multivariate in character and are therefore conveniently expressed and treated algebraically as *vectors* is gaining general acceptance. In studies involving variables of two kinds, the sample observations give rise to *partitioned* vectors. Such vectors consist of a pair of subvectors which may, for example, be made up of p ecological variables and q environmental or other variables, respectively, or *vice versa*. Interest in vector variables of this kind centers on the relationship between the variables comprising each measurement domain. All the available information about linear relationships within and between sets is summarized by the covariance matrix of the vector variables, or by some similar scalar-products matrix. In practice, it is generally convenient to work with the variables in standard form (zero mean, unit variance). In such cases the covariance matrix is identically a correlation matrix.

Suppose we have a sample of N partitioned vectors of the kind described. We shall denote the jth sample by $\mathbf{g}_j=[[\mathbf{z}_j^{(x)}]^t|[\mathbf{z}_j^{(y)}]^t]^t$, $j=1,\ldots,N$, and for concreteness shall suppose that $\mathbf{z}_j^{(x)}$ $(p\times 1)$ and $\mathbf{z}_j^{(y)}$ $(q\times 1)$ represent subvectors of standardized ecological and environmental variables, respectively. Then the sample correlation matrix $\mathbf{R}$ of $\mathbf{g}_j$ may be written as the partitioned matrix

$$\mathbf{R}=(N-1)^{-1}\sum_{j=1}^{N}\mathbf{g}_j\mathbf{g}_j^t=\left[\begin{array}{c|c}\mathbf{R}_{11} & \mathbf{R}_{12}\\ \hline \mathbf{R}_{21} & \mathbf{R}_{22}\end{array}\right],$$

where

$\mathbf{R}_{11}$: $p\times p$ matrix correlations between the ecological variables
$\mathbf{R}_{22}$: $q\times q$ matrix correlations between the environmental variables
$\mathbf{R}_{12}=\mathbf{R}_{21}^t$: $p\times q$ matrix correlations between the variables of each set.

The internal structure of a matrix of this kind is rarely evident on inspection even where there are strong grounds for supposing the existence of an intrinsic structure of some kind corresponding to suspected interrelationships among the variables. The principal task of analysis therefore is to operate on the matrix **R** in some way such that the structure of the matrix becomes manifest. It will perhaps be appreciated that the task of comprehending *between-set* relationships will be made easier if we can at the same time also disentangle and simplify the *within-set* correlations between variables. Thus, the starting point for a canonical analysis may be regarded as the correlation matrix **R**, or some comparable scalar-products matrix. The aim of analysis is to summarize and clarify the internal structure of **R** and in this way throw light on the nature of relationships between the two sets of variables of interest.

The strategy of canonical analysis is in some ways analogous to that of principal component analysis. In each case, analysis entails rotation of the coordinate frame of the sample space determined by the columns of the data matrix to a new position in which some feature or property of the data is emphasized. In components analysis, the feature of interest is the total *variance* associated with a single set of variables; in canonical analysis it is the *covariance* between two sets of variables. In canonical analysis the data can be regarded as generating two sample spaces, of p and q dimensions, respectively, which correspond to the two sets of variables, $\mathbf{Z}^{(x)}$ $(\mathrm{N} \times p)$ and $\mathbf{Z}^{(y)}$ $(\mathrm{N} \times q)$. Nevertheless, it is possible to simultaneously rotate the coordinate frames of each space to new positions in which the covariance or correlation structure of the data is clearly revealed. Moreover, the new coordinate systems are such that relationships within each set of variables are disentangled (much as in components analysis), while at the same time facilitating a reduction in dimensionality (again, much as in components analysis). Algebraically, the rotations are equivalent to finding linear transformations of each set of variables, $u_k = \mathbf{a}_k^t \mathbf{z}^{(x)}$ and $v_k = \mathbf{b}_k^t \mathbf{z}^{(y)}$, such that the simple correlation r_k between the transformed variables u_k and v_k is maximized. The new variables u_k and v_k, which correspond to the rotated axes of the coordinate frames, are called *canonical variates*, while the familiar product-moment correlation coefficient r_k between the canonical variates u_k and v_k is referred to as a *canonical correlation coefficient*. The explanation for this teminology will shortly become apparent. In all there will be s pairs of such linear transformations, $k = 1, \ldots, s$ where s is equal to the smaller of p and q. As in components analysis, successive pairs of canonical variates are required to be uncorrelated with the preceding variates.

The procedure is illustrated diagrammatically in Fig. 1.1 for the first pair of canonical variates. The figure is largely self-explanatory. It is worth remarking, however, that the transformations to u_1 and v_1 entail a rescaling of the new variables. That is, in other words, the dimensions of x-space and y-space are shrunk or stretched to some extent. In canonical analysis it is not the case, as in components analysis, that a simple rotation of axes is involved. We may also mention in passing that certain aspects of canonical analysis are best appreciated by thinking conceptually of the sample as a galaxy of N points in a single space of $(p+q)$-dimensions.

Often, most of the covariation between measurement domains will be concentrated in the first two or three pairs of canonical variates, (u_k, v_k), $k = 1, 2, 3$ (say). The remaining $s-3$ pairs of canonical variates in such cases can then be neglected with little or no loss of useful information. In this way a reduction in the dimensionality of the data may be achieved. Tests of signifi-

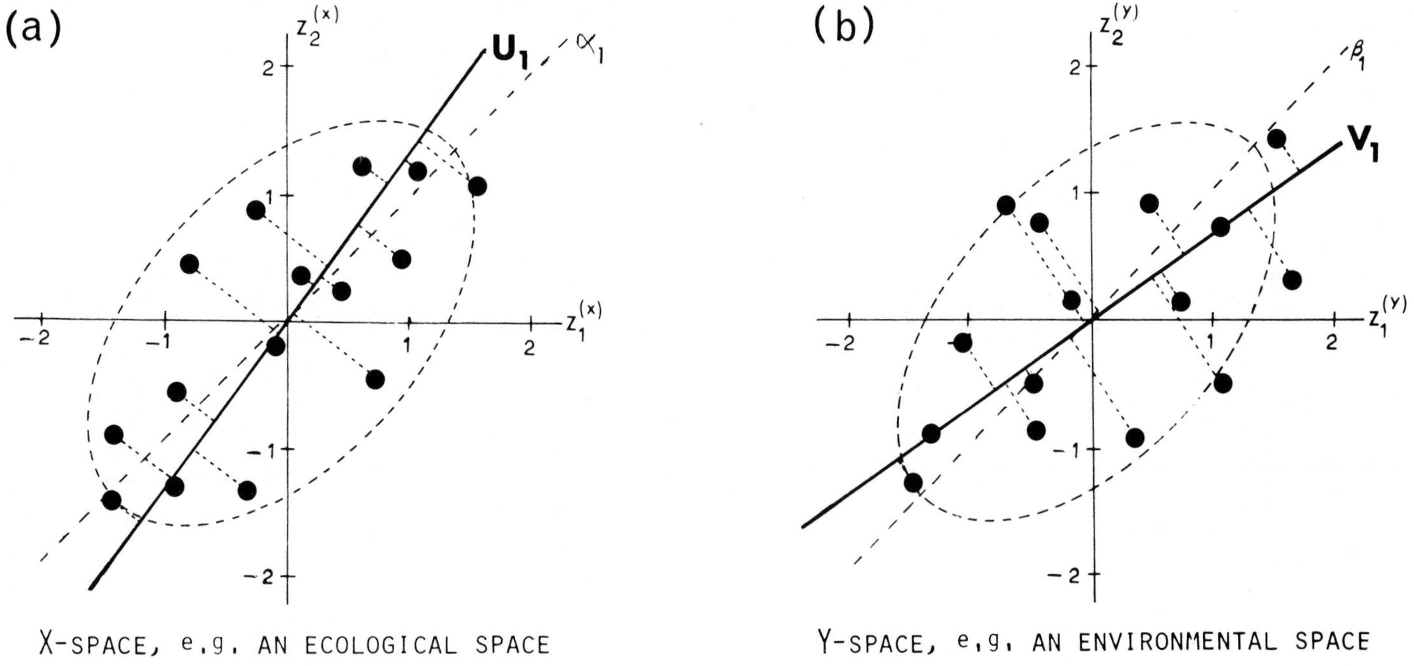

Fig. 1.1 a, b. Diagrammatic illustration of the principle of canonical analysis. A sample (N = 15) and concentration ellipse are shown (**a**) in an ecological space ($p=2$) and (**b**) in an environmental space ($q=2$). The canonical variates u_1 and v_1 are by definition those linear combinations of the variables of each space for which the simple correlation coefficient, r_1, between the projected points is a maximum over all possible choices for u and v; r_1 is the first or largest canonical correlation. For comparison, the first principal components, α_1 and β_1, of each measurement domain are also indicated. In general, there will be $s=\min(p,q)$ pairs of such canonical variates, u_k and v_k, and canonical correlation coefficients, r_k

cance are available under certain conditions and may be helpful in determining dimensionality. Where a reduction in dimensionality is possible, mapping of the sample into the 2- or 3-dimensional space of the retained canonical variates facilitates visual representation and examination of the data. Graphic displays of this kind are generally immensely rewarding in the insight into the covariance structure of the observations which they convey.

It is instructive to consider the correlation matrix $\mathbf{R}_{uv}$ of the transformed variables, u_k and v_k. This matrix has a particularly simple and appealing form. Confining our attention for the moment for simplicity to the case in which $p = q$, $\mathbf{R}_{uv}$ may be represented as follows:

$$\mathbf{R}_{uv} = \begin{array}{c} \\ u \\ v \end{array}\!\!\begin{array}{c} \begin{array}{cc} u & v \end{array} \\ \begin{bmatrix} \mathbf{I} & \mathbf{D} \\ \mathbf{D}^t & \mathbf{I} \end{bmatrix} \end{array},$$

$$\mathbf{I} = \begin{bmatrix} 1 & & 0 \\ & \ddots & \\ 0 & & 1_s \end{bmatrix}, \qquad \mathbf{D} = \begin{bmatrix} r_1 & & 0 \\ & \ddots & \\ 0 & & r_s \end{bmatrix}, \tag{1.1}$$

where $\mathbf{I}$ is the identity matrix of order s while $\mathbf{D}$ is an $s \times s$ diagonal matrix whose non-zero elements, r_k, are the canonical correlation coefficients between the kth pair of canonical variates u_k and v_k for $k = 1, \ldots, s$. It is customary to arrange the correlations r_k so that $r_1 \geq \ldots \geq r_s$. From (1.1) it is apparent that the correlation structure within and between the new variables may be characterized briefly as follows:

(a) all *within-set* correlation is reduced to zero;

(b) all linear correlation *between* sets is channeled through s canonical correlation coefficients r_k.

It is therefore clear that the transformation to canonical variates achieves the joint objectives of disentangling the correlations within each set of variables while emphasizing and clarifying linear relationships between the two measurement domains. The question of the substantive meaning and interpretation of the canonical variates, however, remains. We shall return to this question in Sect. 3.3.

It is the correlation matrix $\mathbf{R}_{uv}$ which gives its name to canonical analysis. In terms of $\mathbf{R}_{uv}$ the entire relationship between domains comes to be expressed in terms of only s non-zero quantities, $r_1, \ldots, r_s$, the diagonal elements of $\mathbf{D}$. The matrix $\mathbf{R}_{uv}$ is said to be a *canonical form* of the correlation matrix $\mathbf{R}$ between the original variables, a canonical form in mathematics being a mathematical entity which allows properties of interest to be perceived readily. In particular, $\mathbf{R}_{uv}$ expresses important or interesting information about the correlation structure between the measurement domains in question. More specifically, $\mathbf{R}_{uv}$ displays those features of the correlation structure which are *invariant* under linear transformations of each set of variables separately, in so far as the invariants can be expressed by correlation coefficients. Canonical analysis therefore does not alter the intrinsic correlation structure of the data; rather, it provides a convenient coordinate system within which the correlation structure of interest is clearly revealed.

We have sketched the derivation of canonical analysis above in an informal way by pointing to analogies between it and the better known procedure of principal component analysis. But canonical analysis is also closely related to regression analysis. In fact canonical analysis can be regarded as a natural generalization of multiple regression from strictly $q=1$ to $q \geq 2$ responses. In other words, canonical analysis involves the regression of a *vector* of response variables on a vector of predictors. A distinction between the two procedures which is worth noticing, however, is that whereas in canonical analysis the two sets of variables are treated symmetrically, this is not the case in multiple regression. In regression analysis it makes a great deal of difference which variable is treated as the response. On the other hand, in canonical analysis we may think of either set of variables being used to predict the other, or, indeed, of both sets of variables simultaneously predicting each other. This point is best appreciated by observing that if the first canonical variate of, say, the x-domain, u_1, is regressed onto the q variables of the y-domain, then the resulting regression coefficients are identical to the elements of the vector of weights $\mathbf{b}_1$ corresponding to v_1, at least up to a constant of proportionality. Similarly, multiple regression of the first canonical variate v_1 of the y-domain onto the set of p x-variables leads to regression weights which are proportional to the elements of the canonical weights vector $\mathbf{a}_1$ corresponding to u_1. Thus canonical analysis may be thought of as an analysis in which two multiple regressions are embedded. A second distinction between the procedures is that while in multiple regression there will be at most one regression relationship, in canonical analysis there are in general s pairs of such relationships, although all s pairs are rarely of equal importance or interest.

Thus we see that canonical analysis shares features in common with both components and regression analyses. Members of a pair of canonical variates may be thought of as components of their respective measurement domains somewhat akin to principal components in nature which are related in a regression sense across domains. It is to be clearly understood, however, that, while on occasion canonical variates may indeed be collinear with principal components, in general collinearity is not to be expected. Corresponding to these shared affinities are differences in emphasis which may be stressed in different applications. In certain studies canonical analysis is used purely descriptively to condense and expose mutual relationships among variables of different kinds, while in other applications it is the predictive aspect which is of most value.

The variables entering into a canonical analysis are not required to be continuous and, moreover, need not be directly measured or observed in the usual sense at all. Various kinds of specialization of the variables of either or both sets are admissible. We shall see that this property enhances the flexibility of canonical analysis very considerably. One of the commonest specializations involves the use of binary-valued dummy variables. Variables of this kind are helpful where a classificatory structure exists or can be superimposed on a p-variate sample in order to utilize the information inherent in the classification and are easily constructed. Where such information is relevant to the problem in hand, it is readily retrieved by the construction of dummy variables and these may then be taken to comprise one set of variables in a canonical analysis. A large class of problems in ecology lends itself to formulation and analysis in this way. Sometimes it is possible to cross-classify a sample with respect to two criteria. In such cases the procedure described can be extended, leading to a canonical analysis in which both sets of variables are binary-valued dummy variables.

Other forms of specialization involve the use of differently coded dummy variables, principal components of either or both sets of observed variables and the spatial coordinates of observed vector-valued variables having a spatial distribution.

As with other multivariate procedures, canonical analysis may be employed as a statistical method with the objectives of estimation and testing, or less formally as a means of exploratory data analysis. In the latter case, tightly specified distributional assumptions of the data are not required. Nevertheless, canonical analysis is specified by a linear, additive model, and, moreover, requires a continuous data structure for its proper use. The realities of ecological data, which frequently embody nonlinearities and discontinuities, should warn us against applying the method where these conditions are not at least approximately satisfied. Fortunately, as with other forms of regression, nonlinearity can frequently be accommodated by the method, while steps can be taken to unfold more complicated nonlinear structures prior to analysis. Thus, while the conditions necessary for the proper use of canonical analysis even as a means of exploratory data analysis are fairly stringent, steps can sometimes be taken to enhance compatibility between the realities of the data and the requirements of the model. The method is best suited to the exploratory phase of research where the need is to detect the presence of certain effects or relationships and to determine their direction. To go beyond this point stronger models based on some concept of the actual processes underlying the data are required. The wider significance of canonical analysis is indeed to provide a starting point for the development of more powerful though less general models of this kind by screening out all but a small number of the most promising variables or linear combinations of variables for consideration.

The terminology associated with canonical analysis is not firmly established. We should therefore take steps to make our own usage plain and to relate it to the terminology of other workers. As employed here, canonical analysis is synonymous with canonical correlation analysis. By canonical correlation analysis we mean a technique of multivariate analysis which seeks linear functions of two sets of variables with special properties in terms of correlations irrespective of the nature of the variables comprising either set. With this understanding in mind, canonical analysis includes as special cases both *canonical variate analysis*, in which one set of variables consists of binary-valued dummy variables designating class membership, and the analysis of *association in* $r \times c$ *contingency tables*, in which both sets of variables are binary in character. Canonical variate analysis in the above sense is formally equivalent to *multiple discriminant analysis*. We observe also the close formal connections between canonical variate analysis and one-way *multivariate analysis of variance*, on the one hand, and between canonical analysis of association and *dual scaling* or *correspondence analysis*, on the other. Finally, we remark that canonical variate analysis is also formally equivalent to a two-stage application of *principal component analysis*.

The term canonical analysis is used by many workers (e.g., Seal, 1964; Pearce, 1969; Goldstein & Grigal, 1972; Kowal, Lechowicz & Adams, 1976) in a narrower sense than understood here which corresponds to canonical variate analysis as defined above. On the other hand, Weinberg and Darlington (1976) and Pielou (1977) use canonical variate analysis in a wider sense than adopted here to refer to what we have called canonical correlation analysis. Comprehensive discussions of canonical analysis are provided by most of the standard works on multivariate analysis. These include the texts of Anderson (1958), Dempster (1969), Kshirsagar (1972), Rao (1973), Bock (1975), Timm (1975), Kendall and Stuart (1976), Morrison (1976), Gnanadesikan (1977), Green (1978), Mardia, Kent and Bibby (1979) and Muirhead (1982). A thorough review of the method and its relationships to other multivariate procedures has been given by McKeon (1965).

The theory of canonical correlation was proposed by Hotelling (1935, 1936) as a means of identifying the most predictable p-variate criterion. We shall see that the theory developed has since proved to have other applications.

Part I
Theory

2. Canonical correlations and canonical variates

2.1 Introduction

In this chapter we shall summarize the essential elements of the theory of canonical correlations and variates. We shall begin by formulating the problem. The derivation of canonical correlations and canonical variates will then be taken up. Canonical analysis can be derived in several ways. Two derivations will be described rather fully – those leading to eigenanalysis and singular value decomposition, respectively. Several other derivations are of interest for the light which they shed on canonical analysis and its relations with other methods, and these will be described more briefly. We shall then consider the properties of canonical correlations, weights and variates. Finally, computational aspects of canonical analysis will be taken up.

Proofs of important results are to be found in the texts by Anderson (1958), Kshirsagar (1972), Mardia, Kent and Bibby (1979) and Muirhead (1982).

2.2 Formulation

Let the partitioned vector $\mathbf{g}=[x_1, \ldots, x_p | y_1, \ldots, y_q]^{\mathrm{t}}$ represent observations on two sets of variables. For convenience and without loss of generality suppose that each variable is measured about its respective mean and that $q \le p$. Furthermore, let $\mathbf{x}=[x_1, \ldots, x_p]^{\mathrm{t}}$ and $\mathbf{y}=[y_1, \ldots, y_q]^{\mathrm{t}}$ be jointly distributed with sample mean vector $\mathbf{m}$ and covariance or other scalar-products matrix $\mathbf{S}$ partitioned as:

$$\mathbf{m}=\begin{bmatrix}\mathbf{0}\\ \mathbf{0}\end{bmatrix}, \quad \mathbf{S}=\begin{bmatrix}\mathbf{S}_{11} & \mathbf{S}_{12}\\ \mathbf{S}_{21} & \mathbf{S}_{22}\end{bmatrix}. \tag{2.1}$$

In the identity on the right $\mathbf{S}_{11}=c\,\mathbf{X}^{\mathrm{t}}\mathbf{X}$, $\mathbf{S}_{12}=\mathbf{S}_{21}^{\mathrm{t}}=c\,\mathbf{X}^{\mathrm{t}}\mathbf{Y}$ and $\mathbf{S}_{22}=c\,\mathbf{Y}^{\mathrm{t}}\mathbf{Y}$ for some $c \le 1$ $(c \neq 0)$, ${}_N\mathbf{X}_p=[\mathbf{x}_1| \ldots |\mathbf{x}_p]$, ${}_N\mathbf{Y}_q=[\mathbf{y}_1| \ldots |\mathbf{y}_q]$. Initially we shall assume that $\operatorname{rank}(\mathbf{X})=p$ and $\operatorname{rank}(\mathbf{Y})=q$.

Relations *between* the x's and the y's are specified by the entries of $\mathbf{S}_{12}$. The internal structure of $\mathbf{S}_{12}$ and hence the connection between the x's and the y's, however, is rarely evident on inspection. Moreover, we shall see that resolution of association between the two sets will be facilitated if account is also taken of the *within-set* covariance matrices $\mathbf{S}_{11}$ and $\mathbf{S}_{22}$.

Hotelling (1935, 1936) proposed the following as a measure of the correlation between the p-dimensional variable $\mathbf{x}$ and the q-dimensional variable $\mathbf{y}$. Consider a linear function of the $x_1, \ldots, x_p$, say $u = a_1 x_1 + \ldots + a_p x_p$, and a linear function of the $y_1, \ldots, y_q$, say $v = b_1 y_1 + \ldots + b_q y_q$. Then determine those values of the coefficients vectors $\mathbf{a}$ $(p \times 1)$ and $\mathbf{b}$ $(q \times 1)$ which maximize the correlation between u and v. To completely describe the relationship between $\mathbf{x}$ and $\mathbf{y}$ further linear functions comparable to those leading to u and v may be required. At most, $s = \min(p, q)$ distinct pairs of solutions $\mathbf{a}_k$ and $\mathbf{b}_k$ $(k = 1, \ldots, s)$ exist. The corresponding linear functions $u_k = \mathbf{a}_k^t \mathbf{x}$ and $v_k = \mathbf{b}_k^t \mathbf{y}$ are required to be maximally correlated subject to their being uncorrelated with all previous pairs.

More precisely, linear combinations of $\mathbf{x}$ and $\mathbf{y}$ are sought

$$u_k = \mathbf{a}_k^t \mathbf{x}, \qquad v_k = \mathbf{b}_k^t \mathbf{y} \tag{2.2}$$

with the $\mathbf{a}_k$ and $\mathbf{b}_k$ scaled so that the linear combinations u_k and v_k have unit variance

$$\mathbf{a}_k^t \mathbf{S}_{11} \mathbf{a}_k = 1, \qquad \mathbf{b}_k^t \mathbf{S}_{22} \mathbf{b}_k = 1,$$

for which the correlation

$$\operatorname{cor}(u_k, v_k) = \mathrm{r}_k \qquad \text{(say)}$$

is maximized subject to the constraints

$$\operatorname{cor}(u_i, u_k) = 0, \qquad \operatorname{cor}(u_i, v_k) = 0, \qquad \operatorname{cor}(v_i, v_k) = 0$$

for $i, k = 1, \ldots, s$ and $i \neq k$.

The problem therefore is to find those matrices of weights

$$_p\mathbf{A}_p = [\mathbf{a}_1 | \ldots | \mathbf{a}_p],$$
$$_q\mathbf{B}_q = [\mathbf{b}_1 | \ldots | \mathbf{b}_q],$$

for which the covariance matrix of the transformed variable $\mathbf{h} = [\mathbf{u}^t | \mathbf{v}^t]^t$ has the simple form

$$\operatorname{var}\begin{bmatrix}\mathbf{u}\\ \mathbf{v}\end{bmatrix} = \begin{bmatrix}\mathbf{A}^t & \mathbf{0}\\ \mathbf{0} & \mathbf{B}^t\end{bmatrix}\begin{bmatrix}\mathbf{S}_{11} & \mathbf{S}_{12}\\ \mathbf{S}_{21} & \mathbf{S}_{22}\end{bmatrix}\begin{bmatrix}\mathbf{A} & \mathbf{0}\\ \mathbf{0} & \mathbf{B}\end{bmatrix} = \begin{bmatrix}\mathbf{I}_p & \boldsymbol{\Gamma}\\ \boldsymbol{\Gamma}^t & \mathbf{I}_q\end{bmatrix}, \tag{2.3}$$

where

$$\boldsymbol{\Gamma} = \begin{bmatrix}\operatorname{diag}(\mathrm{r}_k)\\ \hline \mathbf{0}\end{bmatrix}\}(p-q) \times q, \qquad 1 \geq \mathrm{r}_1 \geq \ldots \geq \mathrm{r}_q \geq 0.$$

The principal task in canonical analysis, therefore, is to obtain the weights matrices $\mathbf{A}$ and $\mathbf{B}$. It so happens that in solving for $\mathbf{A}$ and $\mathbf{B}$ the correlation coefficients r_k will usually be obtained simultaneously. Having calculated the weights, the linear functions u_k and v_k are easily arrived at using (2.2). It is to the derivation of $\mathbf{A}$ and $\mathbf{B}$ that we now turn. Computational aspects of canonical analysis are taken up in Sect. 2.5.

2.3 Derivation of canonical correlation coefficients and canonical variates

Canonical correlations and variates can be derived in several ways. The classical solution (Hotelling, 1935, 1936; Anderson, 1958) uses Lagrange multipliers and eigenanalysis. An alternative derivation based on direct orthogonal decomposition of **X** and **Y** has been given by Lancaster (1958, 1966) and Horst (1961 a). The solution in terms of matrix decomposition is in fact simpler and more intuitive than the classical procedure and has significant computational advantages also. Both derivations are described below. A number of other derivations are briefly outlined principally for the light which they throw on relationships between canonical analysis and other methods.

2.3.1 Eigenanalysis

Let $\mathbf{g}=[\mathbf{x}^t|\mathbf{y}^t]^t$, where $\mathbf{x}$ and $\mathbf{y}$ are p and q component vectors, respectively, with $q \le p$. We seek linear transformations

$$u=\mathbf{a}^t\mathbf{x}, \qquad v=\mathbf{b}^t\mathbf{y} \tag{2.4}$$

from $\mathbf{x}$ to u and from $\mathbf{y}$ to v such that the simple correlation coefficient r between u and v is maximized.

The correlation coefficient r expressed as a function of $\mathbf{a}$ and $\mathbf{b}$ is

$$\begin{aligned} \mathrm{r}(\mathbf{a},\mathbf{b}) &= \mathrm{cov}(\mathbf{a}^t\mathbf{x},\mathbf{b}^t\mathbf{y})/\{\mathrm{var}(\mathbf{a}^t\mathbf{x})\,\mathrm{var}(\mathbf{b}^t\mathbf{y})\}^{\frac{1}{2}} \\ &= \mathbf{a}^t\mathbf{S}_{12}\mathbf{b}/\{(\mathbf{a}^t\mathbf{S}_{11}\mathbf{a})(\mathbf{b}^t\mathbf{S}_{22}\mathbf{b})\}^{\frac{1}{2}}. \end{aligned} \tag{2.5}$$

The coefficient vectors $\mathbf{a}$ and $\mathbf{b}$ in (2.4) are determined only up to a constant of proportionality. For convenience, it is customary to choose $\mathbf{a}$ and $\mathbf{b}$ so that u and v have unit variance. Then

$$\mathrm{var}(u)=\mathbf{a}^t\mathbf{S}_{11}\mathbf{a}=1, \qquad \mathrm{var}(v)=\mathbf{b}^t\mathbf{S}_{22}\mathbf{b}=1. \tag{2.6}$$

Expression (2.5) now reduces to

$$\mathrm{r}(\mathbf{a},\mathbf{b})=\mathbf{a}^t\mathbf{S}_{12}\mathbf{b}. \tag{2.7}$$

The algebraic problem therefore is to find $\mathbf{a}$ and $\mathbf{b}$ to maximize (2.7) subject to (2.6). Writing $\xi(\mathbf{a},\mathbf{b})$ for the function to be maximized under constraints, we have

$$\xi(\mathbf{a},\mathbf{b})=\mathbf{a}^t\mathbf{S}_{12}\mathbf{b}-\tfrac{1}{2}\lambda(\mathbf{a}^t\mathbf{S}_{11}\mathbf{a}-1)-\tfrac{1}{2}\mu(\mathbf{b}^t\mathbf{S}_{22}\mathbf{b}-1),$$

where λ and μ are Lagrange multipliers. Taking the partial derivatives of ξ with respect to first $\mathbf{a}$ and then $\mathbf{b}$ and equating to the null vector we obtain

$$\frac{\partial\xi}{\partial\mathbf{a}}=\mathbf{S}_{12}\mathbf{b}-\lambda\mathbf{S}_{11}\mathbf{a}=\mathbf{0}, \tag{2.8}$$

$$\frac{\partial\xi}{\partial\mathbf{b}}=\mathbf{S}_{12}^t\mathbf{a}-\mu\mathbf{S}_{22}\mathbf{b}=\mathbf{0}. \tag{2.9}$$

Premultiplying (2.8) by $\mathbf{a}^t$ and (2.9) by $\mathbf{b}^t$ gives

$$\begin{aligned} \mathbf{a}^t \mathbf{S}_{12}\, \mathbf{b} - \lambda\, \mathbf{a}^t \mathbf{S}_{11}\, \mathbf{a} = 0, \\ \mathbf{b}^t \mathbf{S}_{12}^t\, \mathbf{a} - \mu\, \mathbf{b}^t \mathbf{S}_{22}\, \mathbf{b} = 0. \end{aligned} \tag{2.10}$$

As $\mathbf{a}^t \mathbf{S}_{11}\, \mathbf{a} = \mathbf{b}^t \mathbf{S}_{22}\, \mathbf{b} = 1$, (2.10) reduces to

$$\mathbf{a}^t \mathbf{S}_{12}\, \mathbf{b} = \lambda = \mu. \tag{2.11}$$

From (2.9) and noting that $\lambda = \mu$ we have

$$\mathbf{S}_{12}^t\, \mathbf{a} = \lambda\, \mathbf{S}_{22}\, \mathbf{b}. \tag{2.12}$$

Multiplying (2.8) on the left by $\mathbf{S}_{21} \mathbf{S}_{11}^{-1}$ gives

$$\mathbf{S}_{21} \mathbf{S}_{11}^{-1} \mathbf{S}_{12}\, \mathbf{b} = \lambda\, \mathbf{S}_{12}^t\, \mathbf{a}. \tag{2.13}$$

Substituting (2.12) into (2.13) yields the equation in $\mathbf{b}$

$$(\mathbf{S}_{21} \mathbf{S}_{11}^{-1} \mathbf{S}_{12} - \lambda^2 \mathbf{S}_{22})\, \mathbf{b} = \mathbf{0}. \tag{2.14}$$

This identity is recognizable as one form of the generalized eigenvalue problem. Hence $\lambda_1^2, \ldots, \lambda_q^2$ and $\mathbf{b}_1, \ldots, \mathbf{b}_q$ are the roots and vectors of the characteristic equation

$$|\mathbf{S}_{21} \mathbf{S}_{11}^{-1} \mathbf{S}_{12} - \lambda^2 \mathbf{S}_{22}| = 0, \tag{2.15}$$

where the determinant is a polynomial in λ^2 of degree q. Non-trivial solutions for the λ_k^2 and $\mathbf{b}_k$ ($k = 1, \ldots, q$) exist provided the determinant in (2.15) is zero. It is convenient to arrange the roots such that $\lambda_1^2 \geq \ldots \geq \lambda_q^2$.

After similar manipulation we can also write

$$(\mathbf{S}_{12} \mathbf{S}_{22}^{-1} \mathbf{S}_{21} - \lambda^2 \mathbf{S}_{11})\, \mathbf{a} = \mathbf{0}, \tag{2.16}$$

giving the nonzero roots $\lambda_1^2, \ldots, \lambda_q^2$ and vectors $\mathbf{a}_1, \ldots, \mathbf{a}_q$. Alternatively, the conjugate weights $\mathbf{a}_k$ can be calculated directly from the $\mathbf{b}_k$ by the relation

$$\mathbf{a}_k = \lambda_k^{-1}\, \mathbf{S}_{11}^{-1}\, \mathbf{S}_{12}\, \mathbf{b}_k \tag{2.17}$$

for $k = 1, \ldots, s$ where $s = \min(p, q)$. For $p > q$ there are $p - q$ further roots λ_k^2 ($k = q+1, \ldots, p$) and vectors $\mathbf{a}_k$, all of which are null.

The coefficients matrices $\mathbf{A}$ and $\mathbf{B}$ are constructed columnwise from the eigenvectors corresponding to the nonnull eigenvalues of (2.16) and (2.14):

$$\begin{aligned} \mathbf{A} &= [\mathbf{a}_1 | \ldots | \mathbf{a}_q], \\ \mathbf{B} &= [\mathbf{b}_1 | \ldots | \mathbf{b}_q]. \end{aligned}$$

From (2.7) and (2.11) it follows that the correlation coefficient r_k between the linear composites u_k and v_k is the square root of the kth eigenvalue of (2.14) or (2.16):

$$\mathrm{r}_k = \sqrt{\lambda_k^2}.$$

Thus the task of clarifying relations between two sets of variables reduces to the solution of a generalized eigenvalue problem. The required coefficients vectors $\mathbf{a}_k$ and $\mathbf{b}_k$ are eigenvectors of (2.16) and (2.14), respectively, and are called the vectors of *canonical weights*. Moreover, the positive square roots of the eigenvalues $\lambda_1^2, \ldots, \lambda_q^2$ of (2.14) or (2.16) are the *canonical correlation coefficients* between the *canonical variates* $u_k = \mathbf{a}_k^t \mathbf{x}$ and $v_k = \mathbf{b}_k^t \mathbf{y}$ for $k = 1, \ldots, q \le p$.

2.3.2 Singular value decomposition

We shall require the following result.

Singular value decomposition. Let $\mathbf{A}$ be a real $p \times q$ matrix, $q \le p$. Then there exist matrices $\mathbf{P}$ $(p \times p)$, $\mathbf{D}$ $(q \times q)$ and $\mathbf{Q}$ $(q \times q)$ such that

$$\mathbf{A} = \mathbf{P} \begin{bmatrix} \mathbf{D} \\ \hline \mathbf{0} \end{bmatrix} \mathbf{Q}^t, \tag{2.18}$$

where

$$\mathbf{D} = \operatorname{diag}(d_1, \ldots, d_q),$$

$$\mathbf{P}^t \mathbf{P} = \mathbf{I}_p, \qquad \mathbf{Q}\mathbf{Q}^t = \mathbf{Q}^t \mathbf{Q} = \mathbf{I}_q,$$

$d_1 \ge \ldots \ge d_q \ge 0$ and $\mathbf{0}$ is the null matrix of order $(p-q) \times q$. Equivalently,

$$\mathbf{P}^t \mathbf{A} \mathbf{Q} = \begin{bmatrix} \mathbf{D} \\ \hline \mathbf{0} \end{bmatrix}. \tag{2.19}$$

Relation (2.18) is known as the *singular value decomposition* of $\mathbf{A}$. Matrices $\mathbf{P}$ and $\mathbf{Q}$ consist columnwise of the orthonormalized eigenvectors of $\mathbf{A}\mathbf{A}^t$ and $\mathbf{A}^t\mathbf{A}$, respectively. The diagonal elements d_k $(k = 1, \ldots, q)$ of $\mathbf{D}$ are the non-negative square roots of the eigenvalues of $\mathbf{A}^t\mathbf{A}$ and are the *singular values* of $\mathbf{A}$. Similarly, the columns of $\mathbf{P}$ and $\mathbf{Q}$ are respectively the left and right *singular vectors* of $\mathbf{A}$. For a proof of this result see Lancaster (1958), Johnson (1963), Lawson and Hanson (1974, p. 18), Green (1976, p. 230) or Ten Berge (1977). We are now in a position to proceed to the derivation of the canonical correlations and variates.

Let $\mathbf{x} = [x_1, \ldots, x_p]^t$ and $\mathbf{y} = [y_1, \ldots, y_q]^t$ be observations on two sets of jointly distributed variables with sample mean vector $\mathbf{m}$ and covariance matrix $\mathbf{S}$ given by (2.1). As in the classical derivation, we seek linear combinations u_k and v_k of the observed x's and y's which are maximally correlated. Provided $\mathbf{X}$ and $\mathbf{Y}$ are of full column rank $\mathbf{S}_{11}$ and $\mathbf{S}_{22}$ will be symmetric positive definite. The Cholesky factorizations of $\mathbf{S}_{11}$ and $\mathbf{S}_{22}$ are therefore

$$\mathbf{S}_{11} = [\mathbf{S}_{11}^{\frac{1}{2}}]^t \mathbf{S}_{11}^{\frac{1}{2}}, \qquad \mathbf{S}_{22} = [\mathbf{S}_{22}^{\frac{1}{2}}]^t \mathbf{S}_{22}^{\frac{1}{2}},$$

where the $\mathbf{S}_{ii}^{\frac{1}{2}}$ $(i = 1, 2)$ are nonsingular, upper-triangular matrices of order p and q, respectively.

Now define

$$\mathbf{K} = [\mathbf{S}_{11}^t]^{-\frac{1}{2}} \mathbf{S}_{12} \mathbf{S}_{22}^{-\frac{1}{2}}, \tag{2.20}$$

where $\mathbf{K}$ is $p \times q$. Then by the singular value decomposition (2.18) orthogonal matrices $\mathbf{L}$ $(p \times p)$ and $\mathbf{M}$ $(q \times q)$ can be found such that

$$\mathbf{L}^t\mathbf{K}\mathbf{M} = \begin{bmatrix}\mathbf{D}\\ \mathbf{0}\end{bmatrix}, \tag{2.21}$$

where $\mathbf{D} = \mathrm{diag}(d_1, \ldots, d_q)$.

Using $\mathbf{L}$ and $\mathbf{M}$, perform orthogonal transformations from $\mathbf{X}$ to $\mathbf{U}$ and from $\mathbf{Y}$ to $\mathbf{V}$:

$$\begin{aligned}[\mathbf{U}\ \mathbf{V}] &= [\mathbf{L}^t\ \mathbf{M}^t]\begin{bmatrix}\mathbf{S}_{11}^{-\frac{1}{2}} & \mathbf{0}\\ \mathbf{0} & \mathbf{S}_{22}^{-\frac{1}{2}}\end{bmatrix}\begin{bmatrix}\mathbf{X}^t\\ \mathbf{Y}^t\end{bmatrix}\\ &= [\mathbf{L}^t\mathbf{S}_{11}^{-\frac{1}{2}}\mathbf{X}^t\ \ \mathbf{M}^t\mathbf{S}_{22}^{-\frac{1}{2}}\mathbf{Y}^t].\end{aligned} \tag{2.22}$$

The covariance matrix of the new variable $[\mathbf{u}^t|\mathbf{v}^t]^t$ is then

$$\begin{aligned}\mathrm{var}\begin{bmatrix}\mathbf{u}\\ \mathbf{v}\end{bmatrix} = \begin{bmatrix}\mathbf{u}\\ \mathbf{v}\end{bmatrix}[\mathbf{u}^t\ \mathbf{v}^t] &= \begin{bmatrix}\mathbf{L}^t\mathbf{S}_{11}^{-\frac{1}{2}}\mathbf{S}_{11}\mathbf{S}_{11}^{-\frac{1}{2}}\mathbf{L} & \mathbf{L}^t\mathbf{S}_{11}^{-\frac{1}{2}}\mathbf{S}_{12}\mathbf{S}_{22}^{-\frac{1}{2}}\mathbf{M}\\ \mathbf{M}^t\mathbf{S}_{22}^{-\frac{1}{2}}\mathbf{S}_{21}\mathbf{S}_{11}^{-\frac{1}{2}}\mathbf{L} & \mathbf{M}^t\mathbf{S}_{22}^{-\frac{1}{2}}\mathbf{S}_{22}\mathbf{S}_{22}^{-\frac{1}{2}}\mathbf{M}\end{bmatrix}\\ &= \begin{bmatrix}\mathbf{I}_p & \mathbf{\Gamma}\\ \mathbf{\Gamma}^t & \mathbf{I}_q\end{bmatrix}, \quad \text{where } \mathbf{\Gamma} = \begin{bmatrix}\mathbf{D}\\ \mathbf{0}\end{bmatrix}\}\,(p-q)\times q\end{aligned} \tag{2.23}$$

and $\mathbf{D} = \mathrm{diag}(d_1, \ldots, d_q)$. Thus, by means of linear transformations on $\mathbf{x}$ and $\mathbf{y}$, the correlation structure implicit in $\mathbf{S}$ has been reduced to a form involving only the quantities $d_1, \ldots, d_q$.

The d_k $(k = 1, \ldots, q)$ are identically the r_k of (2.3). The covariance matrix of $[\mathbf{u}^t|\mathbf{v}^t]^t$ is therefore the same matrix which appears on the right-hand side of (2.3). Thus the required linear functions u_k and v_k have been found; they are respectively the kth row of $\mathbf{U}$ $(p \times N)$ and $\mathbf{V}$ $(q \times N)$ of (2.22)

$$u_k = \mathbf{l}_k^t\mathbf{S}_{11}^{-\frac{1}{2}}\mathbf{x}^t, \quad v_k = \mathbf{m}_k^t\mathbf{S}_{22}^{-\frac{1}{2}}\mathbf{y}^t, \tag{2.24}$$

while the correlations r_k between them are the diagonal elements d_k of $\mathbf{D}$. For a proof of these assertions, see Lancaster (1966, p. 586) or Muirhead (1982, p. 553). The coefficients vectors $\mathbf{a}_k$ and $\mathbf{b}_k$, if required, are, from (2.2) and (2.24):

$$\mathbf{a}_k = \mathbf{S}_{11}^{-\frac{1}{2}}\mathbf{l}_k, \quad \mathbf{b}_k = \mathbf{S}_{22}^{-\frac{1}{2}}\mathbf{m}_k. \tag{2.25}$$

The problem of finding maximally correlated linear functions of $\mathbf{x}$ and $\mathbf{y}$ can therefore be shown to be that of determining the singular value decomposition of the $p \times q$ matrix $\mathbf{K} = [\mathbf{S}_{11}^t]^{-\frac{1}{2}}\mathbf{S}_{12}\mathbf{S}_{22}^{-\frac{1}{2}}$. In other words, the problem is equivalent to finding the canonical form of a *rectangular* matrix under pre- and post-multiplication by orthogonal matrices. The singular values of $\mathbf{K}$ are themselves the *canonical correlation coefficients* while the singular vectors yield the *canonical variates* u_k and v_k and the *canonical weights* vectors $\mathbf{a}_k$ and $\mathbf{b}_k$ by the simple relations (2.24) and (2.25), respectively. Horst (1961a), Lancaster (1966), Golub (1969), Kshirsagar (1972, p. 247), Akaike (1976), Cohen (1977), Mardia, Kent and Bibby (1979, p. 281) and Muirhead (1982) may be consulted for further details.

2.3.3 Other derivations

In this section we mention a number of alternative derivations of canonical correlation coefficients and canonical variates. Though these have little practical significance, they are instructive for the additional light which they throw on canonical analysis.

Least squares. It is well known that in multiple regression maximizing the squared coefficient of multiple correlation, R^2, is equivalent to minimizing the sum of squares of residuals from regression. Indeed, multiple regression analysis is usually derived by explicitly setting out to minimize the squared residuals. In contrast, although canonical analysis involves a straightforward generalization of multiple regression, it is rarely formulated as a minimization problem. Brillinger (1969, 1975, p. 367), however, has shown that canonical variates and correlations can be derived by the principle of least squares. The quantity to be minimized is the trace of the $q \times q$ residual sums of squares and products matrix $\hat{\mathbf{E}}^t \hat{\mathbf{E}}$, where

$$\hat{\mathbf{E}}^t \hat{\mathbf{E}} = [(\mathbf{XA} - \mathbf{YB})^t (\mathbf{XA} - \mathbf{YB})]. \tag{2.26}$$

Matrices $\mathbf{A}$ $(p \times q)$ and $\mathbf{B}$ $(q \times q)$ are sought which minimize the trace of (2.26). The least squares derivation of canonical analysis has been described also by Muller (1980a). Two alternative solutions are given by Muller and in each case are shown to lead to the normal equations (2.8) and (2.9) of the classical solution.

A multivariate general linear model. The term general linear model refers to a family of models characterized by linearity of the parameters of the equation specifying the model. The members of the family are distinguishable by their various assumptions.

The multivariate linear regression model is written

$$\underset{(N \times q)}{\mathbf{Y}} = \underset{(N \times p)}{\mathbf{X}} \; \underset{(p \times q)}{\boldsymbol{\Theta}} + \underset{(N \times q)}{\mathbf{E}} \tag{2.27}$$

where $\mathbf{Y}$ is a matrix of observations on q response variables, $\mathbf{X}$ is a known matrix, $\boldsymbol{\Theta}$ consists of a matrix of unknown parameters to be estimated and $\mathbf{E}$ is a matrix of unobserved random disturbances. The model is of interest where linear functions of $\mathbf{X}$ which in some sense best predict $\mathbf{Y}$ are sought. A principal task in multivariate regression analysis is to obtain an estimate of the parameter matrix $\boldsymbol{\Theta}$ which satisfies this requirement.

Canonical analysis can be represented as special case of (2.27). To see this consider the model

$$\underset{(N \times q)}{\mathbf{Z}^{(y)}} \; \underset{(q \times q)}{\mathbf{B}} = \underset{(N \times p)}{\mathbf{Z}^{(x)}} \; \underset{(p \times q)}{\mathbf{A}} \; \underset{(q \times q)}{\mathbf{D}} + \underset{(N \times q)}{\mathbf{E}} \tag{2.28}$$

where

$\mathbf{Z}^{(y)}, \mathbf{Z}^{(x)}$	are matrices of data in standard score form;
$\mathbf{A}, \mathbf{B}$	are matrices of canonical weights;
$\mathbf{D}$	is a diagonal matrix of canonical correlations;
$\mathbf{E}$	is a matrix of random disturbances.

By identifying the terms $\mathbf{Z}^{(y)}\mathbf{B}$, $\mathbf{Z}^{(x)}$ and $\mathbf{AD}$ of (2.28) with $\mathbf{Y}$, $\mathbf{X}$ and $\boldsymbol{\Theta}$, respectively, of (2.27) it is apparent that canonical analysis can be cast in the form of multivariate regression, and, hence may be conceived of as a special case of the multivariate general linear model (Muller, 1982). In terms of (2.28) the aim of canonical analysis is to obtain an estimate of the matrix $\mathbf{AD}$ of parameters which in the least squares sense best predicts the canonical variates in $\mathbf{V} = \mathbf{Z}^{(y)}\mathbf{B}$ from those of $\mathbf{U} = \mathbf{Z}^{(x)}\mathbf{AD}$. The solution for $\mathbf{AD}$ comparable to the normal equations of multivariate regression is given by $\mathbf{R}_{12}\mathbf{B} = \mathbf{R}_{11}\mathbf{AD}$, where the $\mathbf{R}_{ij}$ are correlation matrices. Muller (1982) provides further details. For an alternative treatment of the regression formulation of canonical analysis in terms of (2.27) see Robinson (1973) and Izenman (1980). An attractive feature of the regression approach is that it provides an explicit *model* for canonical analysis in the sense that a stochastic structure is readily declared.

Minimization of a Euclidean distance. James (1979) has proposed a derivation based on purely geometric reasoning. James's proposal utilizes the equivalence between minimizing the distance between two vectors and maximizing a function of the angle between them. The problem may be formulated as that of finding vectors $\mathbf{a}$ $(p \times 1)$ and $\mathbf{b}$ $(q \times 1)$ which minimize the squared distance

$$d^2 = \|\mathbf{X}\,\mathbf{a} - \mathbf{Y}\,\mathbf{b}\|^2,$$

where $\|\cdot\|$ denotes Euclidean norm. The derivation leads to Eqs. (2.14) and (2.16) of the classical solution. A notable feature of James's derivation is that conditions on the ranks of the data matrices $\mathbf{X}$ and $\mathbf{Y}$ are not required. Where $\operatorname{rank}(\mathbf{X}) < p$ or $\operatorname{rank}(\mathbf{Y}) < q$, James's procedure results in *generalized inverse* forms of (2.14) and (2.16).

We observe finally that Cramer (1973) has described a simple derivation of Eq. (2.14) obtained directly from the relationship between multiple regression and multiple correlation, while Muller (1980a) has provided a useful summary of several novel derivations in addition to those mentioned above.

2.3.4 Concluding remarks

The elements of the coefficients vectors $\mathbf{a}_k$ and $\mathbf{b}_k$ are expressed in the same units as the variables to which they correspond. To compare the magnitude of the weights and to ensure that the dimensions of the composites u_k and v_k are meaningful, it is necessary to re-express the weights in the standardized scale. Re-expression is accomplished by multiplying the elements $a_1, \ldots, a_p$ and $b_1, \ldots, b_q$ by the standard deviation of the corresponding variables. Where the canonical variates u_k and v_k are to be evaluated using (2.4) the adjusted weights are then applied to the respective variables in zero-mean, unit-variance form.

Canonical correlation coefficients are invariant to changes in units of measurement. Consequently, identical canonical correlations will result if in (2.1) the covariance matrix $\mathbf{S}$ is replaced by the corrected sums of squares and products matrix $\mathbf{Q}$ or by the correlation matrix $\mathbf{R}$. The weights vectors $\mathbf{a}_k$ and $\mathbf{b}_k$ nevertheless will in each case differ. Yet it is possible to pass simply and directly between the weights resulting from $\mathbf{Q}$, $\mathbf{S}$ and $\mathbf{R}$. The matrix $\mathbf{Q}$ yields weights which are proportional to those of $\mathbf{S}$, and to which they are converted on multiplication by $\sqrt{N-1}$; the correlation matrix $\mathbf{R}$ yields the standardized weights directly.

Different versions of canonical analysis are distinguished by the scaling convention normally adopted for the canonical variates. Where the variables of both domains are observed or measured directly, it is customary to scale u_k and v_k to unit variance in the sample. In canonical variate analysis, on the other hand, it is usual to scale the canonical variates of the response set, u_k say, such that the average within-groups variance is unity for all k. Furthermore, in the canonical analysis of association the canonical variates u_k and v_k ($k=1, \ldots, s$) are often scaled such that the sum of squares of their elements equals the corresponding root, r_k^2.

Canonical correlations and canonical variates are open to geometric interpretation in terms of angles and vectors in Euclidean space. Let $\mathscr{U}$ and $\mathscr{V}$ be subspaces of an n-dimensional Euclidean space $\mathscr{E}$ generated by the columns of $\mathbf{X}$ ($N \times p$) and $\mathbf{Y}$ ($N \times q$), ($n=p+q$). One may then ask, for example, what is the smallest angle between $\mathscr{U}$ and $\mathscr{V}$ and what is the angle? The smallest angle between $\mathscr{U}$ and $\mathscr{V}$, θ_1, $0<\theta_1<\frac{\pi}{2}$, is defined by

$$\cos\theta_1 = \max_{u\in\mathscr{U}} \max_{v\in\mathscr{V}} u^t v, \qquad \|u\| = \|v\| = 1. \tag{2.29}$$

Assume that the maximum of (2.29) is attained for $u=u_1$ and $v=v_1$. Then $\theta_2(\mathscr{U}, \mathscr{V})$ is defined as the smallest angle between the subspace of $\mathscr{U}$ orthogonal to u_1 and the subspace of $\mathscr{V}$ orthogonal to v_1. Continuing in this way until one of the subspaces is empty, s angles of inclination θ_k ($k=1, \ldots, s$) are obtained between the subspaces $\mathscr{U}$ and $\mathscr{V}$. We summarize this result as follows:

$$\cos\theta_k = \max_{u\in\mathscr{U}} \max_{v\in\mathscr{V}} u^t v = u_k^t v_k, \qquad \|u\| = \|v\| = 1$$

subject to the constraints

$$u_j^t u = 0, \quad v_j^t v = 0, \quad j = 1, \ldots, s-1.$$

The $\theta_1, \theta_2, \ldots, \theta_s$ are called the *canonical angles* between $\mathscr{U}$ and $\mathscr{V}$ and the orthogonal bases $\mathbf{U}=[u_1| \ldots |u_s]$ and $\mathbf{V}=[v_1| \ldots |v_s]$, where the inner-product $(u_i, v_j)=0$, are the *canonical bases*. Dempster (1969, p. 177) has shown that the cosines of the canonical angles θ_k are indeed identically the canonical correlation coefficients r_k and that the canonical bases are identically the canonical variates of $\mathbf{x}$ and $\mathbf{y}$, respectively. For a thorough geometric exposition of canonical analysis Afriat (1957, 1960) and Afriat, Sastry and Tinter (1975) may be consulted while Dempster (1969, pp. 98 and 176), Björck and Golub (1973) and Corsten (1976a) provide shorter accounts.

Further reading. Certain aspects and generalizations of canonical analysis have been the subject of special investigation. Geisser (1965) has approached the question of *estimation* from a Bayesian rather than a sampling theory viewpoint. Estimation has also been considered by Dempster (1966) and by Wold (1966a, b) and Lyttkens (1972), who have described a class of nonlinear iterative partial least squares estimators for the canonical weights. Asymptotic *distribution theory* for the canonical analysis of contingency tables has been investigated by Corsten (1967a), O'Neill (1978a, b) and Haberman (1981). Questions of *inference* have been addressed by Schuenemeyer and Bargmann

(1978), Bargmann (1979), Glynn and Muirhead (1978) and Khatri (1978), while Muirhead and Waternaux (1980) have considered the application of canonical analysis to samples from a nonnormal multivariate distribution. Robinson (1973) has derived asymptotic properties and tests for generalized canonical estimates under assumptions that do not involve normality and independence and which appear to be relatively weak in other respects. Corsten (1976b) has examined in detail the application of canonical analysis to *incomplete block designs*. The canonical analysis of *time series* has been discussed by Brillinger (1969, 1975, p. 367), Robinson (1973) and Akaike (1976) while Miyata (1970) has extended the theory of canonical analysis to *complex-valued* variables. Finally we mention that the generalization of canonical analysis to $m>2$ *sets of variables* has been investigated or discussed by Horst (1961a), McKeon (1965), Kettenring (1971), Gnanadesikan (1977, p. 69) and Van de Geer (1984).

2.4 Properties of canonical correlation coefficients, weights and variates

2.4.1 Properties of canonical correlation coefficients

We have seen that canonical correlation coefficients, r_k, are product-moment correlation coefficients between canonical variates u_k and v_k. It is therefore not surprising that they should share many properties in common with the familiar correlation coefficient. The principal distinction between the two is that a canonical correlation, unlike the familiar correlation coefficient as generally encountered, is a *maximized* quantity. A number of properties of canonical correlation coefficients are summarized below.

(a) Canonical correlations assume values in the range -1 to $+1$. It is, however, customary to work with and report only the absolute value $|r_k|$. The magnitude of r_k expresses the degree of linear correlation between u_k and v_k.
(b) Canonical correlation coefficients are dimensionless. Consequently, they are invariant under nonsingular linear transformations of the variables of either or both sets.
(c) Canonical correlation coefficients are interpretable as multiple correlation coefficients between a particular canonical variate of one domain and the complete set of variables of the other.
(d) Like the square of a simple correlation coefficient, which is a ratio of explained to total variation, a squared canonical correlation coefficient, r_k^2, represents the ratio of two determinants or generalized variances – namely the ratio of the generalized explained or regression variance to the generalized total variance (e.g., see Bartlett, 1947, 1965). The square of a canonical correlation is interpretable also as the proportion of variation in one composite, u_k, say, that is predictable from, or common to, its conjugate v_k.
(e) In statistical applications, the canonical correlation coefficients r_k are re-

garded as sample estimates of corresponding population quantities, ρ_k. In such cases the kth sample canonical correlation coefficient r_k is a biased estimate of the kth population canonical correlation coefficient ρ_k.

We have noted that canonical correlation coefficients are constrained to be ≤ 1. Pathological cases have however been reported in which one or more correlations exceed unity (W.T. Williams, Edye, Burt & Grof, 1973). Williams and his colleagues have suggested that this phenomenon may be analogous to the Heywood case of factor analysis (see Harman, 1967, Sect. 7.3) while Reyment (1976) has remarked that it may be a result of serious deviation of the data from joint normality. The invariance of canonical correlations under changes of origin and scale has some important consequences. In particular, canonical correlation coefficients calculated from the sums of squares and products matrix, the variance-covariance matrix and the correlation matrix for a particular set of data are identical. Furthermore, prior transformation of either or both sets of variables to their respective principal components leaves the canonical correlations unchanged. For a simple proof of these assertions, see Kshirsagar (1972, p. 278) or Bock (1975, p. 390).

The biased nature of sample canonical correlation coefficients may be hinted at as follows. Consider first r_1, the largest sample canonical correlation. As r_1 is the result of selection from among pairs of normalized linear combinations of variables to yield the largest sample value, it seems likely that r_1 will in fact be an upwardly biased estimate of the corresponding population quantity ρ_1 (Dempster, 1966; cf. also Lancaster, 1963, 1969, p. 197; O'Neill, 1978b). Comparable arguments suggest that the smallest sample correlation, r_s, is likely to be downwardly biased, and, further, that canonical correlations closest to the largest or smallest will also similarly be biased. The question of bias correction has been addressed by Dempster (1966), Glynn and Muirhead (1978) and O'Neill (1978b). Using a modified jackknife procedure, Dempster has proposed the following correction for the kth sample canonical correlation r_k as an estimator of ρ_k:

$$\frac{1}{m}\frac{1-r_k^2}{r_k}\left[\tfrac{1}{2}(r_k^2-p-q+2)+(1-r_k^2)\sum_{\substack{j=1\\ j\neq k}}^{q}\frac{r_j^2}{r_j^2-r_k^2}\right],$$

where $m=N-1$, the number of degrees of freedom on which the revised covariance estimate is based. The adjustment is recommended for use with samples of moderately large size.

In small samples the magnitude of the sample canonical correlations r_k depends in part on the relative number of variables to samples. As the number of variables $(p+q)$ approaches sample size N, the value of r_1, in particular, tends rapidly to unity. To illustrate this point it is instructive to consider briefly the special case of multiple regression. It can be shown in multiple regression that if the response variable y and antecendent variables $x_1, \ldots, x_p$ are uncorrelated in the population ($\rho=0$), the expected value of the squared multiple correlation coefficient R^2 is:

$$\mathscr{E}(R^2|\rho=0)=\frac{p}{N-1} \tag{2.30}$$

(Morrison, 1976, p. 108). It is plain from (2.30) that as p tends to N, R^2 and hence also R converge to unity even when the population value ρ is zero. Moreover, in canonical analysis where $(p+q)>N$, one or more canonical correlation coefficients of unity inevitably will arise. This is so because N sample-points can generate a space of at most $N-1$ dimensions. Consequently, the p-space and q-space which correspond to the columns of **X** and **Y**, respectively, will have $[(p+q)-(N-1)]$ dimensions in common. These common dimensions will be represented by canonical correlations of unity. It follows that canonical correlation coefficients computed from small samples with comparatively many variables can be misleading as indices of the extent of linear relationship between the composites in question unless supported by probability statements as to their significance. These connection between sample size, number of variables and the magnitude of canonical correlation coefficients frequently need to be borne in mind in assessing the latter.

An important consequence of the bias of the r_k is that even after systematically arranging the respective sample and population quantities r_k and ρ_k in descending order of magnitude there is no certainty that the kth sample correlation r_k was in fact generated by the kth population quantity ρ_k. Lawley (1959) however has mentioned that where sample size is large, it may roughly speaking be justifiable to assume that the orderings correspond provided that the first j ρ_k are neither very close to one nor very small.

Notice that r_k is a measure of relationship between *linear composites* of each measurement domain rather than of the measurement domains themselves. If u_k and v_k were invariably collinear with the principal components of their respective domains, then u_k and v_k would optimally account for the variation of the variables of which they are linear functions. But in many cases canonical variates are not collinear with principal components and do not account for a high proportion of the total variation of their particular domain – canonical variates are required only to be maximally correlated in pairs across domains. To see this consider as an example an analysis in which just *one* variable of a particular domain is highly correlated with a *single* variable of the second domain. A high canonical correlation will inevitably result, since it is known that the magnitude of at least one canonical correlation must exceed the absolute value of the largest observed zero-order correlation between variables of different domains (e.g., see Bargmann, 1979). In such cases the canonical correlation coefficient could hardly be taken to indicate a high degree of relationship between the measurement domains themselves. It follows that as a canonical correlation coefficient is a measure of correlation between *canonical variates* it may be highly unreliable as an index of relationship between the *measurement domains* themselves. Where a canonical correlation is misconstrued as an index of relationship between the original variables, it will mislead in appearing to overstate the true extent of the relationship. We shall see in Sect. 3.2.4 that interpretive indices which better indicate the extent to which two measurement domains are related have been developed.

A distinction between the first and the remaining canonical correlation coefficients is worth noticing – canonical correlations following the first are actually correlations between *residual* spaces rather than between x-space and

y-space. The r_k $(k>1)$ therefore refer to correlations between complex functions of the original variables rather than to correlations between linear composites of the x's and y's themselves. Nevertheless, the practical consequences of the distinction do not seem to be important. If on statistical or substantive grounds t canonical correlations are judged worthy of further consideration, then t may be called the *rank* of the model for the data. In other words, the dimensionality of the linear relationship between $\mathbf{x}$ and $\mathbf{y}$ is t.

Apart from measuring the relationship between canonical variates, canonical correlations have two uses. First, in providing an indication of rank or dimensionality, and, secondly, in the construction of further interpretive indices.

2.4.2 Properties of canonical weights

The canonical weights $a_{1k}, \ldots, a_{pk}$ and $b_{1k}, \ldots, b_{qk}$ $(k=1, \ldots, s)$ are coefficients which when applied to the observed variables in $\mathbf{x}$ and $\mathbf{y}$, result in new variables u_k and v_k which are maximally correlated. We assume below that the weights are standardized to ensure that they are commensurate and hence can be compared directly. The standardized canonical weights are analogous to the standard partial regression coefficients of multiple regression. Accordingly, the standardized weights share many properties in common with regression weights of this kind, some of which are summarized below.

(a) Standardized weights are scale free.
(b) The magnitude and sign of canonical weights can in principle at least be used as an indication of the presence of certain variables or effects and of their direction.
(c) The numerical value of canonical weights depends on the selection of variables as well as on their scales. Addition or deletion of variables in either set is likely to produce major alterations in the remaining coefficients. Prior standardization of the observed variables to zero mean and unit variance will remove the scaling effects but the inter-dependencies will remain.
(d) Vulnerability to suppression phenomena. Suppression refers to the effect of a predictor variable which is itself positively correlated with a response variable y but which nevertheless receives a negative weight in the canonical relationship; a related occurrence concerns the behavior of a predictor which itself has negligible predictive validity for a criterion y but which in linear combination with other predictors may contribute substantially to prediction of y.
(e) The weights tend to be highly unstable in replicate samples drawn from the same population. Several factors may contribute to instability, notably measurement errors in the observations, collinearity of the variables of either set and inadequate sample size.

As an analog of a standardized partial regression coefficient, a canonical weight reflects association after the influence of all other variables in its set has been removed. Put another way, canonical weights do not pertain to the common parts among the corresponding variables but to their *unique* parts. Thus, two or more sizeable absolute weights do not indicate that the corresponding variables share anything in common which contributes to the variance of a canonical variate of the other domain; rather, such weights indicate that these variables each contribute something distinct, which nevertheless is related to the linear composite of the other domain. Brillinger (1981, p. 373) has given expressions for the asymptotic mean and variance of the sample estimates $\mathbf{a}_k$ and $\mathbf{b}_k$ under the assumption of multivariate normality. The resulting expressions for the variance of the estimates are complicated and in practice the variances are best evaluated by use of the jackknife.

The scaling of the coefficient vectors $\mathbf{a}_k$ and $\mathbf{b}_k$ is entirely a matter of convenience. In canonical correlation analysis it is customary to standardize the weights in such a way that the canonical variates have unit variance. Then $\mathrm{var}(u_k)=\mathbf{a}_k^t\mathbf{S}_{11}\mathbf{a}_k=1$ and $\mathrm{var}(v_k)=\mathbf{b}_k^t\mathbf{S}_{22}\mathbf{b}_k=1$ for $k=1,\ldots,q$. Letting $\mathbf{A}=[\mathbf{a}_1|\ldots|\mathbf{a}_q]$ and $\mathbf{B}=[\mathbf{b}_1|\ldots|\mathbf{b}_q]$ we may therefore write:

$$\mathbf{A}^t\mathbf{S}_{11}\mathbf{A}=\mathbf{I}_q, \qquad \mathbf{B}^t\mathbf{S}_{22}\mathbf{B}=\mathbf{I}_q. \tag{2.31}$$

With this standardization the following relationships hold:

$$\begin{gathered}\mathbf{A}^t\mathbf{S}_{12}\mathbf{B}=\mathbf{D},\\ \mathbf{A}^t\mathbf{S}_{12}\mathbf{S}_{22}^{-1}\mathbf{S}_{21}\mathbf{A}=\mathbf{D}^2, \qquad \mathbf{B}^t\mathbf{S}_{21}\mathbf{S}_{11}^{-1}\mathbf{S}_{12}\mathbf{B}=\mathbf{D}^2,\end{gathered} \tag{2.32}$$

where $\mathbf{D}=\mathrm{diag}(d_1,\ldots,d_q)$ is the matrix whose kth diagonal element $d_k\equiv \mathrm{r}_k$ is the canonical correlation coefficient between the canonical variates u_k and v_k ($k=1,\ldots,q$). Other standardizations, however, may be more appropriate in particular circumstances, as we shall see in connection with canonical variate analysis (Chap. 4) and dual scaling (Chap. 5). A principal distinguishing feature between these special cases of canonical analysis in fact turns out to be the normalization constraint placed on the canonical variates. Moreover, not only is the scaling of the weights at the discretion of the user but so also is the sign attached to the elements of the conjugate vectors $\mathbf{a}_k$ and $\mathbf{b}_k$. This means that the sign of all the elements comprising a particular pair of vectors can be reversed without materially affecting the analysis. Interpretation of the vectors is in fact based on the *pattern* of the signs, not on the signs themselves, and the pattern is preserved under sign reversal. Quite often, reversing the signs does lead to a more natural or meaningful interpretation.

We have mentioned that the sign and magnitude of the canonical weights may indicate the direction and relative importance of the contributions of the variables comprising $\mathbf{x}$ and $\mathbf{y}$ to the canonical correlation r_k. More often, however, substantive interpretation of the overall pattern of the weights is difficult. There are several reasons why this should be so. First, Bock (1975, p. 393) has pointed out that the weights represent a sort of compromise, under the constraint of orthogonality, between maximizing between-set covariation while disentangling within-set covariation. Secondly, as mentioned above, canonical weights, like regression coefficients, depend on the selection of variables for investigation and on their scales. Finally, the weights also depend on sample-specific variation and on a variety of other factors such as those mentioned under (e) above. For these reasons the substantive value of the canonical weights is often small. Fortunately, the correlations between the observed variables and the canonical variates provide a useful alternative to the canonical weights in the interpretation of canonical variates. Furthermore, steps can sometimes be taken to improve the interpretability of the weights themselves. These steps include estimation of the weights by ridge regression rather than least squares procedures (Carney, 1975; Vinod, 1976; Campbell & Reyment, 1978; Campbell, 1980a; Campbell & Palmer, 1984), correcting for measurement errors in the observations (Meredith, 1964), transforming the observed variables to their principal components prior to analysis in order to remove dependencies and the use of large samples. In studies where the canonical weights do prove to have substantive value, graphic display of the weights may be worthwhile. Two or three dimensional plots of the weights against the canonical variates are effective and easily made.

We noted above that canonical correlation coefficients are invariant under nonsingular linear transformations of the original variables. Consequently, canonical correlations calculated for example from such different scalar-products matrices as the sums of squares and products matrix, **Q**, the covariance matrix, **S**, and the correlation matrix, **R**, are identical. The corresponding vectors of weights, however, differ. **Q** and **S** both give rise to weights which are expressed in the metric of the original variables, while **R** leads to canonical weights which are dimensionless. Nevertheless it is possible to pass simply and directly between the weights calculated from any of these matrices (e.g., see Finn, 1974, p. 190).

In view of inherent difficulties in interpretation, the principal use of canonical weights is in the evaluation of the canonical variates. However, the recent introduction of ridge-type estimation procedures promises to enhance the interpretive value of canonical weights.

2.4.3 Properties of canonical variates

Canonical variates are linear combinations of observations $u_k = \mathbf{a}_k^t \mathbf{x}$ and $v_k = \mathbf{b}_k^t \mathbf{y}$, constructed so as to maximize the simple correlation coefficient r_k between the resulting composites u_k and v_k.

In several respects, canonical variates resemble principal components. Both are linear combinations of variables chosen to maximize a particular quantity. In the case of principal components it is the variance of the derived variables which is maximized; in canonical analysis it is the covariance or correlation between corresponding members of a pair of derived variables which is maximized. The solution in the case of components analysis is provided by a singular value decomposition or by an eigenanalysis of a symmetric matrix and in the case of canonical analysis by a singular value decomposition or by an eigenanalysis of a nonsymmetric matrix. In geometric terms, principal components represent a rotation of the coordinate frame of sample space to a new position in which the axes coincide with the directions of maximum scatter in the sample. Similarly, canonical variates represent separate rotations of the coordinate systems of the two sample spaces corresponding to the data. Each system of axes is rotated to a new position in which the angle between corresponding members of a pair of axes across domains is as small as possible. The cosine of this angle is the canonical correlation between a pair of canonical variates.

While a canonical variate may be collinear with a principal component of the same domain, there is no requirement that it must be so. To see this recall that, for a chosen reference inner product, principal components are uniquely defined; the canonical variates of a specified measurement domain on the other hand, depend critically on the composition of the second domain. The choice of variables which might be used for the second domain is almost infinitely large. The canonical variates of the specified domain are correspondingly varied. This fact draws attention to the opportunities provided by canonical analysis for partitioning the total variance of a particular domain in a great variety of ways. Two broad classes of investigation can be distinguished, leading to the analysis of *superimposed* and *extraneous* structure, respectively (see Gnanadesikan, 1977, p. 63). Notice that in the case of *two* specified sets of variables, however, the canonical variates are uniquely defined up to a scale factor.

Canonical variates calculated from the correlation matrix, **R**, are dimensionless. They are assessed accordingly in terms of the standardized variables comprising $\mathbf{z}^{(x)}$ and $\mathbf{z}^{(y)}$. On the other hand, unless the variables of each set are all expressed in the same metric, canonical variates calculated both from the sums of squares and products matrix, **Q**, and the covariance matrix, **S**, have meaningless dimensions. This follows from the nature of their construction as weighted sums of variables expressed in different units. Thus, where either **Q** or **S** is the starting point for analysis, it is usually necessary to multiply the elements of the weight vectors $\mathbf{a}_k$ and $\mathbf{b}_k$ by the standard deviations of their respective variables before calculating the variates. The canonical variates are then evaluated using the standardized observed variables. The resulting canonical variates turn out to be those which would have been obtained had analysis been based on the correlation matrix originally. These considerations lead in practice to canonical analysis being performed most frequently on the correlation matrix **R**. The general reliance on **R**, however, should not allow one to overlook the possibility that analyses based on such alternative scalar-products matrices as **Q** or **S**, or, indeed, on quite different matrices, might not also sometimes be profitable in particular circumstances.

The scaling of canonical variates is arbitrary. It is customary to standardize the variates to zero mean and unit variance. Other scaling procedures, however, may be advantageous in particular circumstances. In dual scaling there is much to be said for scaling the u_k and v_k such that their respective variances are equal to the square of the corresponding canonical correlation coefficient, r_k^2. The canonical variates then reflect the magnitude of the root to which they refer. In applications where the sample is made up of a number of subsamples or treatments, as in canonical variate analysis, it is advantageous to scale the variates so that the average within-groups variance on each canonical variate is unity. This procedure equalizes the within-groups variance on the canonical variates and maximizes the squared distance between group means. Nash (in Falkenhagen & Nash, 1978) has demonstrated the merit of this particular standardization in certain applications of canonical correlation analysis also.

Canonical variates are interpreted in much the same way as principal components. Interpretation therefore involves consideration of both the contribution of the original variables to a canonical variate and the pattern of variation of a variate over the sample as a whole. The contribution of the observed variables to a canonical variate has in the past generally relied on examination of the canonical weights, the magnitude and sign of which were taken to indicate the presence and direction of the effect of particular variables. However, particularly where the variables comprising each domain are even moderately intercorrelated, as they generally are in ecological contexts, the possibility of interpreting the canonical variates by inspection of the canonical weights is practically nil (Meredith, 1964). Realization of this difficulty is increasingly leading to the adoption of ridge-type rather than the usual least squares estimators for the weights, or to the weights being supplanted entirely by the correlations between the original variables and the canonical variates as aids in the interpretation of the latter. The variable/canonical variate correlations tend to be more stable than the usual weights in replicate samples; for this and for other reasons they often prove to be more reliable indicators of the contribution of particular variables to the canonical variates. Inspection of the sample values or scores on a particular canonical variate may in itself sometimes suggest a substantive interpretation for the variate. In other cases graphical displays of the scores may serve the same purpose.

Where the rank of a fitted model does not exceed three, the canonical variates provide a convenient reference frame within which to display and examine the sample. In particular, mappings into the following two- or three-dimensional subspaces defined by the canonical variates may be illuminating:

(a) u_k against v_k; $k=1, \ldots, t$;
(b) u_k against u_m; v_k against v_m; $k, m=1, \ldots, t$ $(k \neq m)$;
(c) difference $(u_k - v_k)$ against difference $(u_m - v_m)$; $k, m=1, \ldots, t$ $(k \neq m)$.

The results are effectively communicated by means of scattergrams or stereograms. Guidelines for the interpretation of displays of this general kind have been given by Coxon (1982, p. 93) and by Chambers, Cleveland, Kleiner and Tukey (1983, p. 319). It is well to be aware that it is the point configuration and its substantive implications rather than the axes which are of primary interest in such displays; while the axes of scattergrams or stereograms of canonical variates may be amenable to interpretation in substantive terms, this is by no means always to be expected. Mappings of the first of the above kinds, that is of u_k against v_k for $k=1, \ldots, t$, each summarize a separate aspect of the correlation structure in question. Mappings of type (b) differ fundamentally from the covariance displays of type (a) in that they involve subspaces of just *one* of the two sample-spaces of interest. Where the canonical variates in question are collinear with the principal components of their respective domains, scattergrams of u_k against u_m and of v_k against v_m may be expected to resemble those resulting from separate components analyses of each domain. Canonical variates, however, in general are unlikely to be collinear with principal components. In such cases graphical displays of the kind in question can be expected to draw attention to quite different features of the sample. The precise nature of these features will depend both on the measurement domains involved and the relationship between them. Mappings of the third kind are informative in drawing attention to the extent and nature of the lack of fit of a model of specified rank. Simultaneous displays of samples and variables with respect to the canonical variates are also feasible. The construction of such joint displays has been described and illustrated by van der Burg and de Leeuw (1983). In connection with scattergrams we observe also that the minimum spanning tree is a further device which can enhance interpretation (Jeffers, 1978; Brown, 1979; Oxnard, 1979).

Where the intrinsic dimensionality of the correlation structure exceeds three some other form of graphical representation is called for. Andrews (1972) has proposed a general method for displaying high-dimensional data which is useful in these cases. The procedure calls for mapping each data point, say $\mathbf{u}_j = [u_{1j}, \ldots, u_{tj}]^t$, where u_{kj} may represent the score of the jth sample on the kth canonical variate of x-space, into the *arbitrary* sine-cosine function

$$f_{\mathbf{u}_j}(t) = u_{1j}/\sqrt{2} + u_{2j}\sin t + u_{3j}\cos t + u_{4j}\sin 2t + u_{5j}\cos 2t + \ldots,$$

and plotting the function over the range $-\pi < t < \pi$. In this way each sample in u_k-space comes to be represented by a *curve* in 2-space. The advantage of so doing is that no constraint on the dimensionality of the data which may be treated arises. The resulting plots have several other useful properties. The most important of these for present purposes is that similarities in location and shape between curves are directly related to the proximity of the corresponding sample-points in the subspace of the retained canonical variates. Put another way, the closer the sample-points in the space of the retained variates the greater the proximity and overall similarity of the corresponding curves on the function-plot.

Where an analysis involves spatially distributed samples, the simple device of plotting the spatial distribution of a canonical variate on a map of the study area can be useful. Such distribution maps are often effective aids in the interpretation of canonical variates.

As mentioned previously, relationships between two measurement domains may be either symmetric or asymmetric. The function of the canonical variates differs in the two cases. Where the relationship is symmetric, the canonical variates provide a parsimonious *summarization* of the data which simultaneously draws attention to the presence and nature of any linear relationship between the domains investigated. It is hoped and expected that by mapping the sample into one or more subspaces defined by the canonical variates, greater insight into the covariance structure of the data will be obtained than by examining either the initial observations themselves or the matrix $\mathbf{R}_{12}$ of interset correlations. Where, on the other hand, the relationship is one of asymmetry, the canonical variates specify those weighted linear combinations of the responses which are most *predictable* from the antecendent variables. For enlightening discussions of canonical variates and their properties the interested reader is referred to Rozeboom (1965) and to Rao (1973, p. 494).

2.5 Computation

2.5.1 Numerical methods

In Sect. 2.3 we saw that canonical correlation coefficients and canonical variates can be derived in several ways. The usual computational procedure is based on the classical derivation described in Sect. 2.3.1. This calls for the eigenanalysis of a cross-products type matrix [cf. Eq. (2.14)]. An alternative derivation based on singular value decomposition was described in Sect. 2.3.2.

While the direct solution of the eigenvalue problem (2.14) is straightforward, there can be serious drawbacks to this form of computation. First, the entries of the data matrices $\mathbf{X}$ and $\mathbf{Y}$ have only a limited number of accurate digits following which the entries are completely uncertain and hence arbitrary. It is important in computation that this fact be taken into account. We observe, in particular that, unless special precautions are taken in computing the covariance matrix $\mathbf{S}$ of (2.1) at the outset of analysis, errors in the data will be compounded. Secondly, the submatrices $c\mathbf{X}^t\mathbf{X}$ and $c\mathbf{Y}^t\mathbf{Y}$ of $\mathbf{S}$ often prove to be ill-conditioned: that is to say, relatively small changes in $\mathbf{X}$ or $\mathbf{Y}$ will induce comparatively large changes in $(c\mathbf{X}^t\mathbf{X})^{-1}$ or $(c\mathbf{Y}^t\mathbf{Y})^{-1}$. Thirdly, the submatrices comprising $\mathbf{S}$ are greatly influenced by round-off errors. Consequently, no matter how (2.14) is solved, errors in the data, the effects of ill-conditioning and round-off errors will all be magnified excessively in the canonical weights. In other words, the direct solution of (2.14) will not necessarily determine a unique set of weights. For this reason it is often advantageous to work with $\mathbf{X}$ and $\mathbf{Y}$ rather than with cross-products matrices derived from them.

Many of the numerical problems inherent in the direct implementation of the classical formulas for evaluating canonical correlations and variates can be avoided by orthogonal decompositions of $\mathbf{X}$ and $\mathbf{Y}$ followed by a singular value decomposition. Though singular value decomposition requires more computer time and storage than eigenanalysis, it does have advantages in dealing effectively with the numerical problems associated with the latter. More specifically, algorithms based on the singular value decomposition have greater numerical stability, especially where ill-conditioning of an extreme kind exists. That is to say, such algorithms will enable solutions to be obtained for a wider range of problems at a fixed precision of computation.

Both the eigenanalysis and singular value decomposition procedures are reviewed below. In view of the prevalence of computational schemes based directly on the formulas of the classical derivation we shall outline the computation of canonical analysis by this means before proceeding to a fuller description in terms of singular value decomposition.

Eigenanalysis. The sought $\{r_k; \mathbf{b}_k\}$ are solutions of the generalized symmetric eigenvalue problem (2.14). Assuming that $\mathbf{S}_{11}$ is nonsingular, Eq. (2.14) may be solved by the simultaneous diagonalization of $\mathbf{S}_{21}\mathbf{S}_{11}^{-1}\mathbf{S}_{12}$ and $\mathbf{S}_{22}$. The conjugate weights $\mathbf{a}_k$ then follow from (2.17) and the canonical variates u_k and v_k from (2.4). Alternatively, provided $\mathbf{S}_{22}$ is also nonsingular, (2.14) may be reduced to the standard nonsymmetric eigenvalue problem

$$(\mathbf{S}_{22}^{-1}\mathbf{S}_{21}\mathbf{S}_{11}^{-1}\mathbf{S}_{12}-\lambda^2\mathbf{I})\,\mathbf{b}=\mathbf{0}, \tag{2.33}$$

where the quadruple matrix product on the left in general is nonsymmetric.

Canonical analysis can also be formulated and solved as a standard symmetric eigenvalue problem. In applications $\mathbf{S}_{22}$ is generally positive definite. Making this assumption we may write

$$\mathbf{S}_{22}=[\mathbf{S}_{22}^{\frac{1}{2}}]^t\,\mathbf{S}_{22}^{\frac{1}{2}}, \tag{2.34}$$

where $\mathbf{S}_{22}^{\frac{1}{2}}$ $(q\times q)$ is upper triangular ($\triangleleft$) and is the Cholesky factor of $\mathbf{S}_{22}$. Using (2.34), Eq. (2.14) can be reduced to the standard symmetric eigenequation

$$(\mathbf{S}_{22}^{-\frac{1}{2}}\mathbf{S}_{21}\mathbf{S}_{11}^{-1}\mathbf{S}_{12}\mathbf{S}_{22}^{-\frac{1}{2}}-\lambda^2\mathbf{I})\,\mathbf{b}^*=\mathbf{0}. \tag{2.35}$$

The roots λ_k^2 of (2.35) are identical to the roots of (2.14), while the eigenvectors $\mathbf{b}_k^*$ are related to the eigenvectors of (2.14) as

$$\mathbf{b}_k=[\mathbf{S}_{22}^{-\frac{1}{2}}]^t\,\mathbf{b}_k^*.$$

The conjugate weights $\mathbf{a}_k$ and canonical variates u_k and v_k then follow as before from (2.17) and (2.4), respectively. These results are readily extended to the case where $\mathbf{S}_{22}$ is rank defective. To allow for the singular case a check is inserted in the factorization (2.34) to detect rows of $\mathbf{S}_{22}$ which are dependent on previous rows. These dependent rows are then deleted. Khatri (1976) has shown that the computational problems which arise where either $\mathbf{S}_{11}$ or $\mathbf{S}_{22}$ are singular can also be dealt with effectively by the use of Rao's g-inverse, $\mathbf{S}_{11}^-$ or $\mathbf{S}_{22}^-$.

Cohen and Ben-Israel (1969) have described an algorithm implementing the standard symmetric eigenvalue solution for canonical analysis. High quality Fortran subroutines for Cholesky decomposition and the symmetric eigenvalue problem are widely available (Smith, Boyle, Dongarra, Garbow, Ikebe, Klema & Moler, 1976; Dongarra, Bunch, Moler & Stewart, 1979). Other useful sources are the collected algorithms of the *International Mathematical and Statistical Library* and the *Numerical Algorithms Group*.

Singular value decomposition. The formulation of canonical analysis in terms of singular value decomposition was described by Lancaster (1958) and Horst (1961a). Attention was first drawn to the numerical advantages of this and related matrix decompositions in statistical computation by Golub (1969). Björck and Golub (1973) subsequently investigated the calculation of canonical correlation coefficients and variates by orthogonal matrix decompositions very thoroughly. Orthogonal matrices have highly desirable computational properties because they do not magnify errors. Another important feature of the singular value decomposition is its ability to detect linear dependence and the lack of uniqueness to which this gives rise. Furthermore, the procedure enables cross-products matrices and their inverses to be avoided entirely.

The first step in calculating canonical correlations and variates by the singular value decomposition is the factorization of suitably scaled forms of **X** and **Y** into two – an orthogonal matrix and an upper triangular matrix. The product of the orthogonal factors of **X** and **Y** is next computed. Finally, singular value decomposition of this matrix product yields the essential quantities for canonical analysis. Provided that the condition numbers $\varkappa(\mathbf{X})$ and $\varkappa(\mathbf{Y})$ of the scaled matrices are small, the canonical correlation coefficients will be well determined. These remarks provide the foundation for the following algorithm:

Step 0: Read $\mathbf{X}(\mathrm{N} \times p)$, $\mathbf{Y}(\mathrm{N} \times q)$. Interchange **X** and **Y** if necessary to preserve the condition $p \geq q$.

Step 1: Adjust and scale the columns of **X** and **Y** to zero mean and unit variance. Denote the preconditioned matrices $\mathbf{Z}^{(x)}$ $(\mathrm{N} \times p)$ and $\mathbf{Z}^{(y)}$ $(\mathrm{N} \times q)$.

Step 2: QR-decompositions of $\mathbf{Z}^{(x)}$ and $\mathbf{Z}^{(y)}$.

$$\underset{(\mathrm{N}\times p)}{\mathbf{Z}^{(x)}} = \underset{(\mathrm{N}\times p)}{\mathbf{Q}^{(x)}} \underset{(p\times p)}{\mathbf{R}^{(x)}}. \quad [\mathbf{Q}^{(x)}]^{\mathrm{t}}\mathbf{Q}^{(x)} = \mathbf{I}_p. \quad \mathbf{R}^{(x)}: \triangledown$$

$$\underset{(\mathrm{N}\times q)}{\mathbf{Z}^{(y)}} = \underset{(\mathrm{N}\times q)}{\mathbf{Q}^{(y)}} \underset{(q\times q)}{\mathbf{R}^{(y)}}. \quad [\mathbf{Q}^{(y)}]^{\mathrm{t}}\mathbf{Q}^{(y)} = \mathbf{I}_q. \quad \mathbf{R}^{(y)}: \triangledown$$

Step 3: Form

$$\underset{(p\times q)}{\mathbf{K}} = \underset{(p\times \mathrm{N})}{[\mathbf{Q}^{(x)}]^{\mathrm{t}}} \underset{(\mathrm{N}\times q)}{\mathbf{Q}^{(y)}}$$

K is identically $[\mathbf{S}_{11}^{\mathrm{t}}]^{-\frac{1}{2}}\mathbf{S}_{12}\mathbf{S}_{22}^{-\frac{1}{2}}$ of Eq. (2.20); see Golub (1969).

Step 4: Singular value decomposition of **K**

$$\underset{(p\times q)}{\mathbf{K}} = \underset{(p\times q)}{\mathbf{W}} \underset{(q\times q)}{\mathbf{D}} \underset{(q\times q)}{\mathbf{Z}}$$

Step 5: The canonical variates u_k and v_k are the columns of **U** and **V**, where

$$\underset{(\mathrm{N}\times q)}{\mathbf{U}} = \underset{(\mathrm{N}\times p)}{\mathbf{Q}^{(x)}} \underset{(p\times q)}{\mathbf{W}}, \qquad \underset{(\mathrm{N}\times q)}{\mathbf{V}} = \underset{(\mathrm{N}\times q)}{\mathbf{Q}^{(y)}} \underset{(q\times q)}{\mathbf{Z}}$$

Step 6: The vectors of canonical weights $\mathbf{a}_k$ and $\mathbf{b}_k$ are the columns of **A** and **B**, where

$$\underset{(p\times q)}{\mathbf{A}} = \underset{(p\times p)}{[\mathbf{R}^{(x)}]^{-1}} \underset{(p\times q)}{\mathbf{W}}, \qquad \underset{(q\times q)}{\mathbf{B}} = \underset{(q\times q)}{[\mathbf{R}^{(y)}]^{-1}} \underset{(q\times q)}{\mathbf{Z}}$$

Remarks

(a) The singular value decomposition of **K** defines canonical analysis:
 (i) the canonical correlation coefficients r_k are the singular values d_k of **K**;
 (ii) the canonical variates and weights are readily obtained from the singular vectors $[\mathbf{w}_1|\ldots|\mathbf{w}_q]$ and $[\mathbf{z}_1|\ldots|\mathbf{z}_q]$ as in *Steps 5* and *6*.

For computational purposes $\mathbf{Q}^{(x)}$, $\mathbf{R}^{(x)}$, $\mathbf{Q}^{(y)}$, $\mathbf{R}^{(y)}$, **W**, **D** and **Z** provide the essential summary of canonical analysis. These matrices are comparatively free from numerical problems and can be used to generate all the relevant statistical information. Some available published subroutines for obtaining the matrices in question are reported in Sect. 2.5.2 below. Björck and Golub (1973) have shown that the orthogonal decompositions $\mathbf{Z}^{(x)}=\mathbf{Q}^{(x)}\mathbf{R}^{(x)}$ and $\mathbf{Z}^{(y)}=\mathbf{Q}^{(y)}\mathbf{R}^{(y)}$ together with the singular value decomposition of $\mathbf{K}=[\mathbf{Q}^{(x)}]^{\mathrm{t}}\mathbf{Q}^{(y)}$ have a simple, elegant interpretation in terms of the canonical angles and canonical bases of Sect. 2.3.4.

(b) To avoid unnecessary numerical inaccuracy in estimating the variances of the variables in $\mathbf{x}$ and $\mathbf{y}$ in connection with *Step 1*, preliminary QR-factorizations of $\mathbf{X}$ and $\mathbf{Y}$ are desirable. The variances of $x_1, \ldots, x_p$, for example, may then be computed from the decomposition $\mathbf{X}=\mathbf{QR}$ as follows. The cross-products matrix $\mathbf{X}^{\mathrm{t}}\mathbf{X}$ is given by:

$$\mathbf{X}^{\mathrm{t}}\mathbf{X}=\mathbf{R}^{\mathrm{t}}\mathbf{Q}^{\mathrm{t}}\mathbf{Q}\mathbf{R}=\mathbf{R}^{\mathrm{t}}\mathbf{R}.$$

The required variances are obtained from the diagonal elements of $\mathbf{C}=\mathbf{R}^{\mathrm{t}}\mathbf{R}$ as $(\mathrm{N}-1)^{-1}c_{ii}$. Similar manipulation yields the variances of $y_1, \ldots, y_q$.

(c) Several methods of computing the QR-decomposition of a matrix have been proposed. Of these, algorithms based on Householder transformations and the modified Gram-Schmidt procedure are often recommended in connection with *Step 2*. Björck and Golub (1973) have shown that these procedures provide results of comparable accuracy. There is, however, an important advantage associated with the modified Gram-Schmidt procedure. This is that $\mathbf{Q}$ is returned explicitly, which is not the case with Householder's method. The significance of $\mathbf{Q}$ is that it is required for the computation of auxiliary quantities. It is essential that for this purpose the matrix be accurately orthogonal. For this reason it is necessary to re-orthogonalize $\mathbf{Q}$ before using the matrix for further computation.

(d) In addition to the quantities already calculated, a number of auxiliary quantities are required. The most important of these are the structure correlations, the variance and redundancy associated with a canonical variate and the variable communalities to be discussed in Chap. 3.
For the within-set structure correlations $\mathbf{S}^{(x)}$ and $\mathbf{S}^{(y)}$, say, we compute

$$\underset{(p\times q)}{\mathbf{S}^{(x)}}=\underset{(p\times p)}{[\mathbf{R}^{(x)}]^{\mathrm{t}}}\,\underset{(p\times q)}{\mathbf{W}}/\mathrm{N}-1, \qquad \underset{(q\times q)}{\mathbf{S}^{(y)}}=\underset{(q\times q)}{[\mathbf{R}^{(y)}]^{\mathrm{t}}}\,\underset{(q\times q)}{\mathbf{Z}}/\mathrm{N}-1.$$

The between-set correlations, $\mathbf{S}^{(xv)}$ and $\mathbf{S}^{(yu)}$, say, are then obtained from $\mathbf{S}^{(x)}$ and $\mathbf{S}^{(y)}$ as

$$\underset{(p\times q)}{\mathbf{S}^{(xv)}}=\underset{(p\times q)}{\mathbf{S}^{(x)}}\,\underset{(q\times q)}{\mathbf{D}}, \qquad \underset{(q\times q)}{\mathbf{S}^{(yu)}}=\underset{(q\times q)}{\mathbf{S}^{(y)}}\,\underset{(q\times q)}{\mathbf{D}},$$

where $\mathbf{D}=\operatorname{diag}(d_1, \ldots, d_q)$. Having obtained the structure correlations, the variances, redundancies and variable communalities are readily obtained using the formulas of Chap. 3.

The indices mentioned provide the basic quantities necessary for the interpretation of canonical analysis. However, a comprehensive analysis would require the availability of programs or subroutines to perform a variety of supporting functions. Convenient access to algorithms implementing the following additional procedures would go an appreciable way towards facilitating incisive analysis:

> assessment of sample homogeneity and distributional properties, projection pursuit, multivariate planing, re-expression of variables, robust estimation of mean vectors and covariance matrices, treatment of incomplete data, canonical ridge estimation, updating, jackknife analysis, rotation of canonical variates, graphic display (scattergrams, stereograms, the minimum spanning tree, biplots, sine-cosine and probability plots), permutation and parametric tests of significance, calculation and analysis of residuals.

2.5.2 Further remarks

Efficient, numerically stable subroutines for singular value decomposition are widely available. These are all based on a procedure originally suggested by Golub (see Golub & Kahan, 1965; Businger & Golub, 1969) and further developed by Golub and Reinsch (1970). The algorithm consists of two main steps. Householder transformations are first used to reduce the matrix to be factored to bidiagonal form. The QR algorithm of Francis (1961, 1962) is then used to find the singular values of the bidiagonal form. The properties and rate of convergence of Francis' algorithm are well understood, it being among the most satisfactory in the whole of numerical analysis. Fortran subroutines for singular value decomposition have been published by Lawson and Hanson (1974, p. 295), Garbow, Boyle, Dongarra and Moler (1977, p. 229), and by Dongarra, Bunch, Moler and Stewart (1979, Chap. 11). Other sources are the widely distributed collected algorithms of the *International Mathematical and Statistical Library* and the *Numerical Algorithms Group*.

The most satisfactory algorithm for the QR-decomposition of *Step 2* is that of Daniel, Gragg, Kaufman and Stewart (1976). The appeal of this algorithm is that it does explicitly return a $\mathbf{Q}$ matrix which is accurately orthogonal. The algorithm is written in Algol 60, however, and no Fortran version appears to be generally available.

Rank deficiency in $\mathbf{X}$ or $\mathbf{Y}$ can be a source of computational difficulty. The commonest cause of rank deficiency is having more variables than samples. The computational problems themselves, however, can now readily be overcome. To see this, consider the matrix $\mathbf{X}$ $(\mathrm{N} \times p)$ of rank $r<p$. Then $\mathbf{X}$ may be approximated by an $\mathrm{N} \times p$ matrix $\mathbf{X}^*$ such that the rank $(\mathbf{X}^*)=r$. The singular value decomposition of $\mathbf{X}$ is first calculated. The quantity r is determined by the number of singular values which exceed some suitable tolerance. $\mathbf{X}^*$ is then constructed from the r leading terms in the decomposition. Alternatively, use may be made of Rao's g-inverses $\mathbf{R}_{11}^{-}$ or $\mathbf{R}_{22}^{-}$ in place of either or both regular inverses in (2.14), as shown by Khatri (1976). Rao (1981) has described a unified approach to cover singular and nonsingular cases of $\mathbf{X}$ and $\mathbf{Y}$ based on singular value decomposition. Just what value is to be placed on the results of analyses having more variables than samples, however, remains unanswered. Where sample size is small it is a considerable assumption that the extraction

of linear composites can be relied upon to condense the data in such a way that important information is not lost.

Further reading. The *computation* of canonical correlation coefficients has been discussed in some detail by Cohen and Ben-Israel (1969). Attention has been drawn to opportunities provided by *matrix decompositions* in statistical computation by Golub (1969) and a detailed account of the calculations for canonical analysis based on orthogonal matrix decomposition given by Björck and Golub (1973). Other accounts of the computation of canonical correlations and variates emphasizing matrix factorization have been presented by Cohen (1972, 1977), Chambers (1977, p. 125) and Kennedy and Gentle (1980, p. 561). A system of Fortran subroutines for *robust statistical estimation* has been developed by Coleman, Holland, Kaden, Klema and Peters (1980). Software for dealing with the problem of *rank degeneracy* has been described by Golub, Klema and Peters (1980). Computational procedures for *updating* the Gram-Schmidt QR-decomposition have been developed by Daniel, Gragg, Kaufmann and Stewart (1976) and for singular value decomposition by Bunch and Nielsen (1978) and Bunch, Nielsen and Sorensen (1978). Procedures for dealing with *incomplete data* have been discussed by Hocking and Marx (1979) and by Smith and Riggs (1982).

For discussions of the computational aspects of *least-squares* problems in a wider context, reference may be made to Golub and Reinsch (1970), Stewart (1973), Lawson and Hanson (1974), Chambers (1974, 1975, 1977), Forsythe, Malcolm and Moler (1977), Björck (1978), Moler (1978) and Kennedy and Gentle (1980).

Computer programs for implementing special cases or certain generalizations of canonical analysis have been described by several authors. Timm and Carlson (1976) have prepared a Fortran program for *part and bipartial* canonical analysis while Young, de Leeuw and Takane (1976) have developed a program for the analysis of *nonmetric* data. The program of Young, de Leeuw and Takane has since been extended (Gifi, 1981) to allow the analysis of $m>2$ sets of variables expressed on a variety of measurement scales. Other programs for *generalized* canonical correlation analysis ($m>2$ sets of variables) have been written by Carroll and Chang (1970) and by Chen and Kettenring (1972). Van den Wollenberg (1977) has prepared a program for *redundancy analysis* while van der Burg and de Leeuw (1983) have described a program for *nonlinear* canonical correlation analysis. A Fortran program for the symmetric and asymmetric orthogonal *rotation* of canonical variates has been written by Huba, Palisoc and Bentler (1982).

An extensive series of programs for *dual scaling*, some permitting the efficient analysis of very large arrays, has been written by Lebart, Morineau and Tabard (1977) and by Lebart and Morineau (1982). Leong, Nishisato and Wolf (1982) have also developed a program for dual scaling. A comprehensive series of programs for *canonical variate analysis* which facilitate robust analysis as well as several other refinements has been developed by SIROMATH (1982). Lastly, we mention that a computer program for *multivariate graphics* of value in canonical analysis has been prepared by Tsianco, Gabriel, Odoroff and Plumb (1981).

3. Extensions and generalizations

3.1 Introduction

In the preceding chapter we introduced the essential ingredients of the theory of canonical correlation. The theory stems from the work of Hotelling in the 1930s. Since Hotelling's pioneering studies the theory has been added to in a number of ways. The present chapter deals with developments of this kind. We shall begin by considering several numerical indices which have been devised in order to further the interpretation of results in addition to the canonical correlations, weights and variates of the classical solution. Other developments are directed towards widening the applicability of the original theory and several generalizations of this kind will be mentioned. Tests of statistical hypotheses are sometimes of interest in canonical analysis and the chapter closes with an account of hypothesis testing.

3.2 Further interpretive devices

Canonical analysis is concerned with the linear relationships between two measurement domains and with the nature of these relationships. We have seen that canonical correlation coefficients go some way towards establishing the number and strength of the relationships present, while canonical weights and variates may be helpful in clarifying their nature. However, canonical correlations and weights each leave something to be desired as interpretive indices; canonical correlation coefficients express relationships between linear composites which, though perhaps highly correlated, may be relatively unimportant components of their respective domains, while the interpretation of canonical weights has all the problems attendant on the interpretation of standard partial regression coefficients in multiple regression. For these reasons a number of auxiliary interpretive indices have been proposed. We turn now to consider some of these additional indices. In particular:

(a) the correlations between canonical variates and the original variables;
(b) the proportion of the total variance of a measurement domain associated with a canonical variate;
(c) the redundancy of a measurement domain;
(d) the communality of a variable.

3.2.1 Correlations between canonical variates and the original variables

The modulus and sign of the variable/canonical variate correlations are sometimes useful in showing (a) which variables contribute most heavily to a canonical variate and the direction of their effect; and (b) the existence of a affinities or contrasts among variables in their relationship with a canonical variate. The correlations may thus contribute towards establishing the nature of relationships which may be present between domains. The correlations tend to be more stable than either the raw or standardized canonical weights under the addition or deletion of variables and in replicate samples drawn from the same population. Moreover, correlations are more readily translated into meaningful terms than are weights such as canonical or regression weights (Dempster, 1969, p. 160). Thus, the square of a variable/canonical variate correlation expresses the proportion of the variance of a variable which is directly associated with a particular canonical variate. For convenience, we shall on occasion refer to variable/canonical variate correlations as *structure correlations*. Variable/canonical variate correlations are of two kinds:

(a) ***intra****set correlations:* correlations between canonical variates and observed variables of the *same* domain;
(b) ***inter****set correlations:* correlations between canonical variates of one domain and the observed variables of the *other*.

Intraset correlation coefficients. Such correlations express the contribution of the variables of a domain to the canonical variates of the same domain. In so doing they help to establish substantive interpretations of these canonical variates. There are two sets of intraset correlations corresponding to the two measurement domains. For the variables and canonical variates of $\mathbf{z}^{(x)}$ the intraset correlations are given by

$$\begin{aligned} \operatorname{cor}(\mathbf{z}_j^{(x)}, u_k) = \operatorname{cor}(\mathbf{z}_j^{(x)}, \mathbf{a}_k^t \mathbf{z}_j^{(x)}) &= \frac{1}{N} \sum_{j=1}^{N} \mathbf{z}_j^{(x)} [\mathbf{z}_j^{(x)}]^t \mathbf{a}_k \\ &= \mathbf{R}_{11} \mathbf{a}_k, \end{aligned} \tag{3.1}$$

where $\mathbf{R}_{11}\mathbf{a}_k$ is the $p \times 1$ vector of correlations between the kth canonical variate of $\mathbf{z}^{(x)}$ and the observed variables of $\mathbf{z}^{(x)}$. We shall denote the ith element of this vector by s_{ik} ($i = 1, \ldots, p$).

In a similar way for the y-domain

$$\operatorname{cor}(\mathbf{z}_j^{(y)}, v_k) = \mathbf{R}_{22} \mathbf{b}_k.$$

The correlation between the hth variable of $\mathbf{z}^{(y)}$ and the kth canonical variate v_k will be denoted s_{hk} ($h = 1, \ldots, q$).

The square of an intraset correlation for a given variable is the proportion of the variance of that variable which is explained by a canonical variate of the same domain.

Interset correlation coefficients. The interset correlations characterize interrelationships between the canonical variates of one measurement domain and the observed variables of the other. There are two sets of interset correlations. For the correlations of the variables of $\mathbf{z}^{(x)}$ with the canonical variates of $\mathbf{z}^{(y)}$ we have

$$\operatorname{cor}(\mathbf{z}_j^{(x)}, v_k) = \operatorname{cor}(\mathbf{z}_j^{(x)}, \mathbf{b}_k^t \mathbf{z}_j^{(y)}) = \frac{1}{N} \sum_{j=1}^{N} \mathbf{z}_j^{(x)} [\mathbf{z}_j^{(y)}]^t \mathbf{b}_k$$
$$= \mathbf{R}_{12} \mathbf{b}_k,$$

where $\mathbf{R}_{12}\mathbf{b}_k$ is the $p \times 1$ vector of interset correlations between the kth canonical variate, v_k, of $\mathbf{z}^{(y)}$ and the observed variables of $\mathbf{z}^{(x)}$.

From (2.8) and recalling that $\lambda = \mathrm{r}$, it is easy to see that

$$\mathbf{R}_{12} \mathbf{b}_k = \mathrm{r}_k \mathbf{R}_{11} \mathbf{a}_k,$$

which shows how the interset correlations between $\mathbf{z}^{(x)}$ and v_k can be obtained directly from (3.1).

Similarly,

$$\operatorname{cor}(\mathbf{z}_j^{(y)}, u_k) = \mathbf{R}_{21} \mathbf{a}_k = \mathrm{r}_k \mathbf{R}_{22} \mathbf{b}_k,$$

where $\mathbf{R}_{21}\mathbf{a}_k$ is the $q \times 1$ vector of correlations between the kth canonical variate, u_k, of $\mathbf{z}^{(x)}$ and the observed variables in $\mathbf{z}^{(y)}$.

Interset correlation coefficients closely resemble multiple correlation coefficients. The distinction between the two is that the composite variable entering into a multiple correlation coefficient is determined under the requirement that it be maximally correlated with the single response y; in the case of an interset correlation the maximization is with respect to a linear *composite* of response variables, $y_1, \ldots, y_q$. This leads to the interset correlation of a specified variable being smaller than its multiple correlation with the same set of predictor variables.

The square of an interset correlation coefficient specifies the proportion of the variance of a variable which is predictable by a canonical variate of the 'other' domain.

While variable/canonical variate correlations are open to interpretation by inspection, graphical analysis may aid in comprehending relationships among them. Scattergrams or stereograms of the correlations are an effective means of obtaining insight into the covariance structure of the data. *Simultaneous* plots of the correlations of both sets of variables against selected canonical variates often prove to be particularly rewarding. Van der Burg and de Leeuw (1983) have described the construction of joint displays of this kind and illustrated their use.

The distinction between the roles of intraset and interset correlations may be worth emphasizing. Intraset correlations are used to *characterize* the canonical variates u_k or v_k in terms of the variables on which they are defined. Interset correlations, on the other hand, or, strictly speaking their squares, are informative as expressions of the extent of the explanatory power of the u_k or v_k *across* domains.

Apart from the uses of variable/canonical variate correlations noted above, we shall see that they are also used in the construction of other interpretive devices, notably in the variance extracted by a canonical variate, redundancy and in variable communalities.

3.2.2 Variance extracted by a canonical variate

The proportion of the total variance of a measurement domain which is associated with a canonical variate is referred to as the variance extracted or accounted for by the canonical variate. This quantity is calculated as the sum of squared structure correlations between the variate and the variables on which it is defined divided by the number of variables in question. To see this, observe that if each variable is standardized to unit variance, the sum of squared correlations yields the total variance explained while the number of variables is equivalent to the total variance which is to be explained. The variance extracted represents the amount of variance common to both a measurement domain and a particular canonical variate; consequently, it is an index of the extent to which a canonical variate accounts for the total variance of the domain of which it is a linear composite.

For the kth canonical variate u_k of $\mathbf{z}^{(x)}$, the variance extracted, U_k^2, is given by

$$\mathrm{U}_k^2 = \sum_{i=1}^{p} s_{ik}^2 / p, \tag{3.2}$$

where U_k^2 is the mean of the p squared intraset correlations, s_{ik}^2, with u_k. Similarly, for the variance extracted by the kth canonical variate v_k of $\mathbf{z}^{(y)}$, V_k^2, we have

$$\mathrm{V}_k^2 = \sum_{h=1}^{q} s_{hk}^2 / q. \tag{3.3}$$

Notice that in general $\mathrm{U}_k^2 \neq \mathrm{V}_k^2$. This is so because, while u_k may be a major dimension of its measurement domain, v_k may correspond to a much smaller dimension of $\mathbf{z}^{(y)}$, or *vice versa*.

3.2.3 Redundancy

Redundancy is the proportion of the total variance of a measurement domain predictable from a linear composite of the *other* domain, given the availability of the second domain. The term redundancy is therefore synonymous with *explained variance*.

The mean sum of squared intraset structure correlations with the kth canonical variate was described above as the variance extracted by the kth variate. If this quantity is multiplied by the squared canonical correlation coefficient, r_k^2, we obtain the proportion of the variance of one measurement domain that is accounted for by the kth canonical variate of the other. To see this, recall that a squared canonical correlation coefficient expresses the proportion of the variance of one of a pair of canonical variates which is accounted for by the other. The quantity arrived at in this way is the explained variance or redundancy. We may therefore write:

$$\mathrm{V}_{\mathbf{x}|v_k}^2 = \mathrm{U}_k^2 \, \mathrm{r}_k^2 \tag{3.4}$$

for the redundancy of $\mathbf{z}^{(x)}$, given the availability of the set $\mathbf{z}^{(y)}$, and

$$U^2_{\mathbf{y}|u_k} = V^2_k r^2_k \tag{3.5}$$

for the redundancy of $\mathbf{z}^{(y)}$, given the availability of $\mathbf{z}^{(x)}$.

Redundancy is open to interpretation in other terms. Thus the redundancy, $V^2_{\mathbf{x}|v_k}$, of $\mathbf{z}^{(x)}$ with respect to the kth canonical variate $v_k = \mathbf{b}^t_k \mathbf{z}^{(y)}$, is given by the mean of the squared *interset* correlations between the elements of $\mathbf{z}^{(x)}$ and v_k. That is

$$V^2_{\mathbf{x}|v_k} = \sum_{i=1}^{p} s^2_{ik}/p \tag{3.6}$$

where s_{ik} is the correlation between the ith variable of $\mathbf{z}^{(x)}$ and the kth canonical variate of $\mathbf{z}^{(y)}$.

The similarity of (3.6) to the expression for the variance extracted by a canonical variate given by (3.2) will be obvious. Evidently, redundancy is the between-set analog of the variance extracted by a canonical variate.

In a similar way, the redundancy, $U^2_{\mathbf{y}|u_k}$, of $\mathbf{z}^{(y)}$, given the availability of $\mathbf{z}^{(x)}$, is given by

$$U^2_{\mathbf{y}|u_k} = \sum_{h=1}^{q} s^2_{hk}/q,$$

where s_{hk} is the correlation between the hth variable of $\mathbf{z}^{(y)}$ and the kth canonical variate of $\mathbf{z}^{(x)}$.

In general $V^2_{\mathbf{x}|v_k} \neq U^2_{\mathbf{y}|u_k}$. This follows on comparing (3.4) and (3.5) and recalling the inequality of $U^2_k \neq V^2_k$. The redundancy associated with a canonical variate is a useful index of the predictive or explanatory power of the canonical variate in relation to the 'other' domain. An alternative measure of this property has been proposed by Finn (1974, p. 191).

3.2.4 Total redundancy

The sum of the redundancy indices for each domain over the retained canonical variates of the rank t model yields an overall or total redundancy index for each domain. For the total redundancy in the set $\mathbf{z}^{(x)}$, given the canonical variates $v_1, \ldots, v_t$ of $\mathbf{z}^{(y)}$, we have

$$V^2_{\mathbf{x}|v_1,\ldots,v_t} = \sum_{k=1}^{t} \left(\sum_{i=1}^{p} s^2_{ik}/p \right) = \sum_{k=1}^{t} V^2_{\mathbf{x}|v_k}.$$

Similarly, the total redundancy in $\mathbf{z}^{(y)}$, given the canonical variates $u_1, \ldots, u_t$ of $\mathbf{z}^{(x)}$, may be written

$$U^2_{\mathbf{y}|u_1,\ldots,u_t} = \sum_{k=1}^{t} \left(\sum_{h=1}^{q} s^2_{hk}/q \right) = \sum_{k=1}^{t} U^2_{\mathbf{y}|u_k}.$$

Total redundancy provides an overall measure of the variance in one variable set accounted for by the variables of the other. Unlike a canonical correlation coefficient, which expresses the relationship between linear composites of each domain, total redundancy is a direct expression of the interrelatedness of the measurement domains themselves. Total redundancy is asymmetric between domains, so that in general $V^2_{\mathbf{x}|v_1,\ldots,v_t} \neq U^2_{\mathbf{y}|u_1,\ldots,u_t}$.

The sampling distribution of redundancy has been investigated by Miller (1975). Arguing partly by analogy from the distribution of the coefficient of multiple determination R^2 in multiple regression and partly on the basis of Monte Carlo studies, Miller (1975) has shown a simple multiple of the ratio $[U^2_{\mathbf{y}|u_1\ldots u_t}]/[1-U^2_{\mathbf{y}|u_1\ldots u_t}]$ to be a ratio of two independent χ^2 variates and hence to have an F distribution. Specifically, in the null case

$$\{[U^2_{\mathbf{y}|u_1\ldots u_t}]/[1-U^2_{\mathbf{y}|u_1\ldots u_t}]\}\cdot\{[N-p-1]\,q/p\,q\}\sim F_{\nu_1,\nu_2}. \tag{3.7}$$

For the model of full rank [$t=s$, where $s=\min(p,q)$]

$$\nu_1=p\,q, \qquad \nu_2=(N-p-1)\,q.$$

For the reduced rank model ($t<s$), the total redundancies of the leading term of the left-hand side of (3.7) are replaced by the redundancies of the reduced-rank model, $U^2_{\mathbf{y}|u_1\ldots u_t}$ and ν_1 and ν_2 by $\nu_1=p\,q'$ and $\nu_2=(N-p-1)\,q'$, respectively, where $q'=t$, the number of retained canonical roots.

The practical importance of this result lies in making a test of the null hypothesis

$$H_0:\ P^2_{\mathbf{y}|u_1\ldots u_t}=0$$

against the alternative

$$H_1:\ P^2_{\mathbf{y}|u_1\ldots u_t}\neq 0$$

possible, where we have written $P^2_{\mathbf{y}|u_1\ldots u_t}$ for the redundancy in the population.

Miller's results extend straightforwardly to the redundancy in the y-domain given the availability of $\mathbf{X}$, $V^2_{\mathbf{x}|v_1\ldots v_t}$.

Total redundancy has application in establishing the rank, t, of the canonical model judged to give the most acceptable fit to the data. The explanatory power of the model for various trial values of t is assessed informally in conjunction with the substantive value of the corresponding canonical variates and the overall reduction in dimensionality which will result. In this way total redundancy can help to establish the number of noteworthy linear relationships between domains.

3.2.5 Variable communalities

The communality of a variable expresses the proportion of the variance of the variable which is accounted for by the retained canonical variates of a fitted model. Two communalities are associated with each variable, corresponding to the two sets of canonical variates, u_k and v_k:

(a) ***intra****set communality:* the proportion of variance accounted for by the retained canonical variates of the variable's *own* set;

(b) ***interset*** *communality:* the proportion of variance accounted for by the retained canonical variates of the '*other*' set.

Intraset communalities. These are conveniently obtained as the sum of squared intraset correlations between a variable and the retained canonical variates. For the ith variable of $\mathbf{z}^{(x)}$, the intraset or within-set communality, h_{wi}^2, is therefore

$$h_{wi}^2 = \sum_{k=1}^{t} s_{ik}^2$$

where s_{ik} is the intraset correlation of variable i with the kth canonical variate of $\mathbf{z}^{(x)}$.

Likewise, for the within-set communality of the hth variable of $\mathbf{z}^{(y)}$, we have

$$h_{wh}^2 = \sum_{k=1}^{t} s_{hk}^2.$$

Interset communalities. There are calculated as the sum of squared interset correlations between a variable and the retained canonical variates. The interset or between-set communality, h_{bi}^2, of the ith variable of $\mathbf{z}^{(x)}$ with the retained canonical variates of $\mathbf{z}^{(y)}$ is therefore given by

$$h_{bi}^2 = \sum_{k=1}^{t} s_{ik}^2,$$

where s_{ik} denotes the *interset* correlation of variable i of $\mathbf{z}^{(x)}$ with the kth canonical variate of $\mathbf{z}^{(y)}$.

In a similar way, the interset communality of variable h of $\mathbf{z}^{(y)}$ with the retained canonical variates of $\mathbf{z}^{(x)}$, is

$$h_{bh}^2 = \sum_{k=1}^{t} s_{hk}^2.$$

Interset communalities may be interpreted as the proportions of variance which the variables of one set have in common with the space spanned by the canonical variates of the other set. The information provided is useful in determining rank and in deciding which variables of either set to retain or delete in efforts to enhance the overall explanatory power or fit of a canonical model.

3.2.6 Concluding remarks

We have seen that part of the interpretive function previously associated with canonical correlation coefficients and canonical weights is now effectively performed by other quantities. Redundancy provides a measure which is in some ways a more satisfactory expression of the interrelatedness of two sets of variables than a canonical correlation coefficient. Similarly, the correlations between a canonical variate and the original variables often prove to be more trustworthy indicators of the variables which contribute most to the variate than canonical weights. This follows because structure correlations are largely free from the instability of the weights and the partialling out of variables which are consequences of near-multicollinearity, as well as from the effects of suppression and related phenomena. Consequently, structure correlations exhibit a simple, direct relationship between a variable and a canonical variate which is free from the effects of other variables and canonical variates. On the other hand, this heavy reliance on the product-moment correlation coefficient could in itself cause concern. Tukey (1969), for example, has expressed scepticism concerning the value of the correlation coefficient in data analysis. The product-moment correlation coefficient is a fragile statistic in the sense that it is exquisitely sensitive to distributional peculiarities in the data (Kowalski, 1972; Devlin, Gnanadesikan & Kettenring, 1975; Gnanadesikan, 1977, p. 275; Thissen, Baker & Wainer, 1981). Consequently, it may be less informative than at first appears if not actually misleading (e.g., see Chambers, Cleveland, Kleiner & Tukey, 1983, p. 76). A robust estimator of the correlation coefficient (Gnanadesikan, 1977, p. 131; Bebbington, 1978; Titterington, 1978), however, may well prove more acceptable than the usual estimator.

We have also suggested that variable communalities may be useful indices of the extent to which the variance of a variable is accounted for by an analysis. Furthermore, total redundancy and variable communalities can be helpful in establishing the rank t of the model judged to best satisfy the joint requirements of explanatory power and parsimony. Under certain conditions, however, the number of statistically significant canonical correlations might be used for this purpose.

Effort continues to be directed towards strengthening and extending canonical analysis. In Sect. 3.3 which follows other developments in this area are reviewed.

Further reading. The task of *interpreting* canonical variates has received considerable attention. The use of *correlation coefficients* between the canonical variates and the observed variables in interpretation was suggested by Bargman (1962) and Meredith (1964). Cooley and Lohnes (1971, p. 33) have emphasized the value of the intraset correlation coefficients in this way, while Bock (1975, p. 393) has remarked that the interset correlations could well prove to be even more informative. The concept of *redundancy* is generally ascribed to Stewart and Love (1968) and to Miller (1969), working independently, but appears to have been arrived at by Glahn (1968, 1969), also working independently. Dawson-Saunders (1982) has shown that estimates of redundancy and total redundancy are biased and has proposed bias correction factors. Darlington, Weinberg and Walberg (1973), Miller (1975), Gleason (1976), Dawson (1977), van den Wollenberg (1977), DeSarbo (1981), Johansson (1981), Muller (1981), Tyler (1982), Dawson-Saunders and Tatsuoka (1983) and Van de Geer (1984) have all discussed or extended the notion of redundancy. Weiss (1972), in a general review of canonical analysis, has given useful accounts of the interpretive functions of structure correlation coefficients and redundancy.

3.3 Extensions and generalizations

The value of canonical analysis is sometimes questioned because of problems of interpretability (e.g., see Dempster, 1969, p. 179; Johnston, 1972, p. 333; Mulaik, 1972, p. 423; Marriott, 1974, p. 31; Kendall, 1975, p. 69; Ottestad, 1975; Kendall & Stuart, 1976, p. 310; Reyment, 1979; Cohen, 1982; DeSarbo, Hausman, Lin & Thompson, 1982). W.T. Williams (1976, p. 66) has gone further in expressing the view that canonical correlation analysis cannot be said to have "been useful in agriculture, or indeed in biology in general." Canonical analysis, however, is the focus of much current research and proposals for strengthening and extending the method continue to be put forward. Activity has been stimulated in particular by difficulties inherent in the interpretation of canonical correlation coefficients and canonical weights. In this section we review attempts to overcome or circumvent these and other difficulties and describe also the generalization of canonical analysis to more than two sets of variables. We first mention, however, an alternative to canonical analysis put forward by van den Wollenberg in 1977.

3.3.1 Redundancy analysis: an alternative to canonical analysis

A canonical correlation coefficient neglects to take account of the within-set variance associated with the canonical variates to which it refers. For certain purposes this omission might be undesirable and led to the development of the concept of redundancy (Stewart & Love, 1968). Unlike a canonical correlation coefficient, redundancy takes account of both the within-set and the between-set variance associated with a given canonical variate. Redundancy therefore is a two-part explained-variance index. Moreover, redundancy is a measure of direct relationship, being asymmetric between domains. Recently, van den Wollenberg (1977) has suggested maximizing redundancy itself, rather than the canonical correlation coefficient or simply the between-set component of explained-variance. The procedure, which is referred to as *redundancy analysis*, is an appealing one. Van den Wollenberg's idea has been further developed by Johansson (1981), while relationships between redundancy analysis, canonical correlation analysis and multivariate regression have been investigated by Muller (1981).

The concept of redundancy has been sharpened and clarified by Tyler (1982) in a unified treatment of the topic. The *most predictable surface mapping* of Lee (1981) embodies a similar concept since it is directed towards maximizing both the canonical correlation coefficient and the variance of one of the associated conjugate pair of canonical variates.

3.3.2 Improving the interpretability of canonical weights

Among the causes of difficulty in the interpretation of canonical weights are dependencies of a linear or almost linear nature among the variables comprising **x** or **y**, errors of measurement in the observations and sample-specific variation and covariation. Steps for dealing directly or indirectly with one or other of these conditions have been put forward by Bargmann (1962), Meredith (1964), Campbell (in Phillips, Campbell & Wilson, 1973), Carney (1975), Vinod (1976), Maxwell (1977), Campbell (1980a), and Campbell and Palmer (1984), amongst others.

Variable selection. The results of analysis may be influenced profoundly by the choice of variables for study. Strong correlation among the variables of either or both sets is a major source of instability in the canonical weights. Instability refers to the sampling variation of the weights over repeated samples, and, indeed, is commonly an indication that one or more variables are redundant. Consequently, the elimination of highly correlated variables in **x** and **y** can be expected to enhance stability and interpretability. For the same reason, the addition of variables which happen to be highly correlated with those originally selected can dramatically affect the magnitude and even the sign of the weights. Moreover, as the number of variables increases relative to sample size, the stability of the weights also deteriorates.

In considering variable selection it will be helpful to distinguish between selection before and following a preliminary analysis. The initial choice of variables is, of course, entirely a matter for substantive judgement. Variables which are merely parallel forms of one another are best considered as candidates for exclusion at this stage. This remark is equally applicable to other pairs of highly correlated variables belonging to the same set. Inspection of the correlation matrices $\mathbf{R}_{11}$ and $\mathbf{R}_{22}$ can be a helpful preliminary step in the identification of such variables. Following analysis it is pertinent to inquire whether some subset of the chosen variables could be eliminated without markedly affecting the overall result. In effect we may wish to determine whether there exist subsets of the variables of each domain on which the remaining variables of the same domain are linearly dependent. In certain circumstances the canonical weights themselves might aid in selection. Variables associated with weights which did not depart appreciably from zero might be regarded as candidates for exclusion. Formal tests of significance are also available. McHenry (1978) has described a procedure for computing a best subset of variables based on a likelihood ratio test to be described shortly (Sect. 3.4.3). The practical value of formal tests of this kind has, however, been questioned by Campbell (1980a) in connection with canonical variate analysis. Alternative criteria for variable selection have been proposed by McCabe (1975), Campbell and Reyment (1978), Campbell (1980a) and McKay and Campbell (1982a).

Collinearity and its treatment. The term collinearity refers to the presence of an exactly or an almost exactly linear relation among two or more variables. Where strong linear relationships exist among the variables comprising **x** and **y** the canonical weights will be ill-determined. That is to say, small changes in the data matrices **X** and **Y** will induce large changes in the weights.

The *detection* of collinearity in itself is not necessarily straightforward. Dependencies between three or more variables, for example, are very difficult to perceive by visual inspection of $\mathbf{R}_{11}$ and $\mathbf{R}_{22}$. The *condition number* $\varkappa(\mathbf{X})$ of a data matrix $\mathbf{X}$ of full rank, however, is a sensitive indicator of such dependencies. The condition number of $\mathbf{X}$ $(N \times p)$ is given by

$$\varkappa(\mathbf{X}) = d_{max}/d_{min},$$

where d_{max}, d_{min} are the largest and smallest singular values of $\mathbf{X}$. The index is interpretable as an error magnification factor, large values warning that an inaccurate solution to the problem in hand may be forthcoming. Thus $\varkappa(\mathbf{X})$ and $\varkappa(\mathbf{Y})$ will be informative in detecting collinearity. Moler (1978) has described the nature and properties of the condition number of a matrix. Where collinearity is diagnosed, the *singular value analysis* of Lawson and Hanson (1974, p. 196) is an effective technique for identifying those columns of the matrix that are nearly linearly dependent.

In multiple regression it is known that the stability and interpretability of regression weights in the presence of collinearity is sometimes improved by the use of ridge estimates rather than their ordinary least squares counterparts (Hoerl & Kennard, 1970; Marquardt & Snee, 1975; Vinod, 1978). Ridge estimates, though slightly biased, have greatly reduced error variances in comparison with least squares estimates. Carney (1975), Vinod (1976), Campbell (1980a), and Campbell and Palmer (1984) have extended the principle of ridge regression to canonical analysis. The resulting weights are referred to as *canonical ridge weights* and may be obtained by replacing the covariance matrix $\mathbf{S}$ of (2.1) by the matrix $\mathbf{S}^*$ where

$$\mathbf{S}^* = \begin{bmatrix} \mathbf{S}_{11} + \mathbf{C}_x & \mathbf{S}_{21} \\ \mathbf{S}_{21} & \mathbf{S}_{22} + \mathbf{C}_y \end{bmatrix},$$

$\mathbf{C}_x = \mathrm{diag}(c_{x1}, \ldots, c_{xp})$, $\mathbf{C}_y = \mathrm{diag}(c_{y1}, \ldots, c_{yq})$ and the c_{xi} $(i = 1, \ldots, p)$ and c_{yh} $(h = 1, \ldots, q)$ are small positive numbers. Then the matrix

$$(\mathbf{S}_{11} + \mathbf{C}_x)^{-\frac{1}{2}} \mathbf{S}_{12} (\mathbf{S}_{22} + \mathbf{C}_y)^{-\frac{1}{2}}$$

rather than (2.20) provides a starting point for analysis. By suitably choosing the c_{xi} and c_{yh} various solutions arise. On setting $c_{xi} = c_1$ for all i and $c_{yh} = c_2$ for all h an *ordinary* ridge solution is obtained. Where, in contrast, the c_{xi} and c_{yh} are in each case not equal to a specified constant a *generalized* ridge solution arises. By letting $c_{xi} = c_{yh} = 0$ for all i, h the usual solution results. Details concerning possible choices for the c_{xi} and c_{yh} and of the evaluation of the ridge weights themselves have been given by Campbell (1980a) and Campbell and Palmer (1984). Canonical ridge weights have been shown to be an improvement over the usual canonical weights, at least in certain cases. The procedure leaves the canonical correlation coefficients and variates substantially unchanged.

The literature on ridge techniques is growing rapidly. There is, however, no consensus as to the merit of the procedure. Alldredge and Gilb (1976) give a comprehensive bibliography up to the mid-1970s. For an introduction to the more recent literature see Campbell and Reyment (1978), Vinod (1978), Willan and Watts (1978), Hoerl and Kennard (1981), Draper and Van Nostrand (1980), Campbell (1980a), Smith (1980), Vinod and Ullah (1981), and Campbell and Palmer (1984).

Collinearity can be entirely eliminated by an *orthogonal transformation* of the observed variables of either or both sets prior to analysis. The observed variables may then be replaced by, for example, their principal components or some subset of them before embarking on canonical analysis (e.g., see Phillips, Campbell & Wilson, 1973). Although the procedure is straightforward, it sometimes has the disadvantage that the substantive meaning of the components themselves is not clear. This inevitably leads to difficulties in the interpretation of the canonical analysis (Kendall, 1975, p. 97). For this reason Ottestad (1975) has proposed the use of an alternative set of orthogonal functions termed *regression functions* which may be more readily interpreted. DeSarbo, Hausman, Lin and Thompson (1982) have suggested *constrained canonical analysis* as a means of combatting collinearity and coefficient instability.

In the presence of collinearity, canonical weights are often highly dependent on the particular sample in hand; addition or deletion of one or more observations then produces marked changes in the value of some or all of the weights, including changes in algebraic sign. One solution which is sometimes practicable in these circumstances is to *increase* sample size. Another possibility is to *discard* one or more of the variables which appear to be responsible for the collinearity. Delection of variables can sometimes be achieved without markedly affecting the overall fit of a model. More is said about sample-specific variation in Sect. 3.3.4 below.

Allowing for measurement error. The effect of errors of measurement on the dependence of two sets of variables has been investigated by Meredith (1964). Provided the error variances are known or can be estimated reliably, then the canonical correlations and variates can be *corrected for attentuation* due to measurement error. The procedure calls for replacing the scalar-products matrix $\mathbf{S}$ of (2.1) by the reliability adjusted matrix $\mathbf{S}^* = \mathbf{S} - \mathbf{\Delta}$, where $\mathbf{\Delta}$ is a $(p+q) \times (p+q)$ diagonal matrix of error variances of the variables in $\mathbf{x}$ and $\mathbf{y}$. Correction for measurement error can dramatically improve the outcome of an analysis. A comparable suggestion which in effect involves adjusting the *off-diagonal* elements of $\mathbf{S}$ has been made by Darlington, Weinberg and Walberg (1973). Even where the error variances are unknown and the sample size too small for their satisfactory estimation, steps can still be taken to correct for the effects of errors of measurement. In situations of this kind, Meredith (1964) has proposed performing a *factor analysis* on each set of variables separately before going on to conduct a canonical analysis across the common-factor spaces of each domain. In so doing, the unique (specific plus error) variance of each system is effectively separated from the non-error variance and removed prior to canonical analysis.

In the context of agricultural experiments, it has been suggested that principal coordinates analysis rather than factor analysis be used for this purpose (W.T. Williams, 1976, p. 76). The transformed variables corresponding to the smaller latent roots are discarded and the remaining principal coordinates used as the input for canonical analysis. The procedure is referred to as *canonical coordinates analysis.* Williams (1976) has remarked that "this version [of canonical analysis] has invariably proved informative" and in practice has been found more profitable than a direct canonical analysis of the original variables. Except perhaps in the rather special case of designed experiments, however, canonical analysis across the common-factor space of each domain would be more appropriate; in general, it would be incorrect to assume, as often seems to be the case, that error variance is reflected only in the smaller roots of a components or comparable dimension-reducing analysis.

The effect of measurement error on ridge estimates in multiple regression has been considered by Maxwell (1977), whose remarks are applicable to canonical ridge estimates also. Maxwell (1977) pointed out that as the elements of a matrix are seldom given exactly, even where a matrix is deficient in rank, it is unlikely that its 'approximation' as realized in a data matrix will similarly be rank deficient. It is assumed here that the preliminary precaution of excluding variables which are merely parallel forms of each other has been exercised. In other words, where errors of measurement are appreciable, poorly conditioned matrices of an extreme kind will occur only infrequently. In such cases multicollinearity will escape detection. The practical value of ridge estimation in these circumstances is plainly nil. Maxwell therefore suggested a prior *factor analysis* of the data before embarking on regression analysis. This approach was shown to be more realistic in its underlying assumptions than ridge regression, as well as being unaffected by nearzero roots in the matrix of intercorrelations between the predictors (Lawley & Maxwell, 1973; Maxwell, 1977). For these reasons, where high unique variances are likely, the use of a factor model to estimate regression or canonical weights may in fact provide a more effective means of dealing with collinearity than ridge estimation.

An alternative to canonical weights. Bargmann (1962) and Meredith (1964) have suggested using the correlations between the original variables and the canonical variates as a means of interpreting the latter. Meredith's results have since been extended by Huynh (1975). Such *structure correlations* have the advantage over canonical weights for this purpose of smaller standard errors and greater stability in replicate samples. Nevertheless, the theoretical advantages of these two interpretive indices has not been explicated adequately. In small or medium size samples, however, or where the variables of either or both sets are highly intercorrelated, it seems likely that the use of structure correlations will be advantageous. Furthermore, correlations are generally regarded as being easier to interpret than regression or canonical weights. Tukey (1969) has, however, questioned the value of the correlation coefficient in data analysis precisely because, as a dimensionless quantity, he considers its interpretation to be far from clear. In addition, Harris (1975, p. 55) and Campbell (1980a) have expressed doubts about the worth of variable/canonical variate correlations in canonical analysis.

As an alternative to both structure correlations and standardized canonical weights, the *semistandardized regression coefficient* of Stavig (1977) and Stavig and Acock (1981) may sometimes be useful. Semistandardized regression coefficients are regression coefficients calculated between a standardized predictor variable and a non-standardized criterion. Unlike structure correlations and standardized canonical weights, semistandardized regression coefficients are open to a functional, rate of change interpretation. Thus, for example, if a (standardized) canonical variate v_k is incremented by one standard deviation unit a criterion variable x will increase or decrease by b'_{xv_k} units, where b'_{xv_k} denotes the semistandardized regression coefficient.

3.3.3 Rotation of canonical variates

Canonical analysis is one of a class of multivariate methods which involves solving a characteristic equation for its latent roots. For some members of this class it is well known that rotation of the composite-variables specified by the latent vectors can improve interpretability while leaving certain optimized properties of the solution unchanged. It has been shown that this is true of canonical analysis also. Cliff and Krus (1976) have demonstrated in particular that *rotation* by the normalized varimax criterion preserves (a) the sum of the canonical correlations; and (b) the total predictable variance of each measurement domain. Furthermore, rotation was found to enhance the interpretability of both the canonical weights and the structure correlations. For these reasons, Cliff and Krus have recommended rotation as a routine procedure in canonical analysis. Identical rotations are applied to both sets of weights or structure correlations. It is of interest to note that, while the sum of the canonical correlations is preserved under rotation, this quantity is more evenly distributed among corresponding pairs of canonical variates than in the unrotated solution. In statistical applications, rotation may be justified by noting that if the original canonical variates are significant, then any linear combination of them is also significant.

Rotation in canonical analysis has been advocated or used by several workers. These include Hall (1969, 1977), Krus, Reynolds and Krus (1976), Skinner (1977, 1978), Scott and Koopman (1977), Wilkinson (1977), Fornell (1979), Huba, Wingard and Bentler (1980), Reynolds and Jackosfsky (1981) and Bentler and Huba (1982). Cohen (1982), on the other hand, has questioned the appropriateness of the rotation of canonical variates. Scott and Koopman (1977) have explored in some detail the possibility of employing *non-identical* transformations of each set of canonical variates, while Huba, Wingard and Bentler (1980) and Huba, Newcomb and Bentler (1981) report applications involving transformations of this sort. The general objective is either to achieve maximum correspondence between the canonical variates and the original variables, or to satisfy some 'simple structure' criterion for either the canonical weights or variates.

3.3.4 Validation

Canonical analysis involves maximization of the canonical correlation coefficients with respect to the variables of a particular sample. The rationale of maximization is that it facilitates prediction. However, as the maximization is performed in relation to a particular sample, it follows that canonical analysis will capitalize on chance variation and covariation specific to the sample in question. Consequently, if canonical weights are applied to a sample other than that on which they were determined, the correlations between the resulting linear composites will be smaller than the original canonical correlations and often appreciably so. In other words, weights which are optimal with respect to one sample may not have good general validity. The smaller the sample and the larger the number of variables, the greater the likelihood of spurious relationships. In order to determine the extent of sample-specific covariation and so have some check on the wider validity of an analysis, *cross validation* and the *jackknife* are useful devices.

Cross-validation. Cross-validation refers to the procedure of dividing a sample randomly into two subsamples, not necessarily of equal size, performing the analysis in question on one subsample and using the second subsample to assess the stability of the results of the analysis. Usually about 25% of the total sample is set aside as the holdout group. Canonical analysis may then be performed on the larger, development sample. The resulting canonical weights are applied to the standardized variables of the holdout group. The correlation coefficients between corresponding pairs of linear composites are calculated and compared with the canonical correlation coefficients of the development subsample. Some diminution is to be expected, but unless reasonable consistency between the two sets of coefficients is found, any generality beyond the sample which the results may possess is unsubstantiated.

Cross-validation makes use of only part of the available data for evaluating the canonical correlations, weights and variates. Thus a serious deficiency of cross-validation is that it is liable to induce the condition of instability which it is designed to detect. However, provided the sample is large, an alternative procedure free from this defect is available. This is *double cross-validation* (Mosier, 1951). Double cross-validation calls for splitting the sample in half and performing independent analyses on each subsample. In the context of canonical analysis, the resulting canonical weights are then each applied to the subsample on which they were not developed. Finally, the ensuing canonical correlation coefficients, weights and variates for both subsamples are compared. Where agreement is close, the wider validity of the results in enhanced. Moreover, a canonical analysis based on the total sample would then be justifiable and would provide the best possible estimates of the population values, given the available data.

The role of cross-validation or double cross-validation in canonical analysis has been discussed by Alpert and Peterson (1972), Thorndike and Weiss (1973, 1983), Thorndike (1976, 1977, 1978), Wood and Erskine (1976), Huba, Wingard and Bentler (1980) and Lee, McCabe and Graham (1983).

The jackknife. Cross-validation calls for the availability of a comparatively large sample. Jackknife methods are based on the principle of sample *re-use* and are independent of sample size. The idea underlying jackknife methods is to perform analyses repeatedly, dropping a subset of the sample at each analysis. The result of a jackknife analysis is an unbiased, or nearly unbiased estimator for the parameter of interest and its associated approximate confidence interval. Bias correction factors and variance estimators are usually arrived at by computing and averaging quantities related to the statistic of interest over *all* subsamples of size $N-1$ out of a sample of size N. An adaptation of this procedure which involves dropping single degrees of freedom rather than individual samples which has application in canonical analysis has been described by Dempster (1966). Bias correction factors for the canonical correlation coefficients r_k were obtained by Dempster and can be evaluated for the $\{\mathbf{a}_k, \mathbf{b}_k; u_k, v_k\}$ also.

Dempster (1969, p. 257) has also described a jackknife method for assessing stability in canonical variate analysis. The method calls for the usual analysis based on all N samples to be performed. The results obtained are summarized graphically by projecting sample-points onto the plane defined by the canonical variates u_1 and u_2. Further canonical analyses are then performed re-using the original sample. In all a total of N such re-analyses is made, each based on $N-1$ samples and involving the omission of a different sample. The canonical variate scores of the omitted samples are then estimated using the weights from the corresponding analysis based on $N-1$ samples. In this way the deleted samples are each treated as if they constituted a future sample observation. The final step involves inserting the omitted samples by means of their estimated scores on the original canonical representation. The point-scatters corresponding to the initial analysis and the re-analyses, together with their centroids and concentration ellipses, are then compared. The similarity of the configurations provides an indication of the extent of the stability of the results of the original analysis in relation to a future sample. Computationally, the method at first sight appears to be demanding, involving as it does a total of N canonical analyses. Dempster (1969, p. 259), however, has developed an efficient computational strategy for its implementation.

A similar validation procedure for use in multiple regression was described by Gollob (1967). There are grounds for supposing therefore that the method might be extended to applications of canonical analysis. Drehmer and Morris (1981) have given an algorithm for implementing Gollob's (1967) procedure. For discussion of cross-validation or the jackknife in a more general setting, reference may be made to Miller (1974), Stone (1974), Geisser (1975), Mosteller and Tukey (1977), Efron (1979b, 1982), and Matloff (1980). A bibliography of the jackknife has been prepared by Parr and Schucany (1980).

3.3.5 Predicting a criterion of maximum utility

One of the principal tasks of canonical analysis is to identify the most predictable criterion. Cronbach (1971) and Cassie (1972) have pointed out, however, that the value of a criterion is not necessarily related to one's ability to predict it; the goal should rather be the most *useful* criterion. Thus a potential conflict may exist between the objective of greatest utility and the mathematics of least-squares prediction. These ideas have resulted in efforts to maximize the criterion of greatest practical value rather than the most predictable criterion in some applications of canonical analysis. For this purpose it is necessary to have a reasonably clear idea of the composition of the criterion composite required or expected based on either theoretical insight or practical experience. Canonical analysis may then be used to find the subset of predictor variables which best approximates the desired criterion. Thorndike (1977) has described a *canonical prediction strategy* for this purpose. The technique involves an empirical stepwise approach to the selection of predictor variables which results in those predictors which best predict the desired criterion being identified and retained.

A more rigorous approach to model construction is available when the criterion variables can be ordered logically. This is the *step-down analysis* of J. Roy (1958). Step-down analysis enables non-informative and redundant dependent variables to be discarded. At each stage the unique contribution of one variable to the analysis is estimated and tested, providing a basis for either retaining or partialling out the variable.

3.3.6 Generalizations of canonical analysis

Several generalizations of canonical analysis to more than two data matrices have been proposed. Broadly, such generalizations are of two kinds, namely those in which (a) one or more of the m ($m>2$) data matrices consists of variables whose effects are to be *removed* before considering relationships between the two measurement domains of primary interest; and those in which (b) the status of all m matrices is alike. Generalizations of quite different kinds have been put forward by Brillinger (1969), Miyata (1970), and Van de Geer (1971).

Partial, part and bipartial canonical analysis. S.N. Roy (1957, p. 26) and Timm and Carlson (1976) have shown how relationships between two sets of variables $\mathbf{x}$ and $\mathbf{y}$ can be investigated after first partialling out the effects of a third set of variables $\mathbf{z}$ from both $\mathbf{x}$ and $\mathbf{y}$. The procedure, which can be regarded as a straightforward extension of the principle of simple partial correlation to canonical analysis, is known as *partial* canonical correlation analysis. The objective is to find linear combinations $u=\mathbf{a}^t\mathbf{e}_x$ and $v=\mathbf{b}^t\mathbf{e}_y$ of unit variance of the variables $\mathbf{e}_x=\mathbf{x}-\hat{\mathbf{x}}$ and $\mathbf{e}_y=\mathbf{y}-\hat{\mathbf{y}}$, where $\mathbf{e}_x$ and $\mathbf{e}_y$ are residual vectors obtained from the regression of $\mathbf{x}$ on $\mathbf{z}$ and of $\mathbf{y}$ on $\mathbf{z}$, such that the simple correlation between u and v is maximized. The resulting partial canonical correlations and variates thus refer to the residuals after the effects of a third set of variables is removed. Two modifications of the technique give rise respectively to *part* and *bipartial* canonical analysis. Part canonical analysis deals with relationships between $\mathbf{x}$ and $\mathbf{y}$ after the linear effects of variables in $\mathbf{z}$ have been removed from $\mathbf{x}$ but not $\mathbf{y}$. Bipartial canonical analysis, on the other hand, provides a means of assessing the linear relationship between $\mathbf{x}$ and $\mathbf{y}$ after first removing the effects of variables in $\mathbf{z}$ from $\mathbf{x}$ and of variables in $\mathbf{w}$ from $\mathbf{y}$. Tests of the significance of the partial, part or bipartial independence of $\mathbf{x}$ and $\mathbf{y}$ can be made provided the usual requirements of independence and multivariate normality are met.

The concepts of partial, part and bipartial canonical correlation analysis have been discussed by B.R. Rao (1969), Cooley and Lohnes (1971), Timm (1975) and Cohen (1982), and further generalized by Lee (1978, 1979).

Relationships among *m* sets of variables ($m > 2$). This problem has been considered by Horst (1961a, b), McKeon (1965), Carroll (1968), Kettenring (1971), Gnanadesikan (1977, p. 69) and Van de Geer (1984). Horst suggested selecting a linear composite of the variables of each set such that the *sum* of the correlations between all the linear composites is maximized. More recently, Carroll has proposed finding sample canonical variates and an auxiliary sample variate such that the *sum* of the squared correlations between the canonical variates and the auxiliary is a maximum. Both methods reduce to Hotelling's classical procedure when $m = 2$. Kettenring (1971) has discussed these and some related proposals for *generalized* canonical analysis.

For further discussion of generalized canonical analysis, Carroll and Chang (1970), Lebart, Morineau and Tabard (1977), Gower (1977), and Green (1978, p. 283) may be consulted.

Maximizing the covariance between two measurement domains. The possibility of optimizing some function other than the correlation between linear transformations of the variables of two measurement domains was encountered above in connection with redundancy analysis (Sect. 3.3.1). A further suggestion of the same kind has been made by Van de Geer (1971, p. 169) and developed somewhat by Muller (1980b). Van de Geer proposed maximizing the *covariance* between linear composites of two sets of variables, rather than the correlation between them. For this purpose it is necessary to scale the vectors of canonical weights such that

$$\begin{aligned} \mathbf{A}^t \mathbf{A} = \mathbf{A}\mathbf{A}^t = \mathbf{I}, \\ \mathbf{B}^t \mathbf{B} = \mathbf{B}\mathbf{B}^t = \mathbf{I}, \end{aligned} \tag{3.8}$$

rather than according to (2.6), as is usual. Relations (2.4) and (3.8) together correspond to an orthogonal rotation of the coordinate axes of x-space and y-space, respectively, to the canonical variates. This is in contrast to the more general transformation specified by (2.4) and (2.6) which, in addition to rotation, also implies shrinking or stretching the dimensions of the respective spaces. The solution has a number of interesting properties which have been explored by Van de Geer (1971), Green (1978, p. 280) and Johansson (1981). See also Brillinger (1981, p. 371). The procedure has application in the context of pattern matching, as for example in the rotation of one matrix towards a fixed 'target' matrix and in mapping the joint trend common to $p \geq 2$ spatially distributed variables (Lee, 1981).

Canonical analysis of complex variables. Miyata (1970) has generalized canonical analysis to *complex-valued* variables in a study of linear relationships between the time histories of two collections of points at sea-level (marigrams). The theory developed is applied in the frequency domain to investigate the power spectra of time series. We note also that the application of canonical analysis to time series has been discussed by Brillinger (1969, 1981, p. 374),

Robinson (1973), and Akaike (1976). Saxena (1978) has prepared a bibliography of complex multivariate statistical analysis.

3.3.7 Concluding remarks

The problem of interpreting canonical variates remains to some extent unresolved. In particular, the relative interpretive value of correlations between canonical variates and observed variables, on the one hand, and of standardized canonical weights on the other, requires clarification. It is frequently the case that these indices do not support consistent interpretations. N.A. Campbell (1981b) has pointed out that where a vector of canonical *weights* reflects a contrast between variables, the correlations between the canonical variates and the original variables will be misleading. For this reason, and because of the well-known instability of the standardized ordinary weights, ridge-type or shrunken weights (Campbell, 1980a; Campbell & Palmer, 1984) may well prove more informative than either structure correlations or standardized weights.

It seems likely that standardized weights, shrunken weights and structure correlations may each offer advantages under different conditions. For example, where analysis is based on variables or linear combinations of variables which are uncorrelated, the usual standardized weights may be perfectly satisfactory. In contrast, where the original variables are correlated to any appreciable extent it is likely that either shrunken weights or structure correlations would be advantageous. In cases where the data are subject to measurement error, however, shrunken estimators may prove to be of limited value. Moreover, the *purpose* of analysis may also influence the choice between canonical weights and structure correlations. Where the aim is prediction, weights in one form or another are likely to be preferred; where the analysis is viewed as an external factor analysis (Bartlett, 1948) to seek underlying factors common to both measurement domains, however, it would be natural to express an interest in structure correlations. Finally, in considering these various indices it is necessary to bear in mind that the distinction between canonical weights and structure correlations is not a sharp one because structure correlations themselves are open to interpretation as regression weights. More precisely, a correlation coefficient between an observed variable and a canonical variate is identically the standardized regression coefficient of the variable on the canonical variate in question (see Browne, 1979).

Until the relative merits of the indices in question are clarified it may be worthwhile to place greater emphasis on the *direct* interpretation of the point configurations which result on mapping the sample into linear subspaces associated with retained canonical variates. Moreover, on occasion an interpretation for a canonical variate may be forthcoming as a result of:

(a) visual inspection of the pattern of variation of sample scores over the sample as a whole; and, in survey applications, by
(b) examining the spatial distribution of the canonical variate on a map of the area surveyed.

3.4 Hypothesis testing

The study of relationships sometimes leads to an interest in hypotheses of one or more kinds. Among these, hypotheses of independence, dimensionality and of the contribution of particular variables to a given relationship may be pertinent in a variety of substantive contexts. The present discussion focuses on hypotheses of these three kinds.

In the context of hypothesis testing the question of underlying distributions must inevitably arise. We have assumed up to this point a sample of jointly distributed $(p+q)$-variate observations but one whose characteristics are otherwise unspecified. Provided either $\mathbf{x}$ or $\mathbf{y}$ can reasonably be regarded as having a multinormal distribution and that the samples are independent, parametric tests are available (see Anderson, 1958, p. 243). Tests developed from likelihood ratio and union-intersection principles are in use and are described below. Where joint normality is lacking, alternative procedures which do not depend on an assumed stochastic model may be feasible. Such tests are considered in Sect. 3.4.4. In connection with the assessment of distributional properties, see Gnanadesikan (1977, p. 150). Other useful sources are Mardia (1974, 1975), Mardia and Zemroch (1975), Cox and Small (1978), Small (1980), Hawkins (1981), Koziol (1982) and Royston (1983).

Where a model of rank $t<s=\min(p,q)$ is found on the basis of tests of dimensionality to adequately account for the data, residuals from the fitted model may be calculated. Study of residuals can be informative in assessing goodness-of-fit and as a check on distributional assumptions. The calculation of residuals is described in Sect. 3.4.5.

3.4.1 Independence

One of the first questions to arise in studies of two sets of variables concerns the linear independence of $\mathbf{x}(p\times 1)$ and $\mathbf{y}(q\times 1)$. Morrison (1976, pp. 91 & 253) has shown that a necessary and sufficient condition for the linear independence of $\mathbf{x}$ and $\mathbf{y}$ is that $\mathbf{\Sigma}_{12}=\mathbf{0}$, where $\mathbf{\Sigma}_{12}$ denotes the population counterpart of the sample scalar-products matrix $\mathbf{S}_{12}$ of (2.1). A test of the hypothesis

$$\mathrm{H}_0\colon \mathbf{\Sigma}_{12}=\mathbf{0} \quad \text{against} \quad \mathrm{H}_1\colon \mathbf{\Sigma}_{12}\neq\mathbf{0} \tag{3.9}$$

therefore can provide a starting point for investigating independence. Hypothesis (3.9) can be re-expressed in several equivalent forms. When $\mathbf{\Sigma}_{12}$ is null, the matrix of population regression coefficients $\mathbf{B}(q\times p)$ in the regression of $\mathbf{y}$ on $\mathbf{x}$, for example, where $\mathbf{B}=\mathbf{\Sigma}_{21}\mathbf{\Sigma}_{11}^{-1}$ will also be null. Thus, the hypotheses

$$\mathrm{H}_0\colon \mathbf{B}=\mathbf{0}, \qquad \mathrm{H}_1\colon \mathbf{B}\neq\mathbf{0} \tag{3.10}$$

are equivalent to (3.9). Similar hypotheses to (3.10) can be written for the regression of $\mathbf{x}$ on $\mathbf{y}$ in terms of $\mathbf{B}^*=\mathbf{\Sigma}_{12}\mathbf{\Sigma}_{22}^{-1}$. Furthermore, when either $\mathbf{B}$ or $\mathbf{B}^*$ is null so are the population canonical correlation coefficients ρ_k $(k=1,\dots,s)$. Hypotheses (3.9) and (3.10) can therefore also be formulated in terms of the ρ_k as

$$H_0\colon \rho_1^2=\rho_2^2=\ldots=\rho_s^2=0$$

$$H_1\colon \rho_k^2 \neq 0 \quad \text{for at least one } k. \tag{3.11}$$

Consequently, the linear independence of **x** and **y** may be examined from several viewpoints. We shall see that equivalent forms of hypothesis (3.9) may be advantageous in connection with the development of different test procedures.

Likelihood ratio test. The principle on which likelihood ratio tests are based involves maximization of the likelihood under the null hypothesis and also under the alternative hypothesis. A criterion for testing hypothesis (3.10) which is closely related to the likelihood ratio λ is Wilks' Λ-statistic, where

$$\begin{aligned}\Lambda &= \frac{|\mathbf{Q}_e|}{|\mathbf{Q}_e+\mathbf{Q}_h|} \\ &= \frac{|\mathbf{S}_{22}-\mathbf{S}_{21}\mathbf{S}_{11}^{-1}\mathbf{S}_{12}|}{|\mathbf{S}_{22}|} = \prod_{k=1}^{s}(1-\mathrm{r}_k^2).\end{aligned} \tag{3.12}$$

Here the $\mathbf{S}_{ij}$ are the scalar-products matrices defined in (2.1) while the r_k^2 are squares of the singular values of (2.20) or, equivalently, are the roots of (2.14). For a derivation of the test, see Mardia, Kent and Bibby (1979, p. 135). Λ is related to the likelihood ratio as $\lambda^{2/\mathrm{N}}$.

The quantity Λ in (3.12) has a Wilks' $\Lambda(u, \nu_h, \nu_e)$ distribution (Kshirsagar, 1972, p. 293; Timm, 1975, p. 147) with parameters:

u: order of $\mathbf{Q}_e$ and $\mathbf{Q}_h$, $\quad \nu_h$: df of $\mathbf{Q}_h$, $\quad \nu_e$: df of $(\mathbf{Q}_e+\mathbf{Q}_h)$.

For testing the regression of **y** on **x** the parameter values are $u=q$, $\nu_h=p$ and $\nu_e=\mathrm{N}-p-1$. Λ varies in the range $0\leq\Lambda\leq 1$. If Λ is sufficiently small, H_0 is rejected and the conclusion drawn that **x** and **y** are linearly dependent. If Λ is close to unity then H_1 is not noticeably more likely than H_0 and the null hypothesis (3.10) is maintained. In this case **x** and **y** may be regarded as independent. The distribution of $\Lambda(u, \nu_h, \nu_e)$ has been tabulated (Wall, 1968; Timm, 1975, p. 624). Critical values of $\Lambda(u, \nu_h, \nu_e)$ at which to reject H_0 can therefore be chosen. However, a satisfactory and often more convenient approximation to the distribution of $\Lambda(u, \nu_h, \nu_e)$, at least for moderately large N, is given by Bartlett's (1938) approximation

$$\chi^2 = -\{(\mathrm{N}-1)-\tfrac{1}{2}(p+q+1)\}\log_e \Lambda. \tag{3.13}$$

Under the null hypothesis χ^2 is distributed approximately as a chi-squared variate on pq degrees of freedom. Hypothesis (3.10) is rejected with confidence $(1-\alpha)$ if

$$\chi^2 > \chi_\alpha^2(pq),$$

where $\chi_\alpha^2(pq)$ is the upper 100α percentage point of the chi-squared distribution with pq degrees of freedom, and maintained otherwise. Bartlett's result (3.13) permits tests of hypotheses (3.9) to (3.11) to be made without the need for special tables.

The matrices $\mathbf{Q}_e(q \times q)$ and $\mathbf{Q}_h(q \times q)$ of (3.12) are known respectively as the *error* and *hypothesis* sums of squares and products (SSP) matrices in the regression of $\mathbf{y}$ on $\mathbf{x}$. These matrices are used in summarizing the test of (3.10) as a *multivariate analysis of variance*, as shown in Table 3.1.

Table 3.1. A test for H_0: $\mathbf{B}=\mathbf{0}$ in the regression of $\mathbf{y}(q \times 1)$ on $\mathbf{x}(p \times 1)$

Source	df	SSP$(q \times q)$	$\Lambda(u, \nu_h, \nu_e)$
Regression of $\mathbf{y}$ on $\mathbf{x}$	p	$\mathbf{Q}_h=\mathbf{S}_{21}\mathbf{S}_{11}^{-1}\mathbf{S}_{12}=\hat{\mathbf{B}}\mathbf{S}_{11}\hat{\mathbf{B}}^t$	$\lvert\mathbf{Q}_e\rvert/\lvert\mathbf{Q}_e+\mathbf{Q}_h\rvert$
Error	$N-p-1$	$\mathbf{Q}_e=\mathbf{S}_{22}-\mathbf{S}_{21}\mathbf{S}_{11}^{-1}\mathbf{S}_{12}$	
Total	$N-1$	$\mathbf{Q}_e+\mathbf{Q}_h=\mathbf{S}_{22}$	

A table corresponding to the regression of $\mathbf{x}$ on $\mathbf{y}$ can be obtained from Table 3.1 by simply interchanging subscripts 1 and 2, and p and q. It is not difficult to show that *both* regressions yield identical values of $\Lambda(u, \nu_h, \nu_e)$ (e.g., see Kshirsagar, 1972, p. 308).

A useful property of $\Lambda(u, \nu_h, \nu_e)$ is that it leads to convenient factorizations which may be employed in examining further hypotheses. Two such factorizations are considered below. The likelihood ratio test nevertheless does have the limitation in rejecting (3.10) of failing to indicate the origin or nature of the significant effect. Furthermore, the likelihood ratio test, unlike the union-intersection test to be considered next, is not defined for singular matrices.

Union-intersection test. A test of (3.9) can be constructed from Roy's (1957) union-intersection principle. The rationale underlying the test can be hinted at as follows.

The multivariate null hypothesis of (3.9) is first expressed as a set of simpler, univariate hypotheses where each component hypothesis can readily be tested. A criterion for these univariate hypotheses then yields a test of the original multivariate hypothesis. The procedure leads to the maximization of a test statistic which is derived from *linear combinations* of the vector-valued variables of interest.

Let $\mathbf{a}(p \times 1)$ and $\mathbf{b}(q \times 1)$ be any arbitrary non-null vectors. When H_0: $\mathbf{\Sigma}_{12}=\mathbf{0}$ is true it is obvious that $\mathbf{a}^t\mathbf{\Sigma}_{12}\mathbf{b}=0$ will be true for all $\mathbf{a}$ and $\mathbf{b}$; conversely, if $\mathbf{a}^t\mathbf{\Sigma}_{12}\mathbf{b}=0$ for every $\mathbf{a}$ and $\mathbf{b}$, H_0 will be true. Similarly, H_0 will be rejected if at least one of the univariate hypotheses:

$$H_{\mathbf{ab}}\colon\ \mathbf{a}^t\mathbf{\Sigma}_{12}\mathbf{b}\neq 0 \tag{3.14}$$

for some choice of $\mathbf{a}$, $\mathbf{b}$.

Acceptance of (3.9) is equivalent to accepting all univariate hypotheses

$$H_0\colon\ \rho(\mathbf{a}, \mathbf{b})=0 \tag{3.15}$$

for different $\mathbf{a}$, $\mathbf{b}$. A simple test statistic for (3.15) is provided by the sample estimate $r(\mathbf{a}, \mathbf{b})$ of $\rho(\mathbf{a}, \mathbf{b})$. From (2.5) we have

$$r(\mathbf{a}, \mathbf{b})=\mathbf{a}^t\mathbf{S}_{12}\mathbf{b}\{(\mathbf{a}^t\mathbf{S}_{11}\mathbf{a})(\mathbf{b}^t\mathbf{S}_{22}\mathbf{b})\}^{-\frac{1}{2}}.$$

The acceptance region for the univariate hypotheses (3.15) would be

$$\mathrm{r}^2(\mathbf{a}, \mathbf{b}) \leq \mathrm{r}_\alpha^2 \tag{3.16}$$

where r_α^2 is an appropriate critical value of the distribution of r. Therefore, the acceptance region for the original multivariate hypothesis (3.9) is the *intersection* of all regions of the kind specified by (3.16):

$$\bigcap_{\mathbf{a}, \mathbf{b}} [\mathrm{r}^2(\mathbf{a}, \mathbf{b}) \leq \mathrm{r}_\alpha^2].$$

Such a region is equivalent to that defined by the maximum for all **a**, **b** of

$$\max_{\mathbf{a}, \mathbf{b}} \mathrm{r}^2(\mathbf{a}, \mathbf{b}) \leq \mathrm{r}_\alpha^2.$$

The test statistic for the multivariate hypothesis (3.9) is this maximum value of r^2. We have seen (Sect. 2.2) that the maximum r^2 is precisely the square of the largest singular value d_1 of (2.23) or, equivalently, the largest root λ_1^2 of (2.14). Kshirsagar (1972, p. 132) has shown that the rejection region for H_0: $\mathbf{\Sigma}_{12} = \mathbf{0}$ is the *union* of the rejection regions for the component hypotheses of (3.15).

The statistic r_1^2 has the greatest characteristic root (gcr) distribution, $\theta_\alpha(s, m, n)$. The parameters s, m and n of the distribution are:

$$s = \min(p, q), \qquad m = (|p - q| - 1)/2, \qquad n = (\mathrm{N} - p - q - 2)/2.$$

The distribution of $\theta_\alpha(s, m, n)$ has been graphed (Morrison, 1976, p. 397) and tabulated (Harris, 1975, p. 300; Timm, 1975, p. 607). Thus the significance of r_1^2 is readily assessed. The null hypothesis (3.9) is accepted in a test of size α if

$$\mathrm{r}_1^2 \leq \theta_\alpha(s, m, n), \tag{3.17}$$

where $\theta_\alpha(s, m, n)$ is the upper 100α percentage point of the distribution of $\theta(s, m, n)$, and rejected otherwise. θ varies in the range $0 < \theta < 1$.

Where the union-intersection test leads to rejection of H_0, one can enquire as to which component rejection regions were responsible for the result obtained. In this way insight into the nature of the deviation from the null hypothesis may be gained.

3.4.2 Dimensionality

Where the overall hypothesis of independence is rejected it is usually worthwhile to establish the rank or dimensionality of association. This quantity is equivalent to the number of non-zero canonical correlation coefficients in the population.

A sequence of null hypotheses is of interest which may be written:

$$\mathrm{H}_0(j)\colon\ \rho_1 \neq 0,\ \rho_2 \neq 0, \ldots, \rho_j \neq 0,\ \rho_{j+1} = \ldots = \rho_s = 0, \tag{3.18}$$

for $j = 1, \ldots, s-1$, where ρ_1 is the largest population canonical correlation, ρ_2 the next largest and so on. Likelihood ratio and union-intersection tests for the sequence of hypotheses (3.18) are available, though they differ fundamentally in rationale. In the likelihood ratio test the largest root, the first two roots and so on are removed while the joint nullity of the *residual* roots is examined at each stage. In the union-intersection test the significance of the individual roots r_k^2 ($k = 2, \dots, s$) is assessed *directly*.

Likelihood ratio test. After removing j non-zero roots, a general criterion for testing the joint nullity of correlations $j+1$ to s is based on a factorization of (3.12). The test criterion Λ_j is

$$\Lambda_j = \prod_{k=j+1}^{s} (1 - r_k^2). \tag{3.19}$$

The significance of Λ_j is conveniently assessed by evaluating

$$\chi^2 = -\{(N-1) - \tfrac{1}{2}(p+q+1)\} \log_e \Lambda_j, \tag{3.20}$$

whose asymptotic null distribution is chi-squared on $(p-j)(q-j)$ degrees of freedom.

The factorization which leads to (3.19) is based on the assumption that the j roots removed do in fact correspond to non-zero population values. Thus Λ_j is strictly a test of the residuals after the effects of the preceding correlations have been removed. E.J. Williams (1967) has pointed out that while hypothesis (3.9) is seldom of overriding importance it is the appropriate null hypothesis when the effects of canonical variates assumed to represent the association between the two sets of variables have been eliminated. In this sense the hypothesis of no relationship is of central importance in hypothesis testing.

By setting $j = 1, \dots, s-1$ in (3.18) sequential tests of all but the first canonical correlation, all but the first two and so on are possible. The results are often summarized as in Table 3.2. It is important to realize that the test of $H_0(j)$ of (3.18) is provided by the *last* χ^2 value of the right-hand column of Table 3.2. The other quantities in that column are not interpretable as chi-squared variates on the degrees of freedom indicated if $H_0(j)$ is true (cf. Lancaster, 1963; Kshirsagar, 1972, p. 328).

Table 3.2. Likelihood ratio test of dimensionality of association between **x** and **y**. Sequential tests of joint nullity of the residual roots r_k^2 ($k = j+1, \dots, s$); $m = \{(N-1) - \tfrac{1}{2}(p+q+1)\}$

k	Roots, r_k^2	df	χ^2
1	r_1^2	$p+q-1$	$-m \log_e(1-r_1^2)$
2	r_2^2	$p+q-3$	$-m \log_e(1-r_2^2)$
$\vdots$	$\vdots$	$\vdots$	$\vdots$
j	r_j^2	$p+q-(2j-1)$	$-m \log_e(1-r_j^2)$
Remaining roots	$\sum_{k=j+1}^{s} r_k^2$	$(p-j)(q-j)$	$-m \log_e \prod_{k=j+1}^{s} (1-r_k^2)$
Total	$r_1^2, r_2^2, \dots, r_s^2$	pq	$-m \log_e \Lambda(u, v_h, v_e)$

For many purposes a more convenient summary of the results is provided by Table 3.3. Table 3.3 has the advantages of unambiguously displaying the sequential nature of the tests while explicitly specifying those residual roots which are the subject of the test at the jth step.

In Tables 3.2 and 3.3 acceptance of $H_0(j)$ is commensurate with concluding that any association is concentrated in the first or preceding j pairs of canonical variates. It may in this way be possible to isolate one or a small number of linear composites of the measures that describe all significant relationship between **x** and **y**. In such cases an indication of dimensionality is obtained as well as of the rank of the model to be fitted. The tests, however, other than for $j=1$ are very conservative unless the roots removed are close to 1.

Table 3.3. Likelihood ratio test of dimensionality of association between **x** and **y**. Sequential tests of joint nullity of the residual roots r_k^2 $(k=j+1, \ldots, s)$; $m=\{(N-1)-\frac{1}{2}(p+q+1)\}$

j	Roots, r_k^2	df	χ^2
0	$r_1^2, r_2^2, r_3^2, \ldots, r_s^2$	pq	$-m \log_e \Lambda(u, \nu_h, \nu_e)$
1	$r_2^2, r_3^2, \ldots, r_s^2$	$(p-1)(q-1)$	$-m \log_e \prod_{k=2}^{s} (1-r_k^2)$
2	$r_3^2, \ldots, r_s^2$	$(p-2)(q-2)$	$-m \log_e \prod_{k=3}^{s} (1-r_k^2)$
$\vdots$	$\vdots$	$\vdots$	$\vdots$
$s-1$	r_s^2	$(p-j)(q-j)$	$-m \log_e \prod_{k=s}^{s} (1-r_k^2)$

The chi-squared approximation (3.20) is due to Bartlett (1938). A slightly more accurate approximation has since been proposed by Lawley (1959):

$$-\left\{(N-1)-j-\tfrac{1}{2}(p+q+1)+\prod_{k=1}^{j} r_k^{-2}\right\} \log_e \prod_{k=j+1}^{s} (1-r_k^2),$$

whose asymptotic null distribution is chi-squared on $(p-j)(q-j)$ degrees of freedom. The accuracy of this approximation has been investigated by Glynn and Muirhead (1978).

Union-intersection test. The union-intersection test of the jth hypothesis $H_0(j)$ of (3.18) is based directly on the kth root r_k^2 $(j=k)$. The procedure is identical to the overall test (3.17) except for adjustments to the degree-of-freedom parameters. For testing the jth root the parameters are $s=\min(p-j-1, q-j-1)$, $m=(|p-q-j|-1)/2$, $n=(N-p-q-j)/2$. Thus a test of $H_0(j)$ is conducted by referring r_k^2 to a table of percentage points of the gcr distribution. The jth null hypothesis is rejected in a test of size α if

$$r_k^2 > \theta_\alpha(s, m, n),$$

where $\theta_\alpha(s, m, n)$ is the upper 100α percentage point obtained from the appropriate chart or table. The hypothesis is accepted if r_1^2 is equal to or less than $\theta_\alpha(s, m, n)$. Like the corresponding likelihood ratio tests, these tests of the r_k^2 for $k \geq 2$ are very conservative (Mardia, Kent & Bibby, 1979, p. 147).

3.4.3 The contribution of particular variables

Rejection of H_0: $\boldsymbol{\Sigma}_{12}=\mathbf{0}$ does not necessarily imply that all p variables in $\mathbf{x}$ contribute to the prediction of $\mathbf{y}$ (or *vice versa* in the case of regression of $\mathbf{x}$ on $\mathbf{y}$). To establish whether a group of variables in $\mathbf{x}$ does not add any further to the relationship between $\mathbf{x}$ and $\mathbf{y}$ once certain variables are already considered we proceed as follows.

Let the variables in $\mathbf{x}(p \times 1)$ be subdivided into two groups, $\mathbf{x}_1(p_1 \times 1)$ and $\mathbf{x}_2(p_2 \times 1)$:

$$\mathbf{x} = [\underset{p_1}{\mathbf{x}_1} | \underset{p_2}{\mathbf{x}_2}]^t \; 1$$

where $p_1 + p_2 = p$. The matrix of regression coefficients $\mathbf{B} = \boldsymbol{\Sigma}_{21}\boldsymbol{\Sigma}_{11}^{-1}$ $(q \times p)$ and the matrix of sample observations on $\mathbf{x}$ are similarly partitioned:

$$\mathbf{B} = [\underset{p_1}{\mathbf{B}_1} | \underset{p_2}{\mathbf{B}_2}] \; q, \qquad \mathbf{X} = [\underset{p_1}{\mathbf{X}_1} | \underset{p_2}{\mathbf{X}_2}] \; \mathrm{N}.$$

Since we are assuming that (3.9) has been rejected, it is clear that the whole of $\mathbf{B}$ cannot be null. It may nevertheless be of interest to establish whether some submatrix $\mathbf{B}_2$ of $\mathbf{B}$ is null. A test of the hypothesis

$$H_0: \; \mathbf{B}_2 = \mathbf{0}$$

will then be informative. A suitable test criterion can be constructed from the hypothesis SSP matrix $\mathbf{Q}_h$ of Table 3.1. The test requires $\mathbf{Q}_h$ to be partitioned into two components, one corresponding to the regression of $\mathbf{y}$ on $\mathbf{x}_1$ alone and the second to the remainder. In Sect. 3.4.1 we saw that in the regression of $\mathbf{y}$ on $\mathbf{x}$ the regression SSP matrix is given by $\mathbf{Q}_h = \mathbf{S}_{21}\mathbf{S}_{11}^{-1}\mathbf{S}_{12}$. In a similar way, in the regression of $\mathbf{y}$ on $\mathbf{x}_1$ the regression SSP matrix, $\mathbf{Q}_h^*$ say, is given by

$$\mathbf{Q}_h^* = \mathbf{S}_{21}^*[\mathbf{S}_{11}^*]^{-1}\mathbf{S}_{12}^*, \tag{3.22}$$

where the $\mathbf{S}_{ij}^*$ are scalar-products matrices computed from the first p_1 variables of $\mathbf{x}$. The matrix $\mathbf{Q}_h^*(q \times q)$ carries p_1 degrees of freedom. The corresponding multivariate analysis of variance for the regression of $\mathbf{y}$ on $\mathbf{x}_1$ is shown in Table 3.4.

Table 3.4. Contribution of particular variables. A test of H_0: $\mathbf{B}_2=\mathbf{0}$ in the regression of $\mathbf{y}$ on $\mathbf{x}$

Source	df	SSP$(q \times q)$	$\Lambda^*(u, \nu_h, \nu_e)$
Regression of $\mathbf{y}$ on $\mathbf{x}_1$	p_1	$\mathbf{Q}_h^* = \mathbf{S}_{21}^*[\mathbf{S}_{11}^*]^{-1}\mathbf{S}_{12}^*$	$\lvert\mathbf{Q}_e\rvert / \lvert\mathbf{Q}_e + \mathbf{Q}_r\rvert$
Remainder	p_2	$\mathbf{Q}_r = \mathbf{Q}_h - \mathbf{Q}_h^*$	
Regression of $\mathbf{y}$ on $\mathbf{x}$	p	$\mathbf{Q}_h = \mathbf{S}_{21}\mathbf{S}_{11}^{-1}\mathbf{S}_{12}$	
Error	$\mathrm{N}-p-1$	$\mathbf{Q}_e = \mathbf{S}_{22} - \mathbf{S}_{21}\mathbf{S}_{11}^{-1}\mathbf{S}_{12}$	
Total	$\mathrm{N}-1$	$\mathbf{Q}_t = \mathbf{S}_{22}$	

The test criterion for (4.13) is

$$\Lambda^* = \Lambda(q, p_2, N-p-1+p_2) = \frac{|\mathbf{Q}_e|}{|\mathbf{Q}_e + (\mathbf{Q}_h - \mathbf{Q}_h^*)|}.$$

Under the null hypothesis Λ^* has the Wilks' $\Lambda(q, p_2, N-p-1+p_2)$ distribution. Accordingly, hypothesis (3.21) may be tested by referring Λ^* to a table of percentage points of $\Lambda(u, \nu_h, \nu_e)$, e.g., Timm (1975, p. 624). Hypothesis (3.21) is rejected with confidence $(1-\alpha)$ if

$$\Lambda^* < \Lambda^\alpha(u, \nu_h, \nu_e),$$

where $\Lambda^\alpha(u, \nu_h, \nu_e)$ is the upper 100α percentage point of the distribution of $\Lambda(u, \nu_h, \nu_e)$. Otherwise H_0 is maintained.

3.4.4 Hypothesis tests for nonnormal data

The assumption of joint normality is often difficult to sustain in practice. Hence there is a need for tests which are free from the restrictive distributional and other assumptions of parametric tests. One class of less-restrictive tests is provided by multivariate *permutation* tests.

A permutation test is conducted by randomly permuting the sample data repeatedly. For each permutation one or more pertinent test statistics are computed. In canonical analysis these quantities could conceivably be the canonical roots r_k^2 or Wilks' criteria Λ or Λ_j. The proportion of data permutations that provide as large a value of the test statistic as that actually realized in the survey or experiment is then evaluated. This proportion itself determines the significance of the obtained result. The frame of reference for a permutation test, therefore, is not a particular distribution but a set of random permutations of the sample values themselves. In other words, statistical significance is determined directly from the observed data, not by reference to statistical tables based on random sampling from known distributions.

Permutation tests are computationally more demanding than their parametric counterparts. As the efficiency of computational methods for calculating the quantities of interest increases and computing power grows, however, multivariate permutation tests are becoming increasingly feasible. While these tests do not yet appear to have been used in applications of canonical analysis, it seems likely that permutation support will come to play a more prominent role in the future.

3.4.5 Residuals from a fitted model

Where a model of rank $t < \min(p, q)$ provides a satisfactory fit to the data, residuals from the fitted model may be calculated. The residuals themselves then provide a starting point for further analysis.

The singular value decomposition of Sect. 2.3.2 provides a convenient means of obtaining a rank t approximation of any $p \times q$ matrix (Johnson, 1963;

Good, 1969). Let us denote the scalar-products type matrix $\mathbf{S}_{11}^{-\frac{1}{2}}\mathbf{S}_{12}\mathbf{S}_{22}^{-\frac{1}{2}}$ of (2.20) by the $p \times q$ matrix $\mathbf{K}$ ($p \geq q$ without loss of generality). From (2.18) recall that $\mathbf{K}$ can be decomposed into the product of three matrices. Moreover, notice that the last $p-q$ columns of $\mathbf{P}$ in (2.18) can be dropped while preserving the equality

$$\underset{(p\times q)}{\mathbf{K}} = \underset{(p\times q)}{\mathbf{P}} \ \underset{(q\times q)}{\mathbf{D}} \ \underset{(q\times q)}{\mathbf{Q}^{\mathrm{t}}}, \tag{3.23}$$

where $\mathbf{P}$ is an orthonormal section, $\mathbf{Q}$ is orthonormal and $\mathbf{D} = \mathrm{diag}(d_1, \ldots, d_q)$ with $d_1 \geq d_2 \geq \ldots \geq d_q$. It is with this compact form of the singular value decomposition (2.18) that we choose, for convenience, to work here. Recall that the singular values d_k $(k=1, \ldots, q)$ of $\mathbf{K}$ are identically the canonical correlation coefficients r_k.

Equation (3.23) can be re-expressed as

$$\mathbf{K} = \mathrm{r}_1 \mathbf{p}_1 \mathbf{q}_1^{\mathrm{t}} + \mathrm{r}_2 \mathbf{p}_2 \mathbf{q}_2^{\mathrm{t}} + \ldots + \mathrm{r}_s \mathbf{p}_s \mathbf{q}_s^{\mathrm{t}}, \tag{3.24}$$

where the column vectors $\mathbf{p}_1, \mathbf{p}_2, \ldots, \mathbf{p}_s$ are orthonormal of length p, the row vectors $\mathbf{q}_1^{\mathrm{t}}, \mathbf{q}_2^{\mathrm{t}}, \ldots, \mathbf{q}_s^{\mathrm{t}}$ are orthonormal of length q and $s = \min(p, q)$. Equation (3.24) expresses $\mathbf{K}$ as the sum of $s p \times q$ matrices, each of rank 1. The sum of the squares of all the elements in each of the products $\mathbf{p}_k \mathbf{q}_k^{\mathrm{t}}$ $(k=1, \ldots, s)$ of (3.24) is 1, so that none of these products is numerically more or less important than any of the others. On the other hand, since $\mathrm{r}_1 \geq \mathrm{r}_2 \geq \ldots \geq \mathrm{r}_s$ it is clear that the major contributions to $\mathbf{K} = \mathbf{S}_{11}^{-\frac{1}{2}}\mathbf{S}_{12}\mathbf{S}_{22}^{-\frac{1}{2}}$ will come from terms with large r_k.

Suppose that a model of rank $t<s$ is fitted. Then the rank t approximation $\mathbf{K}_t$ of $\mathbf{K}$ is given by the sum of the t leading terms in (3.24):

$$\underset{(p\times q)}{\mathbf{K}_t} = \underset{(p\times q)}{\mathbf{P}_t} \ \underset{(q\times q)}{\mathbf{D}_t} \ \underset{(q\times q)}{\mathbf{Q}_t^{\mathrm{t}}}. \tag{3.25}$$

A residual sum of squares and cross products matrix $\hat{\mathbf{E}}$ $(p \times q)$ corresponding to the rank t model is defined as the sum of the remaining $s-t$ terms in (3.24):

$$\hat{\mathbf{E}} = \mathrm{r}_{t+1} \mathbf{p}_{t+1} \mathbf{q}_{t+1}^{\mathrm{t}} + \ldots + \mathrm{r}_s \mathbf{p}_s \mathbf{q}_s^{\mathrm{t}}.$$

By analogy with (3.25), we denote the residual matrix $\hat{\mathbf{E}} = \mathbf{P}_e \mathbf{D}_e \mathbf{Q}_e^{\mathrm{t}}$, where $\mathbf{P}_e$ is $p \times q$, $\mathbf{D}_e$ is $q \times q$ and $\mathbf{Q}_e$ is $q \times q$. The sum of squares of the diagonal elements of $\hat{\mathbf{E}}$ provides a measure of the lack of fit of the rank t model and is equal to the trace $(\hat{\mathbf{E}}^{\mathrm{t}}\hat{\mathbf{E}})$. It is not difficult to show that this index is also given by the sum of the squares of the smallest $s-t$ canonical correlation coefficients $\mathrm{r}_{t+1}^2 + \ldots + \mathrm{r}_s^2$. To see this write

$$\begin{aligned} \mathrm{tr}(\hat{\mathbf{E}}'\hat{\mathbf{E}}) &= \mathrm{tr}(\mathbf{P}_e \mathbf{D}_e \mathbf{Q}_e^{\mathrm{t}} \mathbf{Q}_e \mathbf{D}_e \mathbf{P}_e^{\mathrm{t}}) = \mathrm{tr}(\mathbf{P}_e \mathbf{D}_e^2 \mathbf{P}_e^{\mathrm{t}}) = \mathrm{tr}(\mathbf{D}_e^2) \\ &= \mathrm{r}_{t+1}^2 + \ldots + \mathrm{r}_s^2. \end{aligned}$$

The sum of squared canonical correlations r_k^2 $(k=t+1, \ldots, s)$ therefore yields a convenient index of the goodness-of-fit of the rank t model to the scaled between-set scalar-products matrix $\mathbf{K}$; the smaller is $\Sigma\, \mathrm{r}_k^2$ $(k=t+1, \ldots, s)$ the closer the fit.

The rank t approximation of $\mathbf{K}$ is optimal in the least squares sense of minimizing the sum of squared residuals (Johnson, 1963; Good, 1969; Chambers, 1977, p. 111). The matrix $\hat{\mathbf{E}}$ can itself be a starting point for further analysis and its structure investigated. Analysis of residuals may provide

(a) an indication of the extent and nature of lack of fit;
(b) a check on multinormality assumptions.

For an enlightening discussion of the analysis of multidimensional residuals, see Gnanadesikan (1977, p. 258).

Further reading. *Inference* in canonical correlation analysis has been discussed by Glynn and Muirhead (1978). The derivation of the *likelihood ratio* test of Sect. 3.4.1 has been given by Anderson (1958, p. 230), Kshirsagar (1972, p. 304), Mardia, Kent and Bibby (1979, p. 135) and by Muirhead (1982, p. 291). Timm (1975, p. 352) and Timm and Carlson (1976) have described the extension of this test to *partial, part* and *bipartial* canonical analysis. The *robustness* of the likelihood ratio tests of independence and dimensionality against nonnormality has been studied by Muirhead and Waternaux (1980). These authors have shown the tests to be extremely sensitive to departure from normality; alternative tests are proposed which are applicable when sampling from an elliptical distribution with longer tails than the normal and which involve a simple adjustment of the usual test statistics for nonzero kurtosis. The work of Muirhead and Waternaux has since been extended by Tyler (1983). The robustness of multivariate tests in a wider context has been discussed by Pillai and Hsu (1979) and Kariya (1981). The *adequacy* of sequential tests based on partitions of Wilks' Λ criterion has been assessed by Mendoza, Markos and Gonter (1978). Morrison (1976, p. 258) has discussed the *union-intersection* test of Sect. 3.4.1. A brief account of this test has been given by Mardia, Kent and Bibby (1979, p. 136) while Schuenemeyer and Bargmann (1978) and Bargmann (1979) have described extensions of the test. Olkin and Tomsky (1981) have shown how the union-intersection principle can be used to provide a *unified* treatment for the construction of multivariate tests and have described a *new class* of multivariate tests based on the union-intersection principle. The effect of moderate *nonnormality* on Roy's largest root test has been studied by Davis (1982).

In connection with the *assessment of dimensionality*, Brillinger (1967, p. 236) has suggested the use of gamma probability plots of the residuals from a fitted model for this purpose. Fujikoshi and Veitch (1979) have proposed the use of Mallows' (1973) C_p statistic for assessing dimensionality. Eastment and Krzanowski (1982) have described a cross-validatory technique based on singular value decomposition for the purpose of establishing dimensionality in principal component analysis which may also be useful in canonical analysis. Gupta and Kabe (1979) have investigated goodness-of-fit tests for *hypothesized* canonical variates. *Permutation tests* have been discussed by the Statistical Task Force to the Weather Modification Advisory Board (1978) and by Gabriel (1979). In connection with multivariate permutation tests, a class of computationally feasible procedures based on *normed distances* has been developed by Mielke and his associates (Mielke, Berry & Johnson, 1976; Mielke, 1979; O'Reilly & Mielke, 1980; Mielke, Berry, & Brier, 1981; Mielke, Berry, Brockwell, &

Williams, 1981; Mielke & Iyer, 1982; Berry & Mielke, 1983). Marasinghe and Kennedy (1982) have proposed the use of theoretical results about the distribution of individual characteristic roots to yield *computationally feasible* tests constructed from characteristic roots. *Outlier resistant* permutation tests have been the subject of an investigation by Lambert (1977).

Gleason (1976) has described the construction of *residual matrices* from a viewpoint which differs from that of Sect. 3.4.5. Residual matrices $\mathbf{X}-\hat{\mathbf{X}}$ and $\mathbf{Y}-\hat{\mathbf{Y}}$ were obtained by Gleason from 'optimal' reconstructions of $\mathbf{X}$ and $\mathbf{Y}$, respectively, based on singular value decomposition.

This chapter has been concerned with extensions and generalizations of canonical analysis. In Chaps. 4 and 5, which follow, two special cases of canonical analysis that arise under restrictions on the variables will be considered.

4. Canonical variate analysis

4.1 Introduction

In studies of dependency it is often useful to distinguish between studies of *relationship* and studies of *structure*. The term structure is used here to refer to the way in which the sample is organized or constructed. In studies of relationship it is the connections between two sets of observed or measured *variables* which are of interest. In structural investigations, on the other hand, it is affinities among *subsamples* making up a sample with respect to p response variables considered simultaneously which are of interest. For this purpose a set of q artificial variables may be created to account for the design characteristics of the survey or experiment which gave rise to the data, or to embody hypotheses concerning structural relationships among subsamples. In this chapter we shall be concerned with the use of canonical analysis in structural investigations of these kinds. The topic is a large one and we cannot hope to do more than touch upon a few selected aspects. Many of the theoretical results presented are illustrated by worked, numerical examples in Chaps. 10 and 11.

Considerable freedom exists as to the nature of the variables which may be employed in canonical analysis. While the variables may be continuous (ordinal, interval or ratio-scaled) there is no requirement that they must be so. Claringbold (1958) and Maxwell (1961), for example, have used canonical analysis with binary (0, 1) data. Moreover, it is not necessary for the variables to be observed or measured in the usual sense at all, or, indeed, that they be real-valued. Thus the class of admissible variables in canonical analysis is very wide. We shall see in this chapter and the next that this property greatly enhances the flexibility of the method.

Suppose we have N p-variate observations ${}_N\mathbf{X}_p = [\mathbf{x}_1 | \dots | \mathbf{x}_N]^t$ which fall naturally into g groups or classes. The classes may consist, for example, of plant communities, plant or animal taxa, or to different experimental treatments. In such cases interest often centers on *comparative* relations among the groups with respect to the p observed variables simultaneously. In order to study comparative relationships, it is necessary to make use in some way of the information represented by the group structure or classification. One means of so doing is to create artificial variables especially for the purpose. Such variables are known as *dummy* variables and may be used as one set of variables in a canonical correlation analysis, the observed or measured responses constituting the second set. Canonical analysis of this kind has a number of interesting and useful properties, so much so that it is often regarded as a method in its own right known as *canonical variate analysis*, following Rao (1948, p. 187). The simplest case arises in the context of the one-way design. Here there is a single variable of classification and relationships among classes are unspecified; the classification is purely nominal and the classes arbitrarily arranged. In such cases *binary-valued* dummy variables can be used to clarify relationships among the classes. In relation to more complex designs or where known or suspected relationships among classes can be specified in advance, dummy variables known as *contrasts* or *designed comparisons* are often helpful.

In the following section the nature and construction of binary-valued dummy variables are considered. In Sect. 4.3 the derivation of canonical variate analysis is outlined and in Sect. 4.4 further aspects of the method described. Hypothesis testing is taken up in Sect. 4.5, and attention drawn to connections between canonical variate analysis and other multivariate methods in Sect. 4.6. In Sect. 4.7 the incorporation into the analysis of prior information on group relationships is considered. Finally, in Sect. 48, an introduction to the rather extensive literature dealing with canonical variate analysis is given.

4.2 Binary-valued dummy variables

Variables whose realizations consist of arbitrary values to which meanings are assigned are known as dummy variables. Various coding schemes exist. The simplest scheme leads to *binary-valued* dummy variables, so called because their realization has just two possible states. In practice the alternative states are coded 0 and 1, although other numerical values might be used. An important distinction between a binary variable and a binary-valued dummy variable is that in the former there is no restriction on the number of ones or zeros, whereas in the latter all elements except one must be zero.

Where a sample consists of g distinct groups, $q=g-1$ dummy variables $z_1, \ldots, z_q$ are required to carry the information contained in the classification. A correspondence is then set up between the classes and the z's. A dummy variable, z_1, may be designated to correspond to the 'first' class, c_1, say. Each sample $\mathbf{x}_j^t$ ($j=1, \ldots, N$) would then be assigned a score or value on z_1 according to whether it belonged to c_1 or to some other class; each sample belonging to c_1 would customarily be assigned a value of 1, while the remaining samples (those not belonging to c_1) would receive a value of 0. In a similar way, a second variable, z_2, would be used to characterize the 'second' class, c_2. This would involve giving a score of 1 to each sample belonging to c_2, while non-members would receive a score of 0. The procedure would be continued in the same way until all $g-1$ dummy variables had been generated. Notice that a gth variable, z_{q+1}, is not required because samples not belonging to any of the preceding $g-1$ groups must belong to the gth category. Consequently, samples comprising the gth group are uniquely specified by a score of zero on *all* $g-1$

variables, $z_1, \ldots, z_q$. The procedure may be summarized by observing that the value of the cth dummy variable z_c $(c=1, \ldots, g-1)$ for the jth sample $(j=1, \ldots, N)$ is defined to be:

$$z_{jc}=\begin{cases}1 & \text{if the } j\text{th sample } \mathbf{x}_j^t \text{ belongs to group } c\\ 0 & \text{otherwise.}\end{cases}$$

A simple numerical example is given in Table 4.1. The example refers to N sample-units classified into $g=5$ groups designated $a, b, \ldots, e$. The group structure is expressed in coded form by the four dummy variables z_c. The set of four scores associated with each sample-unit of a given class uniquely specifies the position of the units in the classification. Notice that all n_c members of the cth group receive the same set of scores $(\Sigma_c^g n_c = \mathrm{N})$. It is sometimes instructive to regard the class specified by a score of 0 on all dummy variables as a *reference* group in terms of which all the groups may be compared and related. The choice of group which is signified in this way is unrestricted; it is not in fact necessary that it be the 'last' group.

Table 4.1. Representation of a sample of N observations classified into $g=5$ groups by $g-1$ binary-valued dummy variables z_c. All n_c members of the cth group $(c=1, \ldots, g)$ are identically coded

Group	Dummy variable			
	z_1	z_2	z_3	z_4
a	1	0	0	0
b	0	1	0	0
c	0	0	1	0
d	0	0	0	1
e	0	0	0	0

Having assigned scores for each sample on the variables $z_1, \ldots, z_q$ we now have a sample of size N specified by *two* sets of variables – a set of $q=g-1$ binary-valued dummy variables ${}_{\mathrm{N}}\mathbf{Z}_q=[\mathbf{z}_1|\ldots|\mathbf{z}_{\mathrm{N}}]^t$ corresponding to the partition of the sample and a set of p measured ecological or response variables ${}_{\mathrm{N}}\mathbf{X}_p=[\mathbf{x}_1|\ldots|\mathbf{x}_{\mathrm{N}}]^t$. It will now be apparent that comparative relations among g groups with respect to p responses can be explored by the standard application of canonical analysis.

4.3 Formulation and derivation

The essential features of canonical variate analysis are readily appreciated geometrically. For this purpose it will be helpful to have at our disposal a geometric conceptualization of a sample of N $(p+q)$-valued variates partitioned into g groups.

4.3.1 Point conceptualizations of ${}_{\mathrm{N}}\mathbf{X}_p$ and ${}_{\mathrm{N}}\mathbf{Z}_q$

Let ${}_{\mathrm{N}}\mathbf{X}_p$ denote a sample of N p-variate observations. To arrive at a geometric notion of the sample we proceed as follows.

(a) Establish a correspondence between the ith response variable in $\mathbf{x}^t(1\times p)$ and the ith axis of a rectangular coordinate system in p dimensions.
(b) Use the observations of the jth row, $\mathbf{x}_j=[x_{j1}, x_{j2}, \ldots, x_{jp}]^t$, of ${}_{\mathrm{N}}\mathbf{X}_p$ as coordinate values by means of which to locate a point P_j corresponding to $\mathbf{x}_j$ in p-space ($j=1, \ldots, \mathrm{N}$).

The position of P_j will be uniquely determined by the observed values taken by $x_1, x_2, \ldots, x_p$ in $\mathbf{x}_j$ and carries the sample information corresponding to $\mathbf{x}_j$. The procedure is illustrated diagrammatically in Fig. 4.1a, where for convenience we have taken p to be 3. The procedure is readily extended to all N samples, $\mathbf{x}_j$ ($j=1, \ldots, \mathrm{N}$), leading to a configuration of N points in p-space (Fig. 4.1b). The configuration carries information equivalent to ${}_{\mathrm{N}}\mathbf{X}_p$ itself. The sample mean vector or *centroid*, $\bar{x}=\mathrm{N}^{-1}\Sigma\,\mathbf{x}_j$ and covariance matrix $\mathbf{S}=(\mathrm{N}-1)^{-1}\Sigma(\mathbf{x}_j-\bar{\mathbf{x}})(\mathbf{x}_j-\bar{\mathbf{x}})^t$, where the summations are over $j=1, \ldots, \mathrm{N}$, specify the location and scatter of the sample in p-space and together provide a graphical summary of the sample. The scatter is conveyed by the mean-centered *concentration ellipsoid* which consists of points at most c units distant from the mean, $(\mathbf{x}-\bar{\mathbf{x}})^t\mathbf{S}^{-1}(\mathbf{x}-\bar{\mathbf{x}})\leq c$. Where $p>3$ actual summaries of this kind are not feasible. The principles involved, however, extend conceptually to spaces of any dimension ($p\geq 1$).

Where ${}_{\mathrm{N}}\mathbf{X}_p$ consists of $g\geq 2$ subsamples, the configuration will be made up of g more or less well-defined point clusters, each specified by its centroid and ellipsoid of concentration (Fig. 4.2a). We may also conceive of a separate q-dimensional configuration corresponding to ${}_{\mathrm{N}}\mathbf{Z}_q$. In this space the sample is represented by g point-clusters, the cth of which consists of n_c *coincident* points ($c=1, \ldots, g$). The objective of canonical variate analysis can be grasped from Fig. 4.2a. We seek a graphical representation of the essential differences between groups in terms of the smallest possible number of dimensions. This leads us to map the sample into that particular subspace of p-space which will be as effective as possible in sorting out the groups (see Fig. 4.2b). It so happens that the required projection is given by $u_k=\mathbf{a}_k^t\mathbf{x}$, where u_k ($k=1, \ldots, t\leq p$) is the canonical variate of the responses corresponding to $v_k=\mathbf{b}_k^t\mathbf{z}$ of the dummy variable space. We are therefore led to seek those vectors of coefficients $\mathbf{a}_k$ and $\mathbf{b}_k$ which maximize the correlation coefficient between the linear composites u_k and v_k. We saw in Sect. 2.2 that this is precisely the problem addressed by canonical correlation analysis. The notion of mapping the sample in such a way that it yields the best possible separation of the groups while simultaneously reducing dimensionality is an intuitively appealing one.

Though the axes in Fig. 4.2a are drawn orthogonally this is merely a convention which is not essential in representing the sample graphically. To appreciate this point, see Webster (1977a, Figs. 7.7 and 7.8). Indeed, the response variables $x_1, \ldots, x_p$ usually prove to be correlated in practice. Furthermore, the variances of the responses may also differ. For these reasons, distances in representations such as Fig. 4.2a are properly assessed by a *generalized* dis-

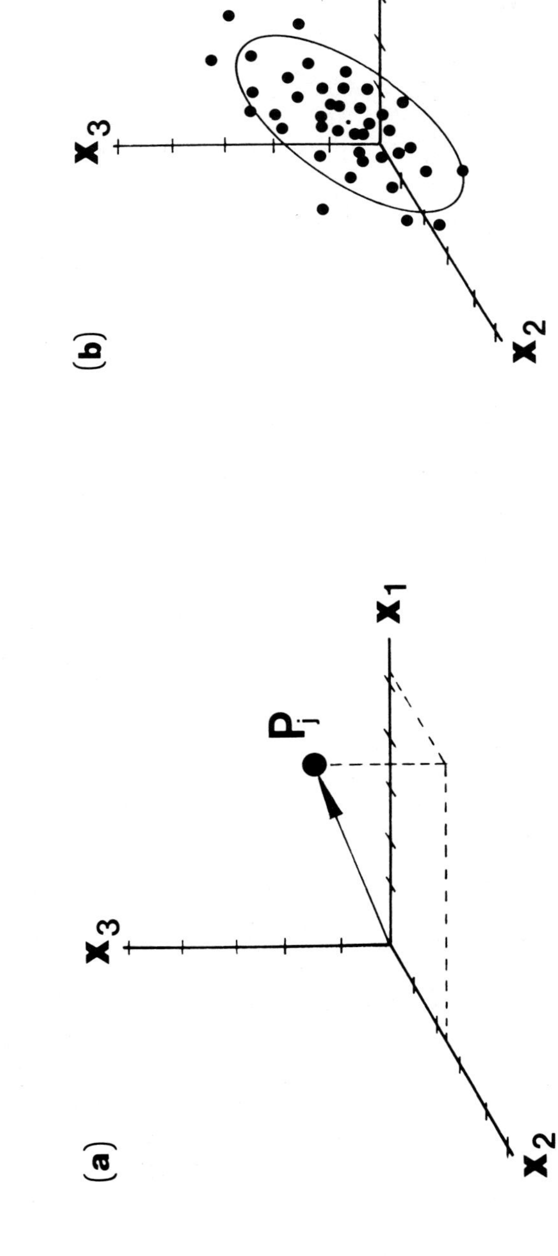

Fig. 4.1. Point conceptualization of **(a)** the *j*th sample, $\mathbf{x}_j$ represented by the point P_j; and **(b)** the data matrix $\mathbf{x}_j$ ($j=1, \ldots, N$) with sample mean (·) and mean-centered concentration ellipsoid

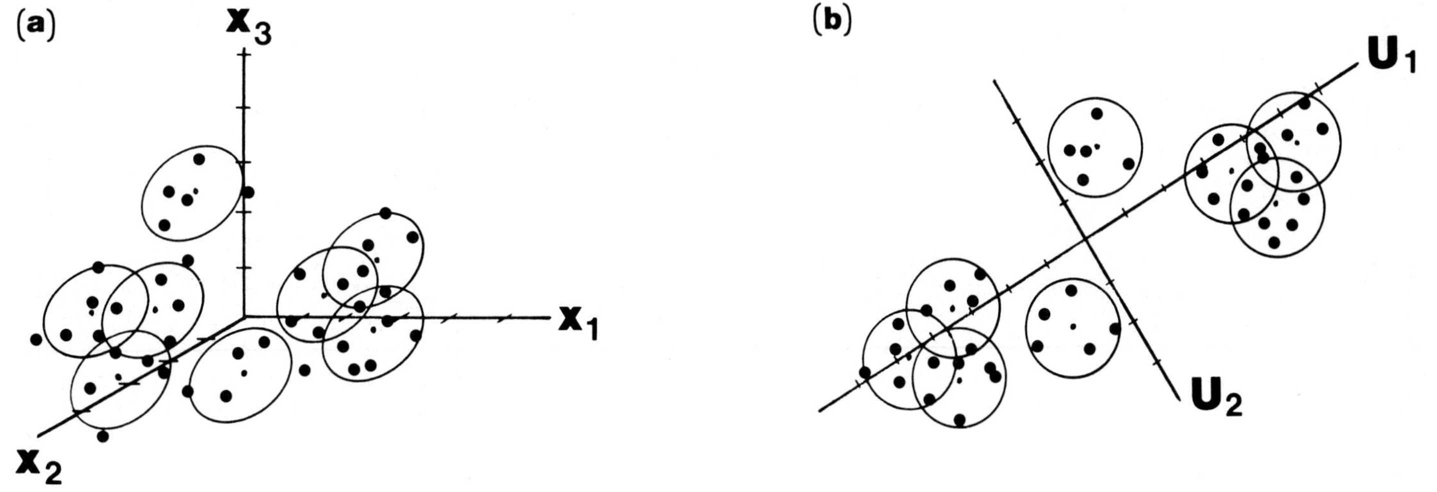

Fig. 4.2. Canonical variate analysis. **a** Point representation and concentration ellipsoids of a sample made up of $g=8$ subsamples; $p=3$, $N=\Sigma_c n_c=40$. **b** The origin of the coordinate frame is first translated to the overall mean. The axes of the new reference frame are then re-scaled and the points simultaneously projected into that subspace (u_1, u_2) of the transformed space which best separates the groups subject to the scatter of each group along each axis being equal. The concentration ellipsoids are now circles. The axes of the new reference frame are the canonical variates

tance function, which takes account of both the variances of the variables and the correlations between them. In other words, the metric structure of the space is not Euclidean and spatial relationships between groups are therefore not open to interpretation in the usual physical way. The canonical variates $u_1, \ldots, u_t$, on the other hand, are not only uncorrelated but are also usually scaled so that the within-groups variance of each is unity. As a consequence, the concentration ellipsoids of Fig. 4.2a become unit ellipses (circles) in the transformed space defined by the canonical variates (Fig. 4.2b). Furthermore, the metric structure of the transformed space is Euclidean. Thus, the mapping effected by canonical variate analysis maps the sample from a p-dimensional non-Euclidean space into a Euclidean space of $t<p$ dimensions. In applications, values for t of 2 or 3 commonly prove acceptable. In such cases the sample can be represented graphically and its structure examined.

4.3.2 Derivation

Suppose that observations on the variables $x_1, \ldots, x_p$ are available for a sample of individuals from each of g groups, the cth group $(c=1, \ldots, g)$ yielding a sample of size n_c. The total sample is then $\Sigma n_c = \mathrm{N}$, where the summation runs from $c=1, \ldots, g$. Define the dummy variables $z_1, \ldots, z_q$, where $q=g-1$, such that $z_{jc}=1$ when the jth individual $(j=1, \ldots, \mathrm{N})$ belongs to the cth group $(c=1, \ldots, q)$ and is zero otherwise. We shall henceforth suppose that the variables of $\mathbf{x}(p \times 1)$ and $\mathbf{z}(q \times 1)$ are expressed as deviations about their respective overall means and that $q \leq p$. We wish to separate the groups as effectively as possible on the basis of the p measurements considered simultaneously. For this purpose a linear transformation from $\mathbf{x}$ to a new variable for which the between-groups sums of squares relative to the total sum of squares is a maximum will be effective.

In Chap. 2 we saw that canonical correlation analysis provides normalized linear combinations of two sets of variables $\mathbf{x}$ and $\mathbf{z}$, say, $u=\mathbf{a}^t\mathbf{x}$ and $v=\mathbf{b}^t\mathbf{z}$ for which the predictable variance r^2 is maximized subject to the constraints $\mathbf{a}^t\mathbf{S}_{11}\mathbf{a}=\mathbf{b}^t\mathbf{S}_{22}\mathbf{b}=1$. Where $\mathbf{z}$ consists of binary dummy variables corresponding to the partition of the sample, it follows that $u=\mathbf{a}^t\mathbf{x}$ represents that linear combination of the response variables which contains the maximum amount of between-groups variance (see Kshirsagar, 1972, p. 360). In other words, the variable which best discriminates the groups is given by the canonical variate

$$u=\mathbf{a}^t\mathbf{x}, \tag{4.1}$$

where $\mathbf{a}$ is the latent vector corresponding to the largest root λ_1^2 of (2.16).

Let us now approach the derivation of u from a rather different viewpoint. For this purpose, it will be helpful to introduce some simplifying notation. Let the regression and total sums of squares and products (SSP) matrices, $\mathbf{S}_{12}\mathbf{S}_{22}^{-1}\mathbf{S}_{21}$ and $\mathbf{S}_{11}$, of (2.16) be denoted by $\mathbf{Q}_h$ and $\mathbf{Q}_t$, respectively. Replacing λ^2 by r^2 for convenience, Eq. (2.16) becomes

$$(\mathbf{Q}_h - \mathrm{r}^2\mathbf{Q}_t)\,\mathbf{a}=\mathbf{0}. \tag{4.2}$$

It can be shown that where $\mathbf{z}$ consists of binary dummy variables, $\mathbf{Q}_h$ is the *between-groups* SSP matrix of $\mathbf{x}$ while $\mathbf{Q}_t$ is the corresponding *total* SSP matrix (e.g., see Kshirsagar, 1972, p. 347; Krzyśko, 1979). Now the sum of squares of a linear transformation, say $u=\mathbf{a}^t\mathbf{x}$ of $\mathbf{x}$, is readily decomposed into separate between-groups and within-groups components. To see this, write

$$\begin{aligned}\mathbf{u}^t\mathbf{u}&=\mathbf{a}^t\mathbf{X}^t\mathbf{X}\,\mathbf{a}=\mathbf{a}^t\mathbf{Q}_t\,\mathbf{a}=\mathbf{a}^t(\mathbf{Q}_h+\mathbf{Q}_e)\,\mathbf{a}\\&=\mathbf{a}^t\mathbf{Q}_h\,\mathbf{a}+\mathbf{a}^t\mathbf{Q}_e\,\mathbf{a},\end{aligned}\tag{4.3}$$

where $\mathbf{X}(\mathrm{N}\times p)$ is the mean-centered matrix of N p-variate observations on the response variables and $\mathbf{Q}_e=\mathbf{Q}_t-\mathbf{Q}_h$ is the *within-groups* SSP matrix of $\mathbf{x}$. The quadratic forms $\mathbf{a}^t\mathbf{Q}_h\mathbf{a}$ and $\mathbf{a}^t\mathbf{Q}_e\mathbf{a}$ are, respectively, the between- and within-groups components of the sum of squares of u. We wish to maximize $\mathbf{a}^t\mathbf{Q}_h\mathbf{a}$ relative to $\mathbf{a}^t\mathbf{Q}_t\mathbf{a}$. The ratio of between-groups to total sums of square in u is

$$\frac{\mathbf{a}^t\mathbf{Q}_h\,\mathbf{a}}{\mathbf{a}^t\mathbf{Q}_t\,\mathbf{a}}=g(\mathbf{a})\ \text{(say)}\qquad |\mathbf{Q}_t|\neq 0.\tag{4.4}$$

The vector $\mathbf{a}$ maximizing (4.4) will be a solution of

$$\frac{\partial g}{\partial\mathbf{a}}=\frac{2[(\mathbf{Q}_h\,\mathbf{a})(\mathbf{a}^t\mathbf{Q}_t\,\mathbf{a})-(\mathbf{a}^t\mathbf{Q}_h\,\mathbf{a})(\mathbf{Q}_t\,\mathbf{a})]}{(\mathbf{a}^t\mathbf{Q}_t\,\mathbf{a})^2}=\mathbf{0}.$$

Dividing the numerator and denominator of the middle term by $\mathbf{a}^t\mathbf{Q}_t\mathbf{a}$ and substituting $g(\mathbf{a})$ from (4.4) we have

$$\frac{2[\mathbf{Q}_h\,\mathbf{a}-g(\mathbf{a})\,\mathbf{Q}_t\,\mathbf{a}]}{\mathbf{a}^t\mathbf{Q}_t\,\mathbf{a}}=\mathbf{0}.$$

Multiplying both sides by the scalar $\mathbf{a}^t\mathbf{Q}_t\mathbf{a}/2$ and simplifying yields

$$(\mathbf{Q}_h-g(\mathbf{a})\,\mathbf{Q}_t)\,\mathbf{a}=\mathbf{0}.\tag{4.5}$$

This equation has a nonzero solution for $\mathbf{a}$ if and only if

$$|\mathbf{Q}_h-g(\mathbf{a})\,\mathbf{Q}_t|=0.\tag{4.6}$$

The nonzero solutions of (4.6) are $\mathrm{r}_1^2\geq\mathrm{r}_2^2\geq\ldots\geq\mathrm{r}_s^2$, the nonzero eigenvalues of $\mathbf{Q}_t^{-1}\mathbf{Q}_h$, where $s=\min(g-1,p)$. Corresponding to the eigenvalue r_k^2 $(k=1,\ldots,s)$ let $\mathbf{a}_k$ be a solution of

$$(\mathbf{Q}_h-\mathrm{r}_k^2\,\mathbf{Q}_t)\,\mathbf{a}_k=\mathbf{0}.\tag{4.7}$$

The required maximum of (4.4) is the largest eigenvalue r_1^2 of $\mathbf{Q}_t^{-1}\mathbf{Q}_h$ and is obtained on setting $\mathbf{a}$ in (4.4) equal to $\mathbf{a}_1$. Notice that Eq. (4.7) is formally identical to Eq. (2.16) of Sect. 2.2.

We have shown that the vector $\mathbf{a}$ which best separates the groups is the eigenvector $\mathbf{a}_1$ corresponding to r_1^2 in (4.7). The remaining eigenvectors $\mathbf{a}_k$ $(k>1)$ will also be of interest if further variables are necessary in order to separate the groups. The function

$$u_k=\mathbf{a}_k^t\,\mathbf{x},\tag{4.8}$$

where $\mathbf{a}_k$ is a solution of (4.7) corresponding to the eigenvalue r_k^2, is the kth *canonical variate* or *discriminant function* $(k=1,\ldots,s)$ for the g-group problem. In canonical correlation analysis, it is usual to determine the weights vectors $\mathbf{a}_k$ under the restriction that the variance of the u_k over the sample as a whole is unity for all k, that is that $(\mathrm{N}-1)^{-1}\mathbf{a}_k^t\mathbf{Q}_t\,\mathbf{a}_k=1$. Some such condition is necessary in order to arrive at a unique solution for the u_k. In canonical variate analysis there are advantages in scaling the u_k to unit *within-groups* variance. This requirement is equivalent to specifying that the $\mathbf{a}_k$ be chosen so that

$$\mathbf{a}_k^t \mathbf{Q}_t \mathbf{a}_k = n_e(1-r_k^2)^{-1}, \qquad (4.9)$$

where $n_e = \Sigma_c^g (n_c - 1)$, rather than $(N-1)^{-1} \mathbf{a}_k^t \mathbf{Q}_t \mathbf{a}_k = 1$. Then $n_e^{-1} \mathbf{a}_k^t \mathbf{Q}_e \mathbf{a}_k = 1$ for $k=1, \ldots, s$. In this scale the between-groups variance of u_k is given by $(g-1)^{-1} \mathbf{a}_k^t \mathbf{Q}_h \mathbf{a}_k$, or, equivalently by

$$f_k/(g-1), \text{ say} \qquad (4.10)$$

where $f_k = r_k^2(1-r_k^2)^{-1}$. The quantity (4.10) is known as the *canonical variance* of the kth canonical variate; the larger f_k, and hence also r_k^2, the greater the separation of groups along the corresponding canonical variate. It is commonly the case that discrimination can for all practical purposes be regarded as effected by a small number $t<s$ of canonical variates which correspond to the larger r_k^2. As a rough guide as to the number of canonical variates to retain, the proportion of the between-groups variance, relative to the within-groups variance, absorbed cumulatively by the first root, the first two roots, and so on may be calculated. This quantity is given by

$$C_t = \sum_{k=1}^{t} f_k/(\mathrm{tr}(\mathbf{Q}_e^{-1} \mathbf{Q}_h)) \qquad (4.11)$$

for $t=1, 2, \ldots, s$, and represents the amount of information retained in the reduction from s to t dimensions. A formal test for the number of statistically significant canonical variates is considered in Sect. 4.5.2.

4.4 Further aspects of canonical variate analysis

Having derived the canonical variates we are in a position to examine some of their properties and to consider the question of their interpretation. Before turning to these matters, however, we first review requirements of the data for analysis.

Assumptions. The original development of canonical variate analysis is attributable to Fisher (1936) and was distribution-free. Inferential aspects dependent on the observations being multivariate normal were added later by Welch (1939). Fisher's results were derived for the two-group case ($g=2$) and were later extended by Rao (1948, 1952) for $g>2$. The fundamental assumption underlying canonical variate analysis is that the distance between any two groups in p-space is properly specified by Mahalanobis' D (Mardia, 1977; Campbell & Atchley, 1981). In relation to a broad class of applications this condition is not a restrictive one, though care is nevertheless required (Gower, 1972). Where the method is used purely for descriptive purposes no tightly specified distributional assumptions are necessary. However, for sensible interpretation, it is desirable that the data not depart too drastically from certain norms. Two properties of importance in this regard are the joint distribution of the variables and the stability of dispersion across groups.

While multinormality is not required in descriptive applications, it is desirable that the joint distribution of the variables be reasonably symmetric and not too long-tailed. These requirements help to ensure that the means, variances and covariances on which the analysis rests will provide a reasonable summary of the data. Extraneous observations, in particular, can exert a profound influence on the results obtained. For this reason, outlying observations are best down-weighted during analysis, or, in extreme cases, removed altogether. With respect to dispersion, it is desirable that the group covariance matrices be relatively homogeneous. This condition is designed to guard against groups possessing aberrant covariance structures unduly affecting the orientation of the canonical axes and the extent of overlap between groups. A rough check of homogeneity is readily made by calculating the within-groups variance of each group on the successive, standardized canonical variates following analysis. The weighted average of these variances on each variate is necessarily unity, but the departure of any group variance from unity would provide a clear warning that the original dispersions cannot have been homogeneous. Heterogeneity of dispersion is by no means uncommon in practice. Where heterogeneity is encountered it is worth examining the effect of the unequal covariance structure on the group configuration. Two complementary procedures have been proposed by Campbell for this purpose (see Phillips, Campbell & Wilson, 1973; Campbell & Mahon, 1974; Campbell, 1978b).

(a) Use of different estimates of the pooled, within-groups covariance. Estimates might, e.g., be based on each within-groups covariance matrix separately, provided the size of groups is not too small, or, alternatively, on a pooled matrix after deleting one or more of the groups. The degree of consistency in the results of analyses based on different estimates will reveal the extent of the effect of any heterogeneity. Dempster (1969, p. 206) and Gnanadesikan (1977, p. 90) have indicated the existence of a wide class of metric matrices which might reasonably be considered as alternatives to the two covariance estimates mentioned above.

(b) Calculation of separate matrices of squared Mahalanobis' distances (Mardia, 1977), D^2, between groups using (i) the usual within-groups covariance matrix; and (ii) the covariance matrix for each pair of groups, ignoring the rest. Principal coordinate analyses (Gower, 1966) of the two D^2 matrices will then reveal the extent of the influence of any heterogeneity.

These strategies depend on the size, n_c, of groups being sufficient to satisfactorily estimate $\mathbf{Q}_e$ for the cth group.

There is some empirical evidence (Campbell, 1980c) that canonical variate analysis is relatively insensitive to differences in covariance structure when used for descriptive purposes or for separation. Cf. also Porebsky (1966), Tatsuoka (1971, p. 233) and Webster (1977a, p. 192).

Diagonalization of $\mathbf{Q}_h$ and $\mathbf{Q}_e$. Let $\mathbf{A}=[\mathbf{a}_1|\ldots|\mathbf{a}_t]$ be the $p\times t$ matrix constructed columnwise from the t leading vectors of canonical weights $\mathbf{a}_k$ scaled as in (4.9). The matrix $\mathbf{A}$ represents a transformation that simultaneously diagonalizes $\mathbf{Q}_h$ and $\mathbf{Q}_e$:

$$(g-1)^{-1}\mathbf{A}^t\mathbf{Q}_h\mathbf{A}=\mathbf{F}, \tag{4.12}$$
$$(N-g)^{-1}\mathbf{A}^t\mathbf{Q}_e\mathbf{A}=\mathbf{I}_t, \tag{4.13}$$

where $\mathbf{F}=\mathrm{diag}(f_1, \ldots, f_t)$. Thus $\mathbf{A}$ diagonalizes $\mathbf{Q}_h$ and reduces $\mathbf{Q}_e$ to an identity matrix. Equation (4.12) shows the between-groups variance of the kth canonical variate $u_k=\mathbf{a}_k^t\mathbf{x}$ to be proportional to the kth root r_k^2 of $|\mathbf{Q}_h-r_k^2\mathbf{Q}_t|=0$. Equation (4.13) shows the u_k to have unit variance within-groups. It follows from the diagonal form of $\mathbf{F}$ and $\mathbf{I}$ that the sample canonical variates are *uncorrelated* between- and within-groups. In other words, within-groups the canonical variates u_k are independent variables with standard deviation 1, while between-groups they are also independent and u_k has variance f_k.

Canonical representation. The descriptive appeal of canonical variate analysis lies in its ability to provide a graphical representation of the sample in a small number of dimensions. It is often the case in applications that the first two or three canonical variates provide a satisfactory representation. The projection of a sample-point $\mathbf{x}_j$ onto the plane or hyperplane of the t retained canonical variates is given by $\mathbf{A}^t\mathbf{x}_j$ $(j=1, \ldots, N)$, where $\mathbf{A}$ is specified in the preceding paragraph. Similarly, the projection of the centroid for the cth group, $\bar{\mathbf{x}}_c=[\bar{\mathbf{x}}_{c1}, \bar{\mathbf{x}}_{c2}, \ldots, \bar{\mathbf{x}}_{cp}]^t$, is $\mathbf{A}^t\bar{\mathbf{x}}_c$ $(c=1, \ldots, g)$. When $t=2$ or $t=3$ the resulting mapping or *canonical representation* provides a visual summary of group relationships. Examples are provided by Hopper and Campbell (1977, Figs. 5–8) and by Campbell and Kitchener (1980, Figs. 4–6). Such mappings require minimal assumptions for their construction – indeed, the only assumption required is that of the appropriateness of the distance function used in specifying group relationships. The displays reveal the shape and extent of the scatter of each group, as well as the extent of the overlap or separation between groups. The latter property is especially useful in assessing the evidence for or against the existence of discrete units such as species or plant communities. In certain circumstances the displays can be enhanced. Where for each group the joint distribution of the responses is approximately p-variate normal, for example, a concentration ellipse centered at the appropriate group-mean can be inscribed to contain some chosen percentage of the individuals of each group. Jolicoeur (1959, Figs. 11 and 12) provides illustrations for the case where $t=2$ and Tsianco, Gabriel, Odoroff and Plumb (1981, Fig. 7.4) for $t=3$. The construction of confidence ellipses requires the orientation and lengths of the principal axes of the corresponding covariance matrix to be determined. Reference may be made to Reyment (1961), Phillips, Campbell and Wilson (1973), Webster (1977a, p. 118) and Altman (1978) for further details. Summary ellipses are particularly valuable where the groups are large and the displays liable to be over-burdened by plotting sample-points individually. Where, in addition to joint normality, group dispersions are substantially homogeneous, the concentration ellipse degenerates to a circle. A circle of radius equal to one standard deviation unit on the canonical variates will contain roughly 68% of the individuals of a particular group. More generally, $100(1-\alpha)\%$ of the points will on average be contained by a circle of radius $[\chi_\alpha^2(2)]^{\frac{1}{2}}$, where $\chi_\alpha^2(2)$ is the upper α percentage point of the chi-squared distribution with two degrees of freedom. See Webster (1977a, Figs. 10.4–10.6), for examples.

The concept of distance underlies the interpretation of these various graphical representations of the sample. We have observed that the canonical variates are uncorrelated and standardized to unit within-groups variance in the sample. This explains why Euclidean distance should be the appropriate measure for interpreting distances in the canonical representation, though not in the original p-dimensional space of the analysis. Moreover, the within-groups variation provides a yardstick against which to judge differences between groups. That is to say, discrimination between groups is made by reference to variation within groups. The transformation to canonical variates has two useful invariance properties. First, distances between groups in the original space are preserved in canonical space. To see this, let $\mathbf{A}$ be the matrix constructed columnwise from the canonical vectors corresponding to all the canonical roots, including null roots if $g-1>p$. Denote the within-groups variance-covariance matrix by $\mathbf{S}_e=(\mathrm{N}-g)^{-1}\mathbf{Q}_e$. The squared Mahalanobis distance between groups c and d with respect to the original coordinate frame, $\mathrm{D}^2(\bar{\mathbf{x}}_c, \bar{\mathbf{x}}_d)$, is then

$$\mathrm{D}^2(\bar{\mathbf{x}}_c, \bar{\mathbf{x}}_d)=(\bar{\mathbf{x}}_c-\bar{\mathbf{x}}_d)^{\mathrm{t}}\mathbf{S}_e^{-1}(\bar{\mathbf{x}}_c-\bar{\mathbf{x}}_d). \tag{4.14}$$

Writing (4.13) in the equivalent form $\mathbf{A}^{\mathrm{t}}\mathbf{S}_e\mathbf{A}=\mathbf{I}$, it is not difficult to show that $\mathbf{A}\mathbf{A}^{\mathrm{t}}=\mathbf{S}_e^{-1}$. Substituting for $\mathbf{S}_e^{-1}$ in (4.14) and denoting the centroid of the cth group with respect to the canonical variates by $\bar{\mathbf{u}}_c=\mathbf{A}^{\mathrm{t}}\bar{\mathbf{x}}_c$, we have

$$\begin{aligned}\mathrm{D}^2(\bar{\mathbf{x}}_c, \bar{\mathbf{x}}_d)&=(\bar{\mathbf{x}}_c-\bar{\mathbf{x}}_d)^{\mathrm{t}}\mathbf{A}\mathbf{A}^{\mathrm{t}}(\bar{\mathbf{x}}_c-\bar{\mathbf{x}}_d)=(\mathbf{A}^{\mathrm{t}}\bar{\mathbf{x}}_c-\mathbf{A}^{\mathrm{t}}\bar{\mathbf{x}}_d)^{\mathrm{t}}(\mathbf{A}^{\mathrm{t}}\bar{\mathbf{x}}_c-\mathbf{A}^{\mathrm{t}}\bar{\mathbf{x}}_d)\\&=(\bar{\mathbf{u}}_c-\bar{\mathbf{u}}_d)^{\mathrm{t}}(\bar{\mathbf{u}}_c-\bar{\mathbf{u}}_d)=\sum_{k=1}^{s}(\bar{u}_{kc}-\bar{u}_{kd})^2.\end{aligned} \tag{4.15}$$

Thus, not only is the squared distance between groups maintained under the transformation but, further, the square of the Mahalanobis distance in the original space becomes the square of an ordinary Euclidean distance in the space of the canonical variates. A second invariance property of the transformation is that relative distances are preserved. That is, if Mahalanobis distances in the original space and in canonical space are defined by

$$\mathrm{D}_c(\mathbf{x})=(\mathbf{x}-\bar{\mathbf{x}}_c)^{\mathrm{t}}\mathbf{S}_e^{-1}(\mathbf{x}-\bar{\mathbf{x}}_c) \quad \text{and} \quad \mathrm{D}_c(\mathbf{u})=(\mathbf{u}-\bar{\mathbf{u}}_c)^{\mathrm{t}}(\mathbf{u}-\bar{\mathbf{u}}_c),$$

respectively, then

$$\mathrm{D}_c(\mathbf{x})-\mathrm{D}_d(\mathbf{x})=\mathrm{D}_c(\mathbf{u})-\mathrm{D}_d(\mathbf{u}). \tag{4.16}$$

The canonical variates always provide the maximum separation between groups relative to variation within groups. Their effectiveness in so doing, however, depends on several factors. Of these, the positions of the group-means themselves are of overriding importance. In addition, the orientation of the canonical variates and the extent of overlap between groups is influenced by the within-groups variances and correlations of the responses. The degree of overlap, in particular, tends to increase as the within-groups correlation decreases. Phillips, Campbell and Wilson (1973) and Albrecht (1980) have examined the effects of these properties of the data on the results provided in some detail. Interpretation of the canonical representation rests largely on inspection of the configuration; distances between groups or between individuals in this transformed space can be used as measures of their relationships. Frequently, the spatial relations among groups in themselves are sufficient to suggest an interpretation in substantive terms. The process may be aided by identifying those variables which contribute most to the separation along a particular direction.

Identifying important variables. Quantities of two kinds are widely used in efforts to obtain some indication of the relative importance of the response variables in separating groups. These are the *weights* $a_{1k}, a_{2k}, \ldots, a_{pk}$ in the kth canonical variate $u_k = \mathbf{a}_k^t \mathbf{x}$, and the *correlation coefficients* between the canonical variates and the responses. It is hoped that the sign and magnitude of the elements of one or other of these sets of indices will reflect the contribution of the corresponding responses, x_i $(i=1, \ldots, p)$. Variables whose contribution is small can often be eliminated with little or no loss of discriminating efficiency.

It is necessary to standardize the canonical weights in order to remove their units of measurement and so provide a uniform basis for comparison before using them in interpretation. The standardized weights are obtained on multiplying the normalized weights $(n_e^{-1} \mathbf{a}_k^t \mathbf{Q}_e \mathbf{a}_k = 1)$ by the standard deviation of the corresponding variable. Writing $\mathbf{A} = [\mathbf{a}_1 | \ldots | \mathbf{a}_t]$ for the matrix constructed columnwise from the t retained vectors of normalized canonical weights, the matrix $\mathbf{A}^*(p \times t)$ of standardized weights is given by

$$\mathbf{A}^* = \mathbf{V}_e^{\frac{1}{2}} \mathbf{A}, \tag{4.17}$$

where $\mathbf{V}_e = \mathrm{diag}(\mathbf{Q}_e/\mathrm{N} - g)$ is the diagonal matrix of within-groups variances of the responses. Interpretation of the separation associated with the kth canonical variate in terms of the standardized weights $\mathbf{a}_k^* = [a_{1k}^*, a_{2k}^*, \ldots, a_{pk}^*]^t$ is nevertheless hazardous. The elements comprising $\mathbf{a}_k^*$, while reflecting the contribution of the responses to discrimination, are also functions of the interrelationships among the responses. As a result, the weights are interdependent and the addition or deletion of a single variable may result in drastic changes in the weights of other variables. Further, where the observed variables are subject to measurement error and are strongly correlated within-groups, variables with the larger absolute weights are not necessarily the more important ones. In such cases the stability of the weights has to be considered – that is to say their sampling variation over repeated samples. Instability may be dealt with by eliminating variables which are shown by high correlations to be redundant or by using shrunken estimators (Campbell & Reyment, 1978) in place of the usual estimates. These adjustments can often be made with little or no loss of discriminatory information.

Correlation coefficients of two kinds may be calculated between the canonical variates and each of the responses. The correlations refer respectively to either the *total* sample or the *within-groups* component of $\mathbf{x}$. The total-sample matrix of canonical variate/variable correlations, $\mathbf{R}^{(t)}(p \times t)$, is calculated as

$$\mathbf{R}^{(t)} = \mathbf{R}_{11} \mathbf{A}_t^*, \tag{4.18}$$

where $\mathbf{R}_{11}$ is the total-sample correlation matrix of the responses and $\mathbf{A}_t^*$ is the matrix of standardized canonical weights normalized to unity over the total sample rather than within-groups as is generally the case. Alternatively, correlation coefficients which help to identify variables contributing to group differences in directions of variation orthogonal to that of the error metric $\mathbf{Q}_e$ may be preferred. The $p \times r$ matrix $\mathbf{R}^{(e)}$ of such coefficients is given by

$$\mathbf{R}^{(e)} = \mathbf{R}_e \mathbf{A}^*, \tag{4.19}$$

where $\mathbf{R}_e$ is the within-groups correlation matrix calculated from $\mathbf{Q}_e$ and $\mathbf{A}^*$ is the matrix of standardized weights specified in (4.17). With respect to correlations of both kinds, it is hoped that the sign and magnitude of the correlation coefficients will guide one to those variables which are important in group separation. Nevertheless, it is possible for important separating variables to have zero correlation with a canonical variate that happens to be a contrast among those variables (Campbell, 1980c; McKay & Campbell, 1982a).

There is scope for improved procedures for establishing the relative importance of variables in group separation. McKay and Campbell (1982a) have recently examined the question in detail. They point out limitations in the procedures described above as well as in widely used *stepwise* methods. Of the existing alternatives, McKay and Campbell regard an all-subsets of variables approach using a simultaneous test procedure most favorably. Interest sometimes attaches to possible substantive interpretations for the canonical variates themselves. In such cases rotation may be found to improve their interpretability.

4.5 Hypothesis testing

Where the response variables have a multinormal distribution, hypotheses about population quantities can be tested.

Let $\mathbf{x}_{hc}$ $(h=1, \ldots, n_c;\ c=1, \ldots, g)$ be $\mathrm{N}=\Sigma_c n_c$ independent observations from g p-variate universes with vector-means $\boldsymbol{\mu}_1 \neq \boldsymbol{\mu}_2 \neq \ldots \neq \boldsymbol{\mu}_g$ and common covariance matrix $\boldsymbol{\Sigma}$. If it is assumed that

$$\mathbf{x}_{hc} \sim \mathrm{N}_p(\boldsymbol{\mu}_c, \boldsymbol{\Sigma}), \tag{4.20}$$

then the statistical tests described in Chap. 3 are applicable. The hypotheses of primary interest are those of the equality of the g population mean vectors and of the dimensionality of the space occupied by the $\boldsymbol{\mu}_c$.

The requirement for normality is not as demanding as might at first appear. For moderately extensive data, a central limit theorem effect ensures that the distribution of sample means is appreciably closer to p-variate normality than the sample observations from which they derive. The same effect is also felt when the observations can be regarded as sums of independent vectors, which is often not unrealistic in practice. The requirement for constant covariance matrices is more serious. It commonly proves difficult to demonstrate convincingly that this condition is not grossly violated. The standard test for homoscedasticity is Bartlett's test (Morrison, 1976, p. 252; Mardia, Kent & Bibby, 1979, p. 140). However, this test is known to be more sensitive to departure from joint normality than to differences in covariance structure. Accordingly, it is wise to apply Bartlett's test only after first examining the joint distribution of the responses.

4.5.1 Equality of g vector-means

We wish to test the hypothesis of the joint equality of the population mean vectors:

$$\mathrm{H}_0\colon \boldsymbol{\mu}_1 = \boldsymbol{\mu}_2 = \ldots = \boldsymbol{\mu}_g \tag{4.21}$$

against

$$\mathrm{H}_1\colon \text{not all } \boldsymbol{\mu}_c \text{ are equal.}$$

Hypothesis (4.21) is equivalent to hypothesis (3.10) of the joint nullity of the matrix of population regression coefficients $\mathbf{B}(q \times p)$ in the regression of $\mathbf{z}$ on $\mathbf{x}$, H_0: $\mathbf{B}=\mathbf{0}$. See Kshirsagar (1972, p. 347) for a proof of this assertion. The criteria used to test (3.10) can therefore be used to test (4.21). Small adjustments in the multiplying factor m and the degrees of freedom for the likelihood ratio statistic (3.13), and for the parameters of Roy's largest-root criterion (3.17), however, are necessary. In canonical variate analysis m is given by $\{(\mathrm{N}-1)-\frac{1}{2}(p+g)\}$ and the degrees of freedom by $p(g-1)$, while the parameter values of $\theta(s, m, n)$ are

$$s=\min(g-1, p), \qquad m=(|g-p-1|-1)/2, \qquad n=(\mathrm{N}-g-p-1)/2,$$

where g is the number of groups. Rejection of (4.21) in a test at some suitable level of α would lead to the conclusion that the data are consistent with the existence of differences between the populations in terms of their vector means. It would then be natural to enquire as to the dimensionality, t, of the space occupied by the means.

4.5.2 Dimensionality

The vector means $\boldsymbol{\mu}_c$ $(c=1, \ldots, g)$ determine a set of g points in p-space. Where the $\boldsymbol{\mu}_c$ are jointly equal the points will be coincident, occupying a space of zero dimension; where there are differences among the $\boldsymbol{\mu}_c$, the points will occupy a space of $0<t\leq s$ dimensions, s being equal to the smaller of $(g-1)$ and p. We wish to establish the minimum value of t necessary to adequately represent group differences. In Sect. 4.3.2 we saw that the kth root r_k^2 of $\mathbf{Q}_t^{-1}\mathbf{Q}_h$ is related to the extent of the separation of groups along u_k. It follows that the number of non-zero roots r_k^2 provides an estimate of the dimensionality of the discriminant space. Tests for the number of non-zero canonical roots were described in Sect. 3.4.2. For our present purpose hypothesis (3.18) of Sect. 3.4.2 may be restated as:

$$\mathrm{H}_0(t)\colon \boldsymbol{\mu}_g \text{ lie in a } t\text{-dimensional hyperplane} \tag{4.22}$$

the alternate being

$$\mathrm{H}_1(t)\colon \boldsymbol{\mu}_g \text{ unrestricted}$$

for $t=1, \ldots, s$. The sequential test procedures of Sect. 3.4.2 provide tests of (4.22) on replacing the arguments j by t and q by $g-1$, respectively. The dimensionality of the space or subspace required to separate the groups is identically the number of non-zero roots. When the dimensionality is given by $t=1$ the means are said to be *collinear*.

For $t=2$ or $t=3$ each group centroid $\boldsymbol{\mu}_c$ can be surrounded by a confidence region with radius given by $(\chi^2_\alpha(2)/n_c)^{\frac{1}{2}}$ or $(\chi^2_\alpha(3)/n_c)^{\frac{1}{2}}$, where $\chi^2_\alpha(\cdot)$ is the upper 100α percentage point of the chi-squared distribution with the corresponding degrees of freedom for some suitable α. As well as indicating the uncertainty of the $\boldsymbol{\mu}_c$ such confidence regions are useful in revealing the extent of group separation.

The assumptions underlying the tests of (4.21) and (4.22) are not always satisfied in practice. A class of distribution-free tests whose assumptions are more realistic and therefore more likely to be satisfied has been developed by Mielke, Berry and Johnson (1976). These tests permit the requirements for multinormality and homogeneity of dispersion to be dispensed with entirely.

4.6 Affinities with other methods

We have seen that canonical variate analysis is a special case of canonical correlation analysis which arises where the sample is partitioned into g groups and one set of variables in the canonical analysis is constructed so as to reflect the group structure. Canonical variate analysis also has affinity with several other multivariate methods. In this section the nature of its connections with multivariate analysis of variance, multiple discriminant analysis and principal component analysis are outlined.

4.6.1 Canonical variate analysis, multivariate analysis of variance and multiple discriminant analysis

The *multivariate single classification* or one-way analysis of variance is concerned with testing the equality of the mean vectors of g p-variate normal distributions with common covariance matrix $\boldsymbol{\Sigma}$, given independent observations from these distributions. The appropriate null hypothesis may be written

$$H_0: \boldsymbol{\mu}_1=\boldsymbol{\mu}_2=\ldots=\boldsymbol{\mu}_g, \tag{4.23}$$

the alternate being

$$H_1: \boldsymbol{\mu}_c \neq \boldsymbol{\mu}_d \text{ for at least one pair of groups } c \text{ and } d.$$

Using matrices, (4.23) can be expressed as the multivariate general linear hypothesis (Timm, 1975, p. 161):

$$H_0: \mathbf{CMA}=\mathbf{0} \tag{4.24}$$

against

$$H_1: \mathbf{CMA}\neq\mathbf{0}.$$

In these expressions a suitable choice for $\mathbf{C}$ has the form

$$\underset{(g-1)\times g}{\mathbf{C}} = \begin{bmatrix} 1 & -1 & 0 & \dots & 0 & 0 \\ 0 & 1 & -1 & \dots & 0 & 0 \\ \vdots & \vdots & \vdots & \vdots & \vdots & \vdots \\ 0 & 0 & 0 & \dots & 1 & -1 \end{bmatrix}, \tag{4.25}$$

M is a $g \times p$ matrix whose rows $\boldsymbol{\mu}_c^t$ are the p-variate group means, $\mathbf{A}(p \times p)$ is an identity matrix and $\mathbf{0}$ is the null matrix of order $(g-1) \times p$. **C** is known as the *hypothesis matrix*. Each row of **C** specifies a contrast whose elements by definition sum to 0 and defines a hypothetical linear relationship among group-means. We shall see in Sect. 4.7 that **C** can be modified to accommodate a wide range of hypotheses in addition to (4.23).

Several criteria for testing (4.23) are available, all of them functions of the roots $f_1, \dots, f_s$ of the determinantal equation $|\mathbf{Q}_h - f\,\mathbf{Q}_e| = 0$. Wilks' likelihood ratio statistic Λ, where

$$\Lambda = \frac{|\mathbf{Q}_e|}{|\mathbf{Q}_e + \mathbf{Q}_h|} = \frac{1}{|\mathbf{I} + \mathbf{Q}_e^{-1}\mathbf{Q}_h|} = \prod_{k=1}^{s} (1 + f_k)^{-1} \sim \Lambda(u, \nu_h, \nu_e) \tag{4.26}$$

is one such criterion. Here $\mathbf{Q}_e(p \times p)$ and $\mathbf{Q}_h(p \times p)$ are within-groups and between-groups SSP matrices, respectively, and f_k is an eigenvalue of $\mathbf{Q}_e^{-1}\mathbf{Q}_h$. The arguments of Λ are $u = p$, $\nu_h = g-1$, $\nu_e = \mathrm{N} - g$. Tables of percentage points of the distribution of $\Lambda(u, \nu_h, \nu_e)$ are not always accessible. A convenient approximation of the distribution is then

$$-\{(\mathrm{N}-1) - \tfrac{1}{2}(p+g)\} \log_e \Lambda = \chi_s^2, \tag{4.27}$$

whose asymptotic null distribution is chi-squared on $p \cdot (g-1)$ degrees of freedom. Hypothesis (4.23) is rejected at the significance level α, if, for the sample value χ_s^2,

$$\chi_s^2 > \chi_\alpha^2,$$

where χ_α^2 is the upper 100α percentage point of the chi-squared distribution with $p \cdot (g-1)$ degrees of freedom.

An alternative criterion for testing (4.23) is readily obtained from the largest root f_1 of $|\mathbf{Q}_h - f\,\mathbf{Q}_e| = 0$. The criterion is

$$\mathrm{r}_1^2 = \frac{f_1}{1+f_1} \sim \theta(s, m, n), \tag{4.28}$$

where $\theta(s, m, n)$ denotes the gcr distribution with arguments

$$s = \min(g-1, p), \qquad m = (|g-p-1|-1)/2, \qquad n = (\mathrm{N}-g-p-1)/2.$$

The null hypothesis is rejected in a test of size α if the sample value of r_1^2, r_s^2, say, is such that

$$\mathrm{r}_s^2 > \theta_\alpha(s, m, n) \tag{4.29}$$

where $\theta_\alpha(s, m, n)$ is the upper 100α percentage point of Roy's largest-root distribution with arguments s, m and n. Otherwise H_0 is maintained.

The test of (4.23) is summarized in the *multivariate analysis of variance* of Table 4.2, where the respective SSP matrices are given by

Table 4.2. Multivariate analysis of variance for the one-way design. ($m=\{(N-1)-\frac{1}{2}(p+g)\}$)

Source of dispersion	df	SSP (symmetric, $p\times p$)	Λ	χ^2
Constant term	1	$[\quad]=\mathbf{Q}_s$		
Between groups	$g-1$	$[\quad]=\mathbf{Q}_h$	$\dfrac{\lvert\mathbf{Q}_e\rvert}{\lvert\mathbf{Q}_e+\mathbf{Q}_h\rvert}$	$-m\log_e\Lambda$
Within groups	$N-g$	$[\quad]=\mathbf{Q}_e$		
Total	N	$[\quad]=\mathbf{Q}_t$		

$$\mathbf{Q}_h=\sum_c n_c(\bar{\mathbf{x}}_c-\bar{\bar{\mathbf{x}}})(\bar{\mathbf{x}}_c-\bar{\bar{\mathbf{x}}})^t,\qquad \mathbf{Q}_e=\sum_c\sum_h(\mathbf{x}_{ch}-\bar{\mathbf{x}}_c)(\mathbf{x}_{ch}-\bar{\mathbf{x}}_c)^t$$

and by

$$\mathbf{Q}_t=\sum_c\sum_h(\mathbf{x}_{ch}-\bar{\bar{\mathbf{x}}})(\mathbf{x}_{ch}-\bar{\bar{\mathbf{x}}})^t,$$

and the summations are over $c=1,\ldots,g$ and $h=1,\ldots,n_c$.

Reference to (4.23) and (4.21) shows the null hypotheses of multivariate analysis of variance and canonical variate analysis to be identical. Further, the respective likelihood ratio and union-intersection tests of these hypotheses can be shown to be equivalent. For the likelihood ratio test, the multivariate analysis of variance criterion is expressed in (4.26) as a function of the roots f_k of $\mathbf{Q}_e^{-1}\mathbf{Q}_h$. The corresponding criterion in canonical variate analysis is given in (3.12) and is a function of the roots r_k^2 of $\mathbf{Q}_t^{-1}\mathbf{Q}_h$:

$$\prod_{k=1}^{s}(1-r_k^2)=|\mathbf{I}-\mathbf{Q}_t^{-1}\mathbf{Q}_h|=\frac{|\mathbf{Q}_t-\mathbf{Q}_h|}{|\mathbf{Q}_t|}$$

$$=\frac{|\mathbf{Q}_e|}{|\mathbf{Q}_e+\mathbf{Q}_h|}\sim\Lambda(u,\nu_h,\nu_e). \tag{4.30}$$

The likelihood ratio test statistics of the two analyses are thus formally identical. Similarly, with respect to the union-intersection test, the multivariate analysis of variance criterion r_1^2 of (4.28) is numerically the same as the quantity r_1^2 in the union-intersection test (3.17) of hypothesis (4.21) of canonical variate analysis.

To complete the comparison it will be helpful to introduce more general notations for Bartlett's transformation of Λ and for the parameters of Roy's largest root statistic, r_1^2. A more general form of Bartlett's approximations (3.13) and (4.27) of the likelihood ratio statistic Λ can be written

$$-\{\nu_e-\tfrac{1}{2}(p+\nu_h+1)\}\log_e\Lambda\sim\chi^2(\nu_h\cdot p). \tag{4.31}$$

In this expression p refers in canonical variate analysis to the number of predictor variables and to the number of responses in multivariate analysis of variance; ν_e and ν_h are the error and hypothesis degrees of freedom, respectively. In both analyses $\nu_e = N-1$. In canonical variate analysis for the hypothesis degrees of freedom $\nu_h = q$, the number of dummy (criterion) variables, and in multivariate analysis of variance $\nu_h = g-1$. With this identification of parameters the equivalence of the test statistics (3.13) and (4.27) is established. More general expressions for the arguments of the largest-root statistic are:

$$s=\min(\nu_h, p), \quad m=(|p-\nu_h|-1)/2, \quad n=(\nu_e-p-1)/2, \tag{4.32}$$

where $\nu_h = g-1$ and $\nu_e = N-g$. The formal equivalence of (3.17) and (4.28) follows immediately.

Suppose hypothesis (4.23) is rejected and the conclusion drawn that there is a difference between mean vectors. Then *multiple discriminant analysis* can be used to characterize the major differences. Multiple discriminant analysis is a procedure for combining variables so that the resulting composites have optimal properties for distinguishing the groups in the smallest possible number of dimensions. More specifically, the objectives are generally to:

(a) identify those linear combinations of the responses which best separate the groups;
(b) establish the dimensionality of the space or subspace of p-space within which the means lie; and to
(c) identify those response variables which contribute most to group separation and the direction of their effect.

The similarity of these objectives to those of interest in canonical variate analysis will be plain. Multiple discriminant analysis also has application in the optimal *allocation* of unidentified samples among several groups from which they are known to derive, though this use of the method is not pursued here. Reference may be made to Lachenbruch (1975, 1982), Geisser (1977), Lachenbruch and Goldstein (1979) or to Fatti, Hawkins and Raath (1982) for accounts of the use of multiple discriminant analysis in allocation. Canonical variate analysis may be used for the same purpose (see Geisser, 1977; Campbell, 1980c).

Multiple discriminant analysis is usually formulated in terms of a search for one or more linear combinations of p mean-corrected response variables, $\mathbf{x}(p \times 1)$, with the property of maximizing the ratio of between- to within-groups sums of squares for the derived composites. Where more than one such composite is calculated, these are required to be uncorrelated. We wish to determine first the $p \times 1$ vector $\mathbf{c}$ for which

$$w = \mathbf{c}^t \mathbf{x} \tag{4.33}$$

maximizes the ratio of between-groups to within-groups sums of squares of w. The function to be maximized with respect to variation in $\mathbf{c}$ is therefore

$$\frac{\mathbf{c}^t \mathbf{Q}_h \mathbf{c}}{\mathbf{c}^t \mathbf{Q}_e \mathbf{c}} = h(\mathbf{c}) \text{ (say)} \quad |\mathbf{Q}_e| \neq 0, \tag{4.34}$$

where $\mathbf{c}^t \mathbf{Q}_h \mathbf{c}$ and $\mathbf{c}^t \mathbf{Q}_e \mathbf{c}$ are scalars corresponding to the between- and within-groups components of the sum of squares of w. Using the standard procedure of Sect. 4.3.2, maximizing (4.34) yields the generalized eigenequation

$$(\mathbf{Q}_h - f_k \mathbf{Q}_e)\,\mathbf{c}_k = \mathbf{0}, \tag{4.35}$$

whose solutions f_k and $\mathbf{c}_k$ $(k=1, \ldots, s)$ are respectively the nonnull eigenvalues and eigenvectors of $\mathbf{Q}_e^{-1}\mathbf{Q}_h$. Let the f_k be arranged in descending order of magnitude so that $f_1 \geq f_2 \geq \ldots \geq f_s$ and order the eigenvectors $\mathbf{c}_k$ corresponding to the f_k similarly. The linear composites $w_k = \mathbf{c}_k^t \mathbf{x}$ are the *discriminant functions*. These new variables are mutually uncorrelated and are generally standardized to unit within-groups variance, $n_e^{-1}\mathbf{c}_k^t \mathbf{Q}_e \mathbf{c}_k = 1$, for all k. The first t discriminant functions w_k $(k=1, \ldots, t)$ may be useful when it is hoped to summarize group differences in $0 < t \leq s$ dimensions.

Equation (4.35) is related to Eq. (4.7) of Sect. 4.3.2. To establish the connection we first rewrite (4.35) as

$$\mathbf{Q}_h \mathbf{c} = f\,\mathbf{Q}_e \mathbf{c}.$$

Adding $f\,\mathbf{Q}_h \mathbf{c}$ to both sides we have

$$(1+f)\,\mathbf{Q}_h \mathbf{c} = f(\mathbf{Q}_e + \mathbf{Q}_h)\,\mathbf{c}$$

or

$$\left(\mathbf{Q}_t^{-1}\mathbf{Q}_h - \frac{f}{1+f}\,\mathbf{I}\right)\mathbf{c} = \mathbf{0}. \tag{4.36}$$

Equation (4.36) shows that if f is an eigenvalue of $\mathbf{Q}_e^{-1}\mathbf{Q}_h$, then $f(1+f)^{-1}$ is an eigenvalue of $\mathbf{Q}_t^{-1}\mathbf{Q}_h$. The eigenvalues r_k^2 of (4.7) and f_k of (4.35) are therefore related as follows:

$$\mathrm{r}_k^2 = \frac{f_k}{1+f_k}, \qquad f_k = \frac{\mathrm{r}_k^2}{1-\mathrm{r}_k^2}. \tag{4.37}$$

The canonical variates $\mathbf{a}_k^t \mathbf{x}$ of $\mathbf{x}$ are solutions of $\mathbf{Q}_t^{-1}\mathbf{Q}_h \mathbf{a}_k = \mathrm{r}_k^2\,\mathbf{a}_k$. Similarly, the discriminant functions $\mathbf{c}_k^t \mathbf{x}$ of $\mathbf{x}$ are solutions of $\mathbf{Q}_e^{-1}\mathbf{Q}_h \mathbf{c}_k = \mathrm{r}_k^2(1-\mathrm{r}_k^2)^{-1}\,\mathbf{c}_k$. The vectors of canonical and discriminant weights $\mathbf{a}_k$ and $\mathbf{c}_k$ therefore stand in the relationship

$$\mathbf{a}_k = (1-\mathrm{r}_k^2)^{-\frac{1}{2}}\,\mathbf{c}_k, \qquad \mathbf{c}_k = (1-\mathrm{r}_k^2)^{\frac{1}{2}}\,\mathbf{a}_k. \tag{4.38}$$

The eigenvectors $\mathbf{a}_k$ and $\mathbf{c}_k$ have arbitrary scale constants which may be absorbed by the vectors. We have seen that in both procedures it is customary to standardize the vectors to a common scale of unit within-groups variance.

Equations (4.37) and (4.38) show the results of canonical variate analysis and multiple discriminant analysis to be interchangeable. In other words, the two procedures are effectively one and the same. It so happens that in deriving canonical variate analysis in Sect. 4.3.2 we chose what may be described as the regression approach to the problem, while in deriving multiple discriminant analysis we did so in terms of maximizing the ratio of between- to within-groups sums of squares in a linear composite of the responses. These are merely alternative solutions to a single problem. In the literature it is by no means uncommon to find canonical variate analysis formulated in terms of maximizing a ratio of between- to within-groups sums of squares (e.g., Rao,

1952; Campbell & Atchley, 1981) or to encounter multiple discriminant analysis treated as a regression problem (e.g., Bartlett, 1965; Kshirsagar, 1972, pp. 206 and 360).

4.6.2 Canonical variate analysis and principal component analysis

We have seen that canonical variate analysis can be derived from the eigenstructure of $\mathbf{Q}_e^{-1}\mathbf{Q}_h$, which is a square, non-symmetric matrix of order p. The eigenvalues f_k $(k=1, \ldots, s)$ of $\mathbf{Q}_e^{-1}\mathbf{Q}_h$ are related to the canonical roots while its eigenvectors, $\mathbf{a}_k$, scaled such that $n_e^{-1}\mathbf{a}_k^t\mathbf{Q}_e\mathbf{a}_k=1$, are the vectors of canonical (or discriminant) weights. In solving for the canonical roots and variates there are computational advantages in working with the symmetric $p \times p$ matrix $\mathbf{Q}_e^{-\frac{1}{2}}\mathbf{Q}_h\mathbf{Q}_e^{-\frac{1}{2}}$, rather than $\mathbf{Q}_e^{-1}\mathbf{Q}_h$. The eigenvalues of $\mathbf{Q}_e^{-\frac{1}{2}}\mathbf{Q}_h\mathbf{Q}_e^{-\frac{1}{2}}$ are identical to those of $\mathbf{Q}_e^{-1}\mathbf{Q}_h$, while its eigenvectors, $\mathbf{a}_k^*$, say, are connected to those of $\mathbf{Q}_e^{-1}\mathbf{Q}_h$ by the relation $\mathbf{a}_k=\mathbf{Q}_e^{-\frac{1}{2}}\mathbf{a}_k^*$. This solution for the canonical roots and vectors can be shown to be equivalent to a two-stage application of principal component analysis. Campbell (1979b) and Campbell and Atchley (1981) have described the geometry and algebra of canonical variate analysis from this viewpoint. The account which follows is based on Campbell and Atchley's enlightening discussion.

The computational scheme outlined above may be performed explicitly as follows. The *first stage* components analysis is accomplished by an eigenanalysis of the pooled, within-groups SSP matrix, $\mathbf{Q}_e$. Writing $\mathbf{Q}_e$ in terms of its eigenvectors and eigenvalues we have

$$\mathbf{Q}_e=\mathbf{PEP}^t, \tag{4.39}$$

where $\mathbf{P}=[\mathbf{p}_1|\ldots|\mathbf{p}_p]$ is the $p \times p$ matrix constructed columnwise from the eigenvectors of $\mathbf{Q}_e$ and $\mathbf{E}=\text{diag}(e_1, \ldots, e_p)$ is the diagonal matrix of eigenvalues of $\mathbf{Q}_e$. The principal components $\mathbf{y}_{ch}(p \times 1)$ of the hth p-variate observation $(h=1, \ldots, n_c)$ of the cth group, $\mathbf{x}_{ch}$, are then

$$\mathbf{y}_{ch}=\mathbf{P}^t\mathbf{x}_{ch}. \tag{4.40}$$

The pooled, within-groups variance of the ith component y_i $(i=1, \ldots, p)$ is $\mathbf{p}_i^t\mathbf{Q}_e\mathbf{p}_i$ and from (4.39) it is easily shown that $\mathbf{p}_i^t\mathbf{Q}_e\mathbf{p}_i=e_i$. The principal components are now standardized to give new variables, z_i, say, whose within-groups variance is unity by dividing y_i by the square root of its variance, $e_i^{\frac{1}{2}}$. For all p components we may write

$$\mathbf{z}_{ch}=\mathbf{E}^{-\frac{1}{2}}\mathbf{P}^t\mathbf{x}_{ch}. \tag{4.41}$$

As the standardized variables $z_1, \ldots, z_p$ have unit variance and are mutually uncorrelated they are referred to as *orthonormal variables*. The elements of $\mathbf{z}_{ch}$ are the coordinates of the hth individual of the cth group with respect to the standardized principal components $z_1, \ldots, z_p$. Similarly, the vector-mean for the cth group in terms of the orthonormal variables, $\bar{\mathbf{z}}_c$, is given by $\bar{\mathbf{z}}_c=\mathbf{E}^{-\frac{1}{2}}\mathbf{P}^t\bar{\mathbf{x}}_c$ $(c=1, \ldots, g)$. The relative positions of the group means are changed by this transformation.

The rotated and scaled axes represented by the orthonormal variables correspond to patterns of *within-groups* variations. They provide the reference coordinate system for the second-stage analysis, which consists of a components analysis of the group means $\bar{\mathbf{z}}_c$ for the orthonormal variables. This stage of the analysis provides an examination of the *between-groups* variation relative to the patterns of within-groups variation defined by the first-stage principal components. Let $\bar{\mathbf{X}}(g \times p)$ be the matrix of group means for the original variables, centered so that the average of the means is zero, with each vector of means weighted by its sample size:

$$\bar{\mathbf{X}}^t = \{n_1^{\frac{1}{2}}(\bar{\mathbf{x}}_1 - \bar{\mathbf{x}}_T) | \dots | n_g^{\frac{1}{2}}(\bar{\mathbf{x}}_g - \bar{\mathbf{x}}_T)\}. \tag{4.42}$$

In this expression $\bar{\mathbf{x}}_T = \Sigma_c^g n_c \bar{\mathbf{x}}_c$. The between-groups SSP matrix $\mathbf{Q}_h$ is then $\bar{\mathbf{X}}^t \bar{\mathbf{X}}$. Further, the matrix of group means for the orthonormal variables is $\bar{\mathbf{Z}} = \mathbf{E}^{-\frac{1}{2}} \mathbf{P}^t \bar{\mathbf{X}}$. The between-groups matrix for the orthonormal variables can now be written:

$$\begin{aligned} \mathbf{Z}^t \mathbf{Z} &= \mathbf{E}^{-\frac{1}{2}} \mathbf{P}^t \bar{\mathbf{X}}^t \bar{\mathbf{X}} \mathbf{P} \mathbf{E}^{-\frac{1}{2}} \\ &= \mathbf{E}^{-\frac{1}{2}} \mathbf{P}^t \mathbf{Q}_h \mathbf{P} \mathbf{E}^{-\frac{1}{2}}. \end{aligned} \tag{4.43}$$

The *second-stage* components analysis is based on an eigenanalysis of the between-groups matrix in (4.43):

$$(\mathbf{E}^{-\frac{1}{2}} \mathbf{P}^t \mathbf{Q}_h \mathbf{P}^t \mathbf{E}^{-\frac{1}{2}} - f\,\mathbf{I})\,\mathbf{a}^* = \mathbf{0}, \tag{4.44}$$

and yields the canonical roots f_k and canonical vectors $\mathbf{a}_k^*$ for the orthonormal variables $\mathbf{z}$. Premultiplying (7.44) by $\mathbf{P}\mathbf{E}^{-\frac{1}{2}}$ we obtain

$$(\mathbf{Q}_e^{-1} \mathbf{Q}_h - f\,\mathbf{I})\,\mathbf{P}\mathbf{E}^{-\frac{1}{2}} \mathbf{a}^* = \mathbf{0}, \tag{4.45}$$

showing that the canonical vectors $\mathbf{a}_k$ for the original variables $\mathbf{x}$ can be obtained from (4.44) by the relation

$$\mathbf{a}_k = \mathbf{P}\mathbf{E}^{-\frac{1}{2}} \mathbf{a}_k^*. \tag{4.46}$$

We have shown canonical variate analysis to be equivalent to a two-stage application of principal component analysis. The connection between the two methods is closer than might at first sight appear from this result. Canonical variate analysis in fact amounts to a principal component analysis of the group mean vectors weighted by sample size in the metric $\mathbf{Q}_e$.

4.7 Imposition of structure

The binary dummy variables introduced in Sect. 4.2 enable the composite nature of a sample to be taken into account where there is just one variable of classification and where relations among classes are unspecified. In the case of more complex designs or where relationships among groups can be postulated before data are collected, it is generally advantageous to employ contrasts which reflect either the experimental design or the relationships in question. Such contrasts are referred to as *designed comparisons*. Analysis will in general be more powerful and readily interpretable where contrasts are chosen to exploit whatever structure may be present. For brevity, we shall confine our attention to contrasts which reflect group relationships arrived at on the basis of *a priori* substantive insight.

4.7.1 Designed comparisons

In constructing dummy variables to reflect postulated relationships, negative as well as positive (including zero) outcome states are allowed. By appropriate selection of values, contrasts which embody the hypotheses of the investigator can be generated. We shall not describe their construction in detail; the principles involved are those governing the construction of contrasts in the analysis of variance. Li (1964, p. 252), Seal (1964, p. 27), Pruzek (1971) or Lewis and Mouw (1978) may be consulted for further details.

Our attention will focus on comparisons of group-means associated with hypotheses of the kind.

$$H_0:\ z_1\boldsymbol{\mu}_1 + z_2\boldsymbol{\mu}_2 + \ldots + z_g\boldsymbol{\mu}_g = \mathbf{0}, \tag{4.47}$$

where $\mathbf{z} = [z_1, z_2, \ldots, z_g]^t$ is a vector of given constants for which $\Sigma_c^g n_c z_c = 0$, n_c being the number of samples in the cth group ($c = 1, \ldots, g$) while the $\boldsymbol{\mu}_c$ are the p-variate group-means of the population from which the sample was drawn. Where there are g groups, $g-1$ contrasts $\mathbf{z}_i$ ($i = 1, \ldots, g-1$) may be constructed, one for each of the $g-1$ between-groups degrees of freedom. Simplicity and interpretability will result if the set of contrasts are mutually orthogonal and our treatment will be restricted to contrasts of this particular kind. In the multivariate analysis of variance formulation of Sect. 4.6.1, the hypothesis matrix $\mathbf{C}$ may be constructed from a set of orthogonal contrasts. Consider, as an example, the following $g = 6$ vegetation types.

c	Vegetation	c	Vegetation
1	Montane grassland	4	Coniferous forest SW aspect
2	Coastal grassland	5	Deciduous forest NE aspect
3	Coniferous forest NE aspect	6	Deciduous forest SW aspect

Structural relationships among these categories might be investigated by means of the orthogonal contrasts of Table 4.3. A suitable matrix $\mathbf{C}$ for the composite hypothesis (4.24) would then be defined by the rows of the table.

Two forms of contrast of wide applicability are *orthogonal polynomial* and *Helmert* contrasts. Orthogonal polynomials may be useful where there are grounds for supposing that groups are equally spaced along some dimension of response-space but perhaps with nonlinear deviations about this line. Helmert

Table 4.3. Orthogonal contrasts for $g=6$ vegetation types

	Vegetation type					
Grassland versus forest	2	2	−1	−1	−1	−1
Montane versus coastal grassland	1	−1	0	0	0	0
Coniferous versus deciduous forest	0	0	1	1	−1	−1
NE versus SW aspect	0	0	1	−1	1	−1
Leaf habit × aspect	0	0	1	−1	−1	1

contrasts are applicable where the groups of interest have a natural order in terms of the responses and it is meaningful to compare the group-means in a systematic way. Dummy variables corresponding to polynomial and Helmert contrasts for the particular case of $g=4$ groups are shown in Table 4.4. General expressions for these contrasts are given by Finn (1974, p. 233) and Bock (1975, p. 300), respectively.

Table 4.4. Dummy variables corresponding to designed comparisons of two kinds. The columns of $\mathbf{P}_4$ and $\mathbf{H}_4$ yield the indicated contrasts

(a) Orthogonal polynomial contrasts of order 4

$$\mathbf{P}_4=\begin{bmatrix} 1 & -3 & 1 & -1 \\ 1 & -1 & -1 & 3 \\ 1 & 1 & -1 & -3 \\ 1 & 3 & 1 & 1 \end{bmatrix}$$

(b) Helmert contrasts of order 4

$$\mathbf{H}_4=\begin{bmatrix} 1 & 1 & 0 & 0 \\ 1 & -\frac{1}{3} & 1 & 0 \\ 1 & -\frac{1}{3} & -\frac{1}{2} & 1 \\ 1 & -\frac{1}{3} & -\frac{1}{2} & -1 \end{bmatrix}$$

The first column of both $\mathbf{P}_4=[\mathbf{1}_4, \mathbf{Z}_{\mathrm{P}}^{(4)}]$ and $\mathbf{H}_4=[\mathbf{1}_4, \mathbf{Z}_{\mathrm{H}}^{(4)}]$ in Table 4.4 consists of unities, the function of which is to account for an overall scale or mean effect. It is readily verified that the remaining columns of $\mathbf{P}_4$ and $\mathbf{H}_4$ sum to zero and that all columns are mutually orthogonal.

4.7.2 Separating the sources of variation

Corresponding to a set of orthogonal contrasts is an additive decomposition of the between-groups SSP matrix, $\mathbf{Q}_h$:

$$\mathbf{Q}_h=\mathbf{u}_0\mathbf{u}_0^{\mathrm{t}}+\mathbf{u}_1\mathbf{u}_1^{\mathrm{t}}+\ldots+\mathbf{u}_q\mathbf{u}_q^{\mathrm{t}}, \tag{4.48}$$

where the $\mathbf{u}_i\mathbf{u}_i^{\mathrm{t}}$ $(i=0, \ldots, q)$ are orthogonal matrices of order p each carrying one degree of freedom. The ith term $(i=1, \ldots, q)$ in (4.48) is a function of the ith contrast $\mathbf{z}_i$, which itself represents a specifiable relationship among the group-means. That is to say, the decomposition (4.48) corresponds to a partition of the between-groups variation among *identifiable* sources. For convenience, it is customary to assemble the $\mathbf{u}_i$ rowwise to form the $q\times p$ matrix $\mathbf{U}^{\mathrm{t}}=[\mathbf{u}_0|\mathbf{u}_1|\ldots|\mathbf{u}_q]$.

The matrix $\mathbf{U}$ is readily constructed from a set of orthogonal contrasts and the vector of group-means for the p responses. To see this, let $\bar{\mathbf{x}}_c$ ($c=1, \ldots, g$) be the p-vector of means for the cth group. Define

$$\bar{\mathbf{X}}^t=[\sqrt{n_1}\,\bar{\mathbf{x}}_1|\sqrt{n_2}\,\bar{\mathbf{x}}_2|\ldots|\sqrt{n_g}\,\bar{\mathbf{x}}_g],$$

and let

$$\mathbf{Z}=[\mathbf{h}|\mathbf{z}_1|\mathbf{z}_2|\ldots|\mathbf{z}_q]=[\underset{1}{\mathbf{h}}|\underset{q}{\mathbf{Z}_0}],$$

where $\mathbf{Z}$ is any $g \times g$ orthogonal matrix whose first column is

$$\mathbf{h}=\left[\left[\frac{n_1}{\mathrm{N}}\right]^{\frac{1}{2}}, \left[\frac{n_2}{\mathrm{N}}\right]^{\frac{1}{2}}, \ldots, \left[\frac{n_g}{\mathrm{N}}\right]^{\frac{1}{2}}\right]^t.$$

Then,

$$\underset{(g\times p)}{\mathbf{U}}=\begin{bmatrix}\mathbf{u}_0^t\\ \mathbf{u}_1^t\\ \vdots\\ \mathbf{u}_q^t\end{bmatrix}=\underset{(g\times g)}{\mathbf{Z}^t}\;\underset{(g\times p)}{\bar{\mathbf{X}}}. \tag{4.49}$$

The ith row of $\mathbf{U}$ generates the ith term in the decomposition (4.48). Suitable choices for $\mathbf{Z}$ in many applications are the polynomial and Helmert matrices $\mathbf{P}_g$ and $\mathbf{H}_g$.

The test of hypothesis (4.47) for the ith contrast, $\mathbf{z}_i$, may be conducted as follows. Denote the SSP matrix $\mathbf{u}_i\mathbf{u}_i^t$ corresponding to $\mathbf{z}_i$ ($i=0, \ldots, q$) by $\mathbf{Q}_q^{(i)}$. Then, the likelihood ratio test criterion for (4.47) is given by

$$\frac{|\mathbf{Q}_e|}{|\mathbf{Q}_e+\mathbf{Q}_h^{(i)}|}=(1+f^{(i)})^{-1}\sim\Lambda(u, \nu_h, \nu_e). \tag{4.50}$$

Bartlett's transformation of Λ is

$$-\{\nu_e+\nu_h-\tfrac{1}{2}(\mathrm{p}+\nu_h+1)\}\log_e\Lambda\sim\chi^2,$$

where $\nu_e=\mathrm{N}-g+1$ and $\nu_h=1$, which for large N is distributed as a chi-squared variate on $\nu_h \cdot p$ degrees of freedom. The complete analysis is conveniently summarized as the *multivariate analysis of variance* of Table 4.5.

What the analysis of Table 4.5 accomplishes is a decomposition of the hypothesis or explained SSP term, $\mathbf{Q}_h$, of Table 4.2, in which each component matrix $\mathbf{Q}_h^{(i)}$ is attributable to an identifiable source and can be individually assessed. In general, it is likely to be advantageous to extend the partitioning of factorial terms such as $\mathbf{Q}_h$ into linear, quadratic, *etc.* or other components as far as possible.

4.7.3 Further comments

The imposition of structure enables analysis to be pursued further than in the case of a standard canonical variate analysis with the promise of greater insight. Designed comparisons allow the objectives of a study to be addressed

Table 4.5. Partition of explained SSP matrix $\mathbf{Q}_h$, for the g-group multivariate analysis of variance

Source of dispersion	df	SSP (symmetric, $p \times p$)	Λ	χ^2
H(0)	1	$\mathbf{u}_0\mathbf{u}_0^t = [\quad] = \mathbf{Q}_h^{(0)}$		
H(1)	1	$\mathbf{u}_1\mathbf{u}_1^t = [\quad] = \mathbf{Q}_h^{(1)}$	$\frac{\lvert\mathbf{Q}_e\rvert}{\lvert\mathbf{Q}_e+\mathbf{Q}_h^{(1)}\rvert}$	$-m\log_e\Lambda^{(1)}$
H(2)	1	$\mathbf{u}_2\mathbf{u}_2^t = [\quad] = \mathbf{Q}_h^{(2)}$	$\frac{\lvert\mathbf{Q}_e\rvert}{\lvert\mathbf{Q}_e+\mathbf{Q}_h^{(2)}\rvert}$	$-m\log_e\Lambda^{(2)}$
⋮	⋮	⋮	⋮	⋮
H(g−1)	1	$\mathbf{u}_{g-1}\mathbf{u}_{g-1}^t = [\quad] = \mathbf{Q}_h^{(g-1)}$	$\frac{\lvert\mathbf{Q}_e\rvert}{\lvert\mathbf{Q}_e+\mathbf{Q}_h^{(g-1)}\rvert}$	$-m\log_e\Lambda^{(g-1)}$
Group means	g	$\sum_c \mathbf{u}_c\mathbf{u}_c^t = [\quad] = \mathbf{Q}_h = \mathbf{U}^t\mathbf{U}$		
Within groups	N−g	$[\quad] = \mathbf{Q}_e = \mathbf{Q}_t - \mathbf{U}^t\mathbf{U}$		
Total	N	$[\quad] = \mathbf{Q}_t$		

in a direct and meaningful way. Not only are comparisons tested on a higher power than the overall test of $\mathbf{Q}_h$ in Table 4.2, but, more importantly, perhaps, examination of the sample in subspaces defined by canonical variates associated with the roots $f^{(i)}$ in successive tests of (4.47) corresponding to different comparisons, $\mathbf{z}_i$, is likely to prove more informative than in terms of the canonical variates of an unstructured analysis. The procedure encourages the investigator to think about the effects of interest and to relate these to some underlying theory bearing on the material in question.

4.8 Concluding remarks

The formulation of the general discriminatory problem as a special case of canonical correlation in which one set of variables is binary in nature is due to Bartlett (1938). Minimal requirements of the data are that the joint distribution of the responses within-groups be approximately symmetric and not too long-tailed, and that the covariance structure be reasonably stable across groups. Where statistical tests are envisaged, requirements are more stringent, namely for independent samples from g multivariate Gaussian universes with constant dispersion.

Techniques for *probing* high-dimensional data for the purpose of obtaining insight into their structure have been described by Cox and Small (1978), Mardia (1980), Morrison (1976, p. 252), Gnanadesikan (1977, pp. 150, 196), Mardia, Kent and Bibby (1979, p. 140), Small (1980), Campbell (1981a), Hawkins (1981), Fatti, Hawkins and Raath (1982, p. 55) and Koziol (1982, 1983).

Gabriel's **h**-plot is an exploratory graphical procedure which may be used for the informal comparison of g covariance matrices (e.g., see Gabriel, 1981, p. 155). Campbell (1981a) has also developed graphical methods for comparing covariance matrices. The detection of multivariate *outliers* has been considered by Gnanadesikan (1977, p. 258), Campbell (1978a) and Campbell and Reyment (1980), while the use of *robust M-estimation* procedures to reduce the weight or influence of atypical observations in distribution-free and parametric applications of canonical variate analysis has been investigated by Campbell (1979a, Sect. 4.1; 1980b), Campbell and Reyment (1980) and Harner and Whitmore (1981). Accounts of the *geometry* of canonical variate analysis in the distribution-free spirit of the original Fisher-Rao derivations have been given by Jolicoeur (1959), Lubischew (1962), Rempe and Weber (1972), Green (1976, p. 285), Gnanadesikan (1977, p. 90), Albrecht (1980), Campbell (1980c) and Campbell and Atchley (1981). The effect of the *variances and covariances* of the original variables on group separation has been examined by Lubischew (1962), Phillips, Campbell and Wilson (1973) and Albrecht (1980). Methods for assessing the effects of *covariance heterogeneity* have been proposed by Dempster (1969, p. 206), Campbell (in Phillips, Campbell & Wilson, 1973), Campbell and Mahon (1974), Gnanadesikan (1977, p. 90) and Campbell (1978b, 1981a). Extensions of canonical variate analysis which are useful where the covariance matrices differ have been described by Campbell (1984).

The consequences of moderate *non-normality* on the likelihood ratio and union-intersection tests of hypotheses (4.21) and (4.22) of Sect. 4.5 have been studied by Davis (1980, 1982). Effects of differences in covariance structure have been considered by Campbell (see Phillips, Campbell & Wilson, 1973; Campbell & Mahon, 1974; Campbell, 1978b) and B.K. Williams (1981, 1983); cf. also Lachenbruch (1982). There is evidence that canonical variate analysis may be *robust* against moderate heteroscedasticity, particularly where groups are well-separated (Webster, 1977a, p. 192; Campbell, 1980c; Webster & McBratney, 1981; Campbell & Woodings, 1981; cf. also Porebsky, 1966, p. 228). Furthermore, a *distribution-free* generalization of canonical variate analysis in which the requirement for a relatively stable covariance structure is relaxed has been proposed by Digby and Gower (1981). Misra (1980) has described a method for estimating and testing the significance of discriminant functions where the covariance matrices are known to be *unequal*, while Mielke, Berry and Johnson (1976) have developed a class of *multivariate permutation tests* based on distance statistics whose underlying assumptions are more realistic than those of parametric tests and hence are more likely to be satisfied in practice. The estimation of covariance matrices in canonical variate analysis where there are *missing observations* has been investigated by Huseby, Schwertman and Allen (1980), Campbell (1982) and Smith and Riggs (1982). *Transformations* to enhance symmetry of distribution or to stabilize the covariance structure have been discussed by Gnanadesikan (1977, p. 137), Dunn and Tubbs (1980) and Dunn (1981). The *formulation* of canonical variate analysis as to two-stage principal component analysis has been discussed by Campbell (1979b) and by Campbell and Atchley (1981). An alternative formulation in terms of optimal, low-rank matrix approximation has been given by Greenacre and Underhill (1982, p. 205), while Green (1976, pp. 251, 285) has discussed canonical variate analysis in terms of the *simultaneous diagonalization* of two quadratic forms. Gower (1966) has described an efficient computational procedure for evaluating canonical variates when the number of groups, g, is less than the number of variables, p, and has also shown (Gower, 1968) how a point may be added to a graphical display of the canonical variates.

Several proposals to aid interpretability have been made. *Generalized ridge* or *shrunken* estimators for the canonical weights have been developed by Campbell and Reyment (1978), Campbell (1980a), DiPillo (1979), Reyment (1980) and Vinod and Ullah (1981, p. 298). The question of *variable selection* in canonical variate analysis has been studied by McKay and Campbell (1982a) while the *rotation* of canonical variates has been considered by Hall (1977), Tukey and Tukey (1981, p. 227) and Bentler and Huba (1982). Canonical variate analysis, like other maximization procedures, capitalizes on chance features of the data to which it is applied. A need accordingly arises following analysis to protect against an overly optimistic assessment of the results obtained. A *jackknife* procedure has been proposed by Lachenbruch and Mickey (1968), Dempster (1969, p. 257), Geisser (1977) and by Fatti, Hawkins and Raath (1982, p. 15) for this purpose. Graphical displays are unsurpassed for communicating results and guiding their interpretation. Where the dimensionality t of a solution is such that $t=2$ or $t=3$ *scattergrams* and *stereograms* are invaluable. Such displays can be supplemented or specialized in several ways. The *minimum spanning tree*, for example, may be used to incorporate information on the proximities of group-centroids in the full s-dimensional space of an analysis. Examples are provided by Jeffers (1978, p. 178), Oxnard (1979) and Gelvin (1980). Gabriel's $\mathbf{GH}^t$- and $\mathbf{JK}^t$-*biplots* (Gabriel, 1981, p. 152; Cox & Gabriel, 1982, p. 52) permit both the response variables and summary ellipses or ellipsoids for the sample-groups to be displayed simultaneously, as Tsianco, Gabriel, Odoroff and Plumb (1981) have shown. An alternative simultaneous display of several p-variate groups is given by the *generalized Procrustes analysis* outlined by Gower and Digby (1981, p. 110). Where $t \geq 3$, Andrews' (1972) *sine-cosine plot* for displaying high-dimensional data is available. For a discussion and examples of the application of sine-cosine plots in canonical variate analysis, see Andrews (1972) and Oxnard (1973, 1983).

The applicability of canonical variate analysis can be widened in a number of ways. We have seen that *designed comparisons* enable research objectives to be addressed in a meaningful way. The opportunities provided have been described and illustrated by E.J. Williams (1967), Hope (1968, p. 129), Pruzek (1971), Kshirsagar (1972, p. 366), Madsen (1977) and Merembeck and Turner (1980), while the extension of designed comparisons from one-way to *higher-order classifications* has been treated by Dempster (1969, p. 222), Pruzek (1971), Kshirsagar (1972, p. 373), Corsten (1976b) and by Seif, Evans and Balaam (1979). The same concepts have been discussed in the terminology of the multivariate analysis of variance by Tatsuoka (1971, p. 200), Finn (1974, p. 219), Bock (1975, p. 236) and Timm (1975, pp. 374, 414). Robson (1959) has shown how orthogonal polynomials can be applied to unequally spaced treatments while Lancaster (1965) has given an account of the Helmert matrices and their application. *Graphical methods* for assessing single-degree-of-freedom contrasts have been discussed by Gnanadesikan (1977, p. 227). The availability of g pre-existing groups is normally presupposed in canonical variate analysis. The requirement for pre-existing groups can, however, be relaxed. Hawkins and his co-workers (Hawkins & Merriam, 1974; Hawkins & Ten Krooden, 1979) have generalized the applicability of canonical variate analysis to situations where the samples from an *ordered sequence*, as they do for example along a transect, rather than discrete groups. Their method is based on identifying the $g-1$ changepoints which in a defined sense optimally divide the sequence into g relatively homogeneous segments. This proves to be a problem in cluster analysis with constraints on cluster membership (Gordon, 1981, p. 61; Hawkins, Muller & Ten Krooden, 1982, p. 349). The approach has been further extended to the analysis of *spatial* data (see Hawkins, Muller & Ten Krooden, 1982, p. 345). Switzer (1980) has described a refinement of linear discriminant analysis which adjusts for *spatial autocorrelation* in the analysis of spatial data. Finally, we observe that a *growth-free* version of canonical variate analysis which enables growth or other time-dependent effects to be partialed-out has been proposed by Gower (1976).

Further reading. Applications of canonical variate analysis in ecology or related areas have been made by several workers. Among these, attention is drawn to studies by Rao (1948), Mahalanobis, Majumdar and Rao (1949), Ashton, Healy and Lipton (1957), Jolicoeur (1959), Nair and Mukerji (1960), Delany and Healy (1966), Dempster (1969, pp. 193 and 222), Phillips, Campbell and Wilson (1973), Campbell and Mahon (1974), Lisowski, Albrecht and Oxnard (1974, 1976), Ashton, Flinn and Oxnard (1975), Ashton, Flinn, Oxnard and Spence (1976), Campbell and Saunders (1976), Hopper and Campbell (1977), Oxnard (1979, 1983), Campbell (1979b), Campbell and Dearn (1980), Campbell and Kitchener (1980), Campbell and Reyment (1980) and Merembeck and Turner (1980). *Growth-free* canonical variate analysis has been used by Reyment and Banfield (1976) and Reyment (1982). Campbell (1976) has discussed relationships between canonical variate analysis and *principal component analysis* using the empirical results of parallel analyses of one and the same set of data. A novel comparison of canonical variate analysis and *canonical correlation analysis* has been made by Nash (see Falkenhagen & Nash, 1978). Nash has shown that in investigations of geographical variation, where in addition to observations on p responses certain auxiliary information is also available, canonical correlation analysis can lead to greater group separation than canonical variate analysis provided the canonical variates are appropriately standardized. Finally, in this section we remark that Webster and McBratney (1981) have described an application of canonical variate analysis to *transect data* – that is to data from an ordered series of samples in which groups are not initially defined.

5. Dual scaling

5.1 Introduction

In the previous chapter we saw that canonical analysis can be applied to a sample of p-variate observations partitioned into $g \geq 2$ classes for the purpose of obtaining insight into relationships among the classes. Canonical variate analysis was introduced as a special case of canonical analysis in which the data matrix for one of the two sets of variables is specialized so as to account for the partition of a sample. Suppose now that we have a sample of individuals which are simultaneously classified with respect to two categorical variables. Dual scaling may be set up as a double canonical variate analysis in which both data matrices are specialized so that each corresponds to one of the two variables of classification for the analysis of data of this kind. Doubly-classified observations arise frequently. In palynology, for example, the data generally consist of counts of fossil pollen taxa at a number of stratigraphic levels. In a similar way, studies of natural communities often lead to estimates of the density of different animal species in a variety of plant communities. The data in such cases are conveniently organized in the form of a 2-way array or table. We may then enquire as to the nature of the relationships within and between the row and column categories of the table. This is the question addressed by dual scaling. Dual scaling is in fact applicable to an m-way classification of individuals ($m \geq 2$), though for simplicity we shall confine our attention to the case where m is strictly equal to two.

The term dual scaling is used here in a generic sense to refer to a family of related methods. The family includes *optimal scaling* (Bock, 1960; Young, 1981), *reciprocal averaging* (Horst, 1935; Hill, 1973), *correspondence analysis* (Benzécri, 1973; Hill, 1974; Greenacre, 1981; Greenacre & Underhill, 1982), and the *canonical analysis of contingency tables* (Fisher, 1938; Williams, 1952; Lancaster, 1958; Good, 1965; Corsten, 1976a). Broadly speaking, these are all procedures for quantifying qualitative data. Two separate though related themes are contained within the general approach. One theme, apparent in the work of Fisher (1938, 1940) and developed further by E.J. Williams (1952) and others, centers on the analysis of association in contingency tables. In this approach, numerical values or *scores* are sought for the row and column categories of the original table which have certain optimal properties in terms of correlations. The second theme is more recent and is directed towards obtaining a *low-rank approximation* of the original table, which, in the least squares sense, minimizes the approximation error. This approach leads to a graphical representation of the original table in which rows and columns are simultaneously displayed as points in a low-dimensional vector space. The graphical component of dual scaling has been emphasized by Benzécri (1973), Corsten (1976a), Greenacre (1981), Greenacre and Underhill (1982) and Heiser and Meulman (1983a). The correlational and graphical approaches are indeed substantially equivalent, the difference between them being principally one of emphasis. The term dual scaling is due to Nishisato (1980), who has traced the historical development of the method and also given a comprehensive account of the procedure itself. The important contributions of the French School of mathematicians to the subject, however, are not covered by Nishisato's review. Reference may be made to Benzécri (1977) for an account of this work. Here we shall not attempt to do more than provide a general introduction to the method. In particular, we shall emphasize those features of dual scaling which reflect its affiliation with canonical analysis.

We saw in Chap. 4 that $g-1$ dummy variables can be used to carry information on the partition of a sample into g groups. In canonical variate analysis, linear functions of such dummy variables may be sought which are maximally correlated with linear functions of p response variables in order to sort out relationships among the groups. An alternative formulation of canonical variate analysis leads to a search for linear functions of the responses which maximize the ratio of between-groups to within-groups variation and best discriminate the groups. Both formulations extend readily to a sample of doubly-classified individuals. In cases involving a doubly-classified sample, *two* sets of dummy variables may be constructed to represent the joint classification. Linear functions of both sets may then be sought for which the correlation coefficient is as large as possible. Alternatively, the ratio of between-groups to within-groups variation in linear compounds of the row and column entries might simultaneously be optimized with the aim of separating the row categories and the column categories to the greatest possible extent. Dual scaling can therefore be regarded as both:

(a) a special case of canonical correlation analysis in which both sets of variables are binary dummy variables; and as
(b) a generalization of canonical variate analysis from a one-way to a two-way classification.

In the correlational formulation, scores are sought for row and column categories of the initial table which are maximally correlated; in the discriminative formulation, scores which maximally distinguish between row categories and between column categories are sought. We shall see that both formulations lead to one and the same solution. In addition, dual scaling can be formulated in several other ways.

The derivation of dual scaling is taken up in Sect. 5.2. Properties of the solution are described in Sect. 5.3 and in Sect. 5.4 hypothesis testing is addressed. Section 5.5 is devoted to a review of relationships between dual scaling and a number of other statistical methods. The next section of the chapter, Sect. 5.6, is not primarily concerned with dual scaling. At several places in the text reference has been made to connections between canonical analysis and other analytical procedures and Sect. 5.6 is devoted to a general review of relationships of this kind. A worked, numerical example of dual scaling based on the theory developed in the present chapter forms the substance of Chap. 12.

5.2 Formulation and derivation

Consider a sample of N observations simultaneously classified with respect to two attributes whose states define the rows and columns of a two-dimensional array. Interest might be expressed in obtaining indices for the rows and columns of the array with the aim of clarifying relationships within and between the two sets of attribute-states. Numerous criteria exist whose optimization might be expected to yield the desired values. Of these criteria we shall consider just two. First, the correlation coefficient between linear functions of two sets of dummy variables generated to correspond respectively to the row and column classifications; secondly, a ratio of quadratic forms in linear functions of the row or column categories. Initially we shall confine our attention to a particular kind of two-dimensional array – the $r \times c$ contingency table. Later, we will see that dual scaling is not confined to arrays of this kind. Before proceeding we shall require some notation.

Table 5.1. $r \times c$ contingency table. N observations simultaneously classified with respect to two characteristics, A and B

A \ B	B_1	…	B_j	…	B_c	Total
A_1	x_{11}	…	x_{1j}	…	x_{1c}	r_1
⋮	⋮	⋱	⋮	⋱	⋮	⋮
A_i	x_{i1}	…	x_{ij}	…	x_{ic}	r_i
⋮	⋮	⋱	⋮	⋱	⋮	⋮
A_r	x_{r1}	…	x_{rj}	…	x_{rc}	r_r
Total	c_1	…	c_j	…	c_c	N

Let $\mathbf{X}(r \times c)$ denote a contingency table with N individuals classified by reference to two attributes, A and B (see Table 5.1). Attribute-states are represented by the $r=p+1$ rows $\{A_i\}$ and the $c=q+1$ columns $\{B_j\}$, respectively. There are x_{ij} individuals possessing the ith row and jth column attribute-states. Letting $\mathbf{1}$ be a vector of ones, the row totals are given by the $r \times 1$ vector $\mathbf{r} = \mathbf{X1}$, the column totals by the $c \times 1$ vector $\mathbf{c} = \mathbf{X}^t\mathbf{1}$ and the total sample by $N = \mathbf{r}^t\mathbf{1} = \mathbf{c}^t\mathbf{1}$. Finally, let the diagonal matrix of row totals, diag($\mathbf{r}$), be denoted by $\mathbf{R}(r \times r)$ and the corresponding matrix of column totals, diag($\mathbf{c}$), by $\mathbf{C}(c \times c)$. We wish to explore the relationship between the characteristics A and B.

5.2.1 Maximizing the correlation between rows and columns

An indication of the extent of linear correlation between A and B might be expected to be informative in view of our declared interest. One plausible avenue of approach would be to derive scores for the rows and columns so as to maximize the correlation for the table. This is the general problem addressed by canonical correlation analysis. Where the data of interest form a contingency table, it is convenient to re-express the table as a *partitioned* data matrix – that is, as a matrix whose rows correspond to samples and whose columns correspond to two sets of distinct though associated variables – in order to perform a canonical analysis. Re-expression is accomplished by defining two sets of binary-valued dummy variables as described below.

Let $\mathbf{z}_1^t(1 \times p)$ and $\mathbf{z}_2^t(1 \times q)$, where $p = r-1$ and $q = c-1$, be vectors of binary dummy variables corresponding respectively to the first $r-1$ rows and first $c-1$ columns of $\mathbf{X}$. Now set the elements of $\mathbf{z}_1^t$ and $\mathbf{z}_2^t$ for the hth sample ($h = 1, \ldots, \mathrm{N}$) to:

$$z_{hi} = \begin{cases} 1 & \text{if the } h\text{th sample belongs to } \mathrm{A}_i \; (i=1, \ldots, p) \\ 0 & \text{otherwise} \end{cases}$$
$$z_{hj} = \begin{cases} 1 & \text{if the } h\text{th sample belongs to } \mathrm{B}_j \; (j=1, \ldots, q) \\ 0 & \text{otherwise.} \end{cases} \tag{5.1}$$

Table 5.2 illustrates the procedure. $\mathbf{X}$ is a 3×4 array containing $\mathrm{N} = 10$ doubly-classified individuals. $\mathbf{Z}_r(\mathrm{N} \times r)$ is the *incidence* matrix for the rows of $\mathbf{X}$, that is the matrix whose columns $\mathbf{z}_r^{(i)}$ ($i = 1, \ldots, r$) register the occurrence of individuals in the rows of $\mathbf{X}$. To see how $\mathbf{Z}_r$ was constructed, consider the four samples belonging to the first row of $\mathbf{X}$. Each of these samples has been assigned a value of 1 on $\mathbf{z}_r^{(1)}$ and of 0 on $\mathbf{z}_r^{(2)}$ and $\mathbf{z}_r^{(3)}$ (Table 5.2b). Proceeding to the second row of $\mathbf{X}$, each of the four samples has been assigned values of 1 on $\mathbf{z}_r^{(2)}$ and of 0 on $\mathbf{z}_r^{(1)}$ and $\mathbf{z}_r^{(3)}$; and so on for the remaining row. The $\mathrm{N} \times c$ matrix $\mathbf{Z}_c$ is the incidence matrix for the columns of $\mathbf{X}$. In other words, the columns $\mathbf{z}_c^{(j)}$ ($j = 1, \ldots, c$) of $\mathbf{Z}_c$ register the occurrence of individuals in the columns of $\mathbf{X}$. $\mathbf{Z}_c$ was obtained from $\mathbf{X}$ rowwise, one row at a time. We find, for example, one individual in the first row and the first column of $\mathbf{X}$; this individual has been given a score of 1 on $\mathbf{z}_c^{(1)}$ and of 0 on $\mathbf{z}_c^{(2)}$, $\mathbf{z}_c^{(3)}$ and $\mathbf{z}_c^{(4)}$. The two individuals in the first row and second column of $\mathbf{X}$ have each been assigned scores of 1 on $\mathbf{z}_c^{(2)}$ and of 0 on $\mathbf{z}_c^{(1)}$, $\mathbf{z}_c^{(3)}$ and $\mathbf{z}_c^{(4)}$. There is one individual in the first row and third column of $\mathbf{X}$ and this has been given a score of 1 on $\mathbf{z}_c^{(3)}$ and scores of 0 on the variables $\mathbf{z}_c^{(j)}$ ($j = 1, 2, 4$). The process is continued in the same way until all the individuals have been accounted for. Finally, a partitioned data matrix is constructed from the row and column incidence matrices. A suitable matrix $\mathbf{Z} = [\mathbf{Z}_1 | \mathbf{Z}_2]$, of order $\mathrm{N} \times (p+q)$, is produced by merging the leading $r-1$ columns of $\mathbf{Z}_r$ and the leading $c-1$ columns of $\mathbf{Z}_c$, as shown in Table 5.2c. $\mathbf{Z}_1$ denotes the $\mathrm{N} \times p$ matrix given by the first $r-1$ columns of $\mathbf{Z}_r$ and $\mathbf{Z}_2$ the $\mathrm{N} \times q$ matrix given by the first $c-1$ columns of $\mathbf{Z}_c$. Notice that in $\mathbf{Z}$ samples belonging to the rth row of $\mathbf{X}$ are uniquely specified by scores of 0 on each $\mathbf{z}_1^{(i)}$

Table 5.2. Re-expression of a doubly-classified table $\mathbf{X}(r \times c)$ as a partitioned data matrix $\mathbf{Z}(\mathrm{N} \times (p+q))$ suitable for canonical analysis

(a) $r \times c$ table $\mathbf{X}$ with marginal row and column totals.

$$\mathbf{X} = \begin{array}{|cccc|c} \hline 1 & 2 & 1 & 0 & 4 \\ 2 & 1 & 0 & 1 & 4 \\ 1 & 0 & 0 & 1 & 2 \\ \hline 4 & 3 & 1 & 2 & 10 \end{array}$$

(b) Incidence matrices $\mathbf{Z}_r$ and $\mathbf{Z}_c$ registering the occurrence of individuals in the rows and columns of $\mathbf{X}$, respectively.

$$\mathbf{Z}_r = \begin{bmatrix} 1 & 0 & 0 \\ 1 & 0 & 0 \\ 1 & 0 & 0 \\ 1 & 0 & 0 \\ 0 & 1 & 0 \\ 0 & 1 & 0 \\ 0 & 1 & 0 \\ 0 & 1 & 0 \\ 0 & 0 & 1 \\ 0 & 0 & 1 \end{bmatrix}, \qquad \mathbf{Z}_c = \begin{bmatrix} 1 & 0 & 0 & 0 \\ 0 & 1 & 0 & 0 \\ 0 & 1 & 0 & 0 \\ 0 & 0 & 1 & 0 \\ 1 & 0 & 0 & 0 \\ 1 & 0 & 0 & 0 \\ 0 & 1 & 0 & 0 \\ 0 & 0 & 0 & 1 \\ 1 & 0 & 0 & 0 \\ 0 & 0 & 0 & 1 \end{bmatrix}$$

Note the following relationships:

$$\mathbf{Z}_r'\mathbf{Z}_r = \mathrm{diag}(\mathbf{r}) = \begin{bmatrix} 4 & 0 & 0 \\ 0 & 4 & 0 \\ 0 & 0 & 2 \end{bmatrix} \qquad \mathbf{Z}_c'\mathbf{Z}_c = \mathrm{diag}(\mathbf{c}) = \begin{bmatrix} 4 & 0 & 0 & 0 \\ 0 & 3 & 0 & 0 \\ 0 & 0 & 1 & 0 \\ 0 & 0 & 0 & 2 \end{bmatrix} \qquad \mathbf{Z}_r'\mathbf{Z}_c = \mathbf{X} = \begin{bmatrix} 1 & 2 & 1 & 0 \\ 2 & 1 & 0 & 1 \\ 1 & 0 & 0 & 1 \end{bmatrix}$$

(c) Partitioned data matrix $\mathbf{Z} = [\mathbf{Z}_1 | \mathbf{Z}_2]$ where $\mathbf{Z}_1$ consists of the leading $r-1$ columns of $\mathbf{Z}_r$ and $\mathbf{Z}_2$ of the leading $c-1$ columns of $\mathbf{Z}_c$.

$$\mathbf{Z} = [\mathbf{Z}_1 | \mathbf{Z}_2] = \left[\begin{array}{cc|ccc} 1 & 0 & 1 & 0 & 0 \\ 1 & 0 & 0 & 1 & 0 \\ 1 & 0 & 0 & 1 & 0 \\ 1 & 0 & 0 & 0 & 1 \\ 0 & 1 & 1 & 0 & 0 \\ 0 & 1 & 1 & 0 & 0 \\ 0 & 1 & 0 & 1 & 0 \\ 0 & 1 & 0 & 0 & 0 \\ 0 & 0 & 1 & 0 & 0 \\ 0 & 0 & 0 & 0 & 0 \end{array}\right]$$

$(i=1, \ldots, p)$ and that samples belonging to the cth column of $\mathbf{X}$ are similarly specified by scores of 0 on each $\mathbf{z}_2^{(j)}$ $(j=1, \ldots, q)$.

The relationship between A and B is now the relationship between $\mathbf{z}_1$ and $\mathbf{z}_2$. The theory developed in Chap. 2 and later chapters can therefore be applied. This leads us to seek linear transformations from $\mathbf{z}_1$ to $u=\mathbf{a}^t\mathbf{z}_1$ and from $\mathbf{z}_2$ to $v=\mathbf{b}^t\mathbf{z}_2$ for which the correlation coefficient r_{uv} is as large as possible.

Let the incidence matrix registering the occurrence of samples in the first $r-1$ rows of $\mathbf{X}$ be denoted by $\mathbf{Z}_1(\mathrm{N}\times p)$ and the incidence matrix for the first $c-1$ columns by $\mathbf{Z}_2(\mathrm{N}\times q)$. Define the $\mathrm{N}\times(p+q)$ matrix $\mathbf{Z}=[\mathbf{Z}_1|\mathbf{Z}_2]$. The sample matrix of mean-corrected sums of squares and products (2.2) is

$$\underset{(p+q)\times(p+q)}{\mathrm{N}\mathbf{S}}=\mathbf{Z}^t\mathbf{H}\mathbf{Z}=\mathbf{Z}^t\mathbf{Z}-\mathrm{N}\bar{\mathbf{z}}^t\bar{\mathbf{z}}=\mathrm{N}\begin{bmatrix}\mathbf{S}_{11} & \mathbf{S}_{12}\\ \mathbf{S}_{21} & \mathbf{S}_{22}\end{bmatrix}, \tag{5.2}$$

where $\mathbf{H}=\mathbf{I}-(\mathrm{N}^{-1})\mathbf{1}\,\mathbf{1}^t$ is the centering matrix of order N. Now, following equation (2.20), write

$$\underset{(p\times q)}{\mathbf{K}}=\underset{(p\times p)}{[\mathbf{S}_{11}^t]^{-\frac{1}{2}}}\;\underset{(p\times q)}{\mathbf{S}_{12}}\;\underset{(q\times q)}{\mathbf{S}_{22}^{-\frac{1}{2}}}. \tag{5.3}$$

A convenient form of the singular value decomposition of $\mathbf{K}$ may be written

$$\underset{(p\times q)}{\mathbf{K}}=\underset{(p\times s)}{\mathbf{P}}\;\underset{(s\times s)}{\mathbf{D}}\;\underset{(s\times q)}{\mathbf{Q}^t}, \tag{5.4}$$

where $\mathbf{P}^t\mathbf{P}=\mathbf{Q}^t\mathbf{Q}=\mathbf{I}_s$, $\mathbf{D}=\mathrm{diag}(d_1,\ldots,d_s)$ and $s=\min(p,q)$. The matrix $\mathbf{P}$ consists of the s orthonormalized eigenvectors associated with the s largest eigenvalues of $\mathbf{K}\mathbf{K}^t$, the matrix $\mathbf{Q}$ of the s corresponding orthonormalized eigenvectors of $\mathbf{K}^t\mathbf{K}$ and the d_i are squareroots of the eigenvalues of either of the symmetric, positive-semidefinite matrices $\mathbf{K}\mathbf{K}^t$ or $\mathbf{K}^t\mathbf{K}$.

The matrices of canonical weights $\mathbf{A}$ and $\mathbf{B}$ are obtained from the singular value decomposition (5.4) as

$$\underset{(p\times s)}{\mathbf{A}}=\underset{(p\times p)}{\mathbf{S}_{11}^{-\frac{1}{2}}}\;\underset{(p\times s)}{\mathbf{P}}\;\underset{(s\times s)}{\mathbf{D}^{\beta}} \tag{5.5}$$

$$\underset{(q\times s)}{\mathbf{B}}=\underset{(q\times q)}{\mathbf{S}_{22}^{-\frac{1}{2}}}\;\underset{(q\times s)}{\mathbf{Q}}\;\underset{(s\times s)}{\mathbf{D}^{\beta}}, \tag{5.6}$$

where the parameter β determines the scaling. Choices of $\beta=0$, $\frac{1}{2}$ or 1 are common. The columns of $\mathbf{A}=[\mathbf{a}_1|\ldots|\mathbf{a}_s]$ and $\mathbf{B}=[\mathbf{b}_1|\ldots|\mathbf{b}_s]$ contain the required numerical indices for the row and column categories of $\mathbf{X}$. The canonical variates u_k and v_k $(k=1,\ldots,s)$ might now be calculated as $u_k=\mathbf{a}_k^t\mathbf{z}_1$ and $v_k=\mathbf{b}_k^t\mathbf{z}_2$. In practice, however, this step is not required. To see this, notice from Table 5.2c that for a given sample the p-vector of dummy variables $\mathbf{z}_1=[z_1,z_2\ldots z_p]^t$ contains only *one* non-zero component. Consequently, $u_k=a_{1k},a_{2k},\ldots,$ or a_{pk}. In other words, the canonical variate score of a sample belonging to the ith row of $\mathbf{X}$ is precisely the weight attached to the dummy variable z_i corresponding to the ith row. Similarly, $v_k=b_{1k},b_{2k},\ldots,$ or b_{qk}. The score (u_i,v_j) of a sample belonging to the ith row category and the jth column category of $\mathbf{X}$ is therefore identically (a_i,b_j). There will be x_{ij} samples with the score (u_i,v_j), or, equivalently, (a_i,b_j). Observe also that the scores of samples belonging to the rth row or cth column of $\mathbf{X}$ will be zero because of the definition of $\mathbf{z}_1$ and $\mathbf{z}_2$ in (5.1). As a consequence, the distribution of scores along a given canonical variate is not centered about an origin located at the mean, which is the custom usually adopted. To obtain scores which are mean-centered, first calculate the mean of u_k and v_k $(k=1, \ldots, s)$, $\bar{u}_k = N^{-1} \Sigma u_{jk}$ and $\bar{v}_k = N^{-1} \Sigma v_{jk}$, respectively, where the summations are over $j=1, \ldots, N$. The required scores are then the deviations $u_{jk} - \bar{u}_k$ and $v_{jk} - \bar{v}_k$.

It may be recalled from Sect. 2.3.2 that the kth singular value, d_k, of $\mathbf{K}$ is the canonical correlation coefficient r_k between the variates u_k and v_k. There are $s = \min(p, q)$ such non-zero correlation coefficients. An interesting relationship exists among the canonical correlation coefficients, namely that the kth coefficient, r_k, is, apart from sampling effects, the kth power of the largest coefficient, r_1. Put somewhat differently, the correlation between the kth pair of canonical variates, u_k and v_k $(k>1)$, is given by $|r_1|^k$.

Frequently, the relationship between A and B can be described satisfactorily by the canonical variates corresponding to some subset $t<s$ of these canonical correlations. An informal procedure for choosing t can be arrived at as follows. The trace of $\mathbf{K}\mathbf{K}^t$ expresses the total predictable or shared variance of the data. The proportion of this quantity which is accounted for by t retained roots is therefore

$$C_t = \sum_{k=1}^{t} r_k^2 / \mathrm{tr}(\mathbf{K}\mathbf{K}^t). \tag{5.7}$$

In other words, C_t represents the amount of information retained in the reduction from s to t dimensions. This quantity may be evaluated for several trial values of t, while noting the extent to which meaningful substantive interpretation of the corresponding score vectors is possible as t tends to s. The required value will be the smallest value of t for which C_t is judged acceptably large. Values of $t=2$ and $t=3$ are not uncommon in practice. In these cases the resulting solution has the advantage of yielding a graphical representation of $\mathbf{X}$. Formal test procedures may also sometimes be useful in assessing dimensionality. Such tests are described in Sect. 5.4.2.

The scaling of canonical variates is arbitrary. The usual practice in canonical correlation analysis is to standardize the variates to unit variance. This is the condition specified in Eq. (2.6). Setting the scale parameter β in Eqs. (5.5) and (5.6) to $\beta=0$ leads to canonical variates which are scaled in this way. In dual scaling, however, there are advantages in standardizing the kth pair of variates, u_k and v_k, such that their variances $(N-1)^{-1} \Sigma_j (u_{jk} - \bar{u}_k)^2$ and $(N-1)^{-1} \Sigma_j (v_{jk} - \bar{v}_k)^2$ are equal to the canonical root r_k^2 rather than unity. In this case we require

$$\mathrm{var}(u_k) = \mathbf{a}_k^t \mathbf{S}_{11} \mathbf{a}_k = r_k^2, \tag{5.8}$$

$$\mathrm{var}(v_k) = \mathbf{b}_k^t \mathbf{S}_{22} \mathbf{b}_k = r_k^2, \tag{5.9}$$

for $k=1, \ldots, s$. The appropriately scaled vectors of weights $\mathbf{a}_k$ and $\mathbf{b}_k$ are obtained from (5.5) and (5.6), respectively, on setting the parameter β to $\beta=1$. The resulting canonical variates are appealing in that the contribution of the root r_k^2 to the overall relationship between A and B is absorbed by the corresponding variates, u_k and v_k. As $r_1^2 \geq r_2^2 \geq \ldots \geq r_t^2$, those aspects of the relationship corresponding to the larger roots are emphasized. Consequently, the results of analysis are summarized and communicated by the graphical procedures of Sect. 2.4.3 in a highly effective way. A refinement, due to Nash (1983), leads to r_k^2 on the right-hand side of (5.8) and (5.9) being replaced by $(1-r_k^2)^{-1}$, or by a simple multiple of this quantity which takes account of degrees of freedom. Vectors of weights $\mathbf{a}_k'$ and $\mathbf{b}_k'$ standardized so as to yield canonical variates scaled in accordance with this scheme can be obtained directly from the vectors $\mathbf{a}_k$ and $\mathbf{b}_k$ comprising the columns of $\mathbf{A}$ and $\mathbf{B}$ in (5.5) and (5.6) with $\beta=0$ by the relations

$$\mathbf{a}_k' = \left\{\frac{(N-c)}{(N-1)(1-r_k^2)}\right\}\mathbf{a}_k, \quad \mathbf{b}_k' = \left\{\frac{(N-r)}{(N-1)(1-r_k^2)}\right\}\mathbf{b}_k. \tag{5.10}$$

5.2.2 Maximizing the separation between rows and columns

The preceding derivation was motivated by a desire to maximize the correlation between linear functions of variables representing the row and column classifications of $\mathbf{X}$. Suppose now that in the spirit of canonical variate analysis we set out to discriminate between the rows *and* between the columns of the contingency table to the maximum possible extent. One possibility would be to seek scores for the rows with the property that a linear function of them will best differentiate the columns; and, in a similar way, to seek scores for the columns such that a linear function of them will differentiate the rows to the maximum possible extent.

Consider first the rows. We shall sketch a two-stage procedure, obtaining first a set of indices for the columns from which the optimal row scores will be derived as linear functions. The rows of $\mathbf{X}$ can be regarded as constituting a one-way classification of the sample, $\mathbf{X}^t=[\mathbf{x}_1|\dots|\mathbf{x}_r]$. The elements $x_{i1}, x_{i2}, \dots, x_{ic}$ of the ith row $\mathbf{x}_i^t$ specify the number of individuals belonging to the jth column attribute-state $(j=1, \dots, c)$. We shall obtain a linear transformation from the column attribute-states to a new variable for which the variation between rows is maximized relative to the total variation. Nishisato (1980, p. 28) has shown that the criterion to be optimized is

$$\frac{\mathbf{b}^t \mathbf{Q}_h \mathbf{b}}{\mathbf{b}^t \mathbf{Q}_t \mathbf{b}} = f(\mathbf{b}), \qquad |\mathbf{Q}_t| \neq 0, \tag{5.11}$$

where $\mathbf{Q}_h$ and $\mathbf{Q}_t$ are SSP matrices of order c and $\mathbf{b}(c\times 1)$ is a vector of coefficients to be determined. The derivation will be simplified by requiring $\mathbf{b}$ to satisfy the condition $\mathbf{b}^t\mathbf{c}=0$, which asserts that the weighted sum of column scores shall be zero. Then, the matrix $\mathbf{Q}_h$ in the numerator of (5.11) may be computed directly from $\mathbf{X}$ as $\mathbf{X}^t\mathbf{R}^{-1}\mathbf{X}$, where $\mathbf{R}(r\times r)$ is the diagonal matrix of row totals, while $\mathbf{Q}_t$ in the denominator is identically the diagonal matrix $\mathbf{C}$ $(c\times c)$ of column totals. Under the restriction $\mathbf{b}^t\mathbf{c}=0$, the criterion to be optimized is therefore

$$\frac{\mathbf{b}^t[\mathbf{X}^t\mathbf{R}^{-1}\mathbf{X}]\,\mathbf{b}}{\mathbf{b}^t\mathbf{C}\mathbf{b}} = f(\mathbf{b}). \tag{5.12}$$

The required vector of column indices is the vector $\mathbf{b}$ which maximizes $f(\mathbf{b})$. To restrict (5.12) to a unique solution a convenient side condition is to require the variance of the transformed variable to be unity, $(\mathrm{N}-1)^{-1}\mathbf{b}^t\mathbf{C}\mathbf{b}=1$. Maximizing (5.12) with respect to $\mathbf{b}$ then yields the generalized eigenequation

$$(\mathbf{X}^t\mathbf{R}^{-1}\mathbf{X}-\mathrm{r}^2\mathbf{C})\,\mathbf{b}=\mathbf{0}, \tag{5.13}$$

where r^2 is a latent root of $\mathbf{C}^{-1}\mathbf{X}^t\mathbf{R}^{-1}\mathbf{X}$. Now $\mathbf{C}$, $\mathbf{R}$ and $\mathbf{X}$ are matrices of (uncorrected) sums of squares and cross-products, as may be verified by reference to Table 5.2b, so that the quadruple matrix product $\mathbf{C}^{-1}\mathbf{X}^t\mathbf{R}^{-1}\mathbf{X}$ has the form of the matrix $\mathbf{S}_{22}^{-1}\mathbf{S}_{21}\mathbf{S}_{11}^{-1}\mathbf{S}_{12}$ encountered implicitly in Eq. (2.14). The latent roots r_k^2 $(k=0, \ldots, s)$ of $\mathbf{C}^{-1}\mathbf{X}^t\mathbf{R}^{-1}\mathbf{X}$ are therefore *canonical* roots. The nontrivial solutions of (5.13) are the required stationary points and values of (5.12). In applications, $\mathbf{C}$ generally proves to be positive definite. Equation (5.13) is therefore conveniently solved after a transformation to an equivalent standard symmetric form. Writing $\mathbf{C}^{\frac{1}{2}}$ for the Cholesky factor of $\mathbf{C}$, (5.13) is re-expressed as the symmetric eigenequation

$$(\mathbf{C}^{-\frac{1}{2}}\mathbf{X}^t\mathbf{R}^{-1}\mathbf{X}\mathbf{C}^{-\frac{1}{2}}-\mathrm{r}^2\mathbf{I})\,\mathbf{b}^*=\mathbf{0}, \tag{5.14}$$

where $\mathbf{b}^*=\mathbf{C}^{\frac{1}{2}}\mathbf{b}$. Once $\mathbf{b}^*$ is obtained, $\mathbf{b}$ is readily computed as

$$\mathbf{b}=\mathbf{C}^{-\frac{1}{2}}\mathbf{b}^*. \tag{5.15}$$

The condition $\mathbf{b}^t\mathbf{c}=0$ is responsible for a solution of (5.13) which has the form $\mathrm{r}_0^2=1$, $\mathbf{b}_0=\mathbf{1}(c\times 1)$ but which does not correspond to a variance ratio. We shall henceforth disregard this *extraneous* solution, confining our attention to the remaining roots r_k^2 and vectors $\mathbf{b}_k$ $(k=1, \ldots, s)$. With this understanding in mind, the maximum of (5.12) is given by the largest root, r_1^2, of $\mathbf{C}^{-\frac{1}{2}}\mathbf{X}^t\mathbf{R}^{-1}\mathbf{X}\mathbf{C}^{-\frac{1}{2}}$.

The elements $b_1, \ldots, b_c$ of $\mathbf{b}$ are indices for the columns of $\mathbf{X}$ from which the $r\times 1$ vector of most discriminative row scores, $\mathbf{a}$, is yielded by the relation

$$\mathbf{a}=\mathrm{r}^{-1}\mathbf{R}^{-1}\mathbf{X}\mathbf{b}. \tag{5.16}$$

As it is the maximum possible separation of rows which is required, it follows that it is the largest root r_1^2 and associated vector $\mathbf{b}_1$ of (5.13) which are needed in (5.16). However, disregarding r_0^2 and $\mathbf{b}_0$, Eq. (5.13) has a total of $s=\min(p, q)$ solutions $\{\mathrm{r}_k^2; \mathbf{b}_k\}$ for $k=1, \ldots, s$, some or all of which will be of interest if further separation of the row categories is possible. The complete solution is therefore

$$\mathbf{a}_k=\mathrm{r}_k^{-1}\mathbf{R}^{-1}\mathbf{X}\mathbf{b}_k, \tag{5.17}$$

where r_k and $\mathbf{b}_k$ are solutions of (5.13) when $\mathrm{r}^2=\mathrm{r}_k^2$ $(k=1, \ldots, s)$. It is of interest to observe that the sum of the row scores weighted by the corresponding row totals is zero, that is $\mathbf{r}^t\mathbf{a}_k=0$. To see this write

$$\begin{aligned}\mathbf{r}^t\mathbf{a}_k&=\mathbf{r}^t\mathbf{R}^{-1}\mathbf{X}\mathbf{b}_k/\mathrm{r}_k=\mathbf{1}^t\mathbf{X}\mathbf{b}_k/\mathrm{r}_k\\&=\mathbf{c}^t\mathbf{b}_k/\mathrm{r}_k=0,\end{aligned} \tag{5.18}$$

as $\mathbf{c}^t\mathbf{b}_k=0$.

Let us now suppose that we had set out initially to obtain scores which in some sense best separate the columns. For this purpose we shall regard the columns of **X** as constituting a one-way classification of the sample, $\mathbf{X} = [\mathbf{x}_1| \dots |\mathbf{x}_c]$. We shall first obtain indices $a_1, \dots, a_r$ for the row categories and proceed to derive the column scores as linear functions of these. As in deriving **b** above, it will be advantageous to require the weighted sum of row scores to be zero, $\mathbf{r}^t\mathbf{a} = 0$. The criterion to be optimized corresponding to (5.12) is then

$$\frac{\mathbf{a}^t[\mathbf{X}\mathbf{C}^{-1}\mathbf{X}^t]\,\mathbf{a}}{\mathbf{a}^t\mathbf{R}\,\mathbf{a}} = g(\mathbf{a}), \qquad |\mathbf{R}| \neq 0. \tag{5.19}$$

Imposing the condition $(\mathrm{N}-1)^{-1}\mathbf{a}^t\mathbf{R}\mathbf{a} = 1$ and following the standard procedure we find that maximization of (5.19) amounts to solving the eigenequation

$$(\mathbf{R}^{-\frac{1}{2}}\mathbf{X}\mathbf{C}^{-1}\mathbf{X}^t\mathbf{R}^{-\frac{1}{2}} - \mathrm{r}^2\,\mathbf{I})\,\mathbf{a}^* = \mathbf{0}, \tag{5.20}$$

where r^2 is a latent root of $\mathbf{R}^{-\frac{1}{2}}\mathbf{X}\mathbf{C}^{-1}\mathbf{X}^t\mathbf{R}^{-\frac{1}{2}}$. The nonzero roots r_k^2 $(k = 0, \dots, s)$ are identical to the r_k^2 of (5.14) and are therefore also canonical roots. The condition $\mathbf{a}^t\mathbf{r} = 0$ however gives rise to an extraneous root $\mathrm{r}_0 = 1$ and corresponding vector $\mathbf{a}_0^* = \mathbf{R}^{\frac{1}{2}}\mathbf{1}(r \times 1)$, which we shall again disregard. The maximum of (5.19) is given by the largest retained root r_1^2 of (5.20). Further, the vector **a** which maximizes (5.19) is obtained from $\mathbf{a}^*$ as

$$\mathbf{a} = \mathbf{R}^{-\frac{1}{2}}\,\mathbf{a}^*. \tag{5.21}$$

Neglecting $\mathrm{r}_0^2 = 1$ and $\mathbf{a}_0 = \mathbf{1}$, Eqs. (5.20) and (5.21) have $s = \min(p, q)$ solutions, r_k^2 and $\mathbf{a}_k$ $(k = 1, \dots, s)$, in all. Once the vectors of optimal row weights $\mathbf{a}_k$ $(k = 1, \dots, s)$ have been obtained the s sets of most discriminative scores for the columns, $\mathbf{b}_k$, are given by

$$\mathbf{b}_k = \mathrm{r}_k^{-1}\,\mathbf{C}^{-1}\mathbf{X}^t\,\mathbf{a}_k. \tag{5.22}$$

The separate derivations of row and column scores each yield identical sets of values for *both* rows and columns. To show the identities we begin by writing

$$\underset{(r\times c)}{\mathbf{G}} = \underset{(r\times r)}{\mathbf{R}^{-\frac{1}{2}}}\;\underset{(r\times c)}{\mathbf{X}}\;\underset{(c\times c)}{\mathbf{C}^{-\frac{1}{2}}}. \tag{5.23}$$

Substituting **G** into (5.12) and recalling from (5.14) that $\mathbf{b}^* = \mathbf{C}^{\frac{1}{2}}\,\mathbf{b}$, we obtain

$$\frac{[\mathbf{b}^*]^t\,\mathbf{G}^t\,\mathbf{G}\,\mathbf{b}^*}{[\mathbf{b}^*]^t\,\mathbf{b}^*} = \mathrm{r}^2. \tag{5.24}$$

Similarly, substituting **G** into (5.19) and noting from (5.21) that $\mathbf{a}^* = \mathbf{R}^{\frac{1}{2}}\,\mathbf{a}$, we have

$$\frac{[\mathbf{a}^*]^t\,\mathbf{G}\,\mathbf{G}^t\,\mathbf{a}^*}{[\mathbf{a}^*]^t\,\mathbf{a}^*} = \mathrm{r}^2. \tag{5.25}$$

Now, for any arbitrary $r \times c$ matrix **G**, the matrices $\mathbf{G}^t\mathbf{G}$ and $\mathbf{G}\mathbf{G}^t$ are symmetric and have the same eigenvalues. Mardia, Kent and Bibby (1979, p. 468) provide a proof of this assertion. On comparing (5.24) and (5.25) it follows immediately that both derivations yield the same distribution of r^2. Moreover, as the two derivations employ the same scale unit for the vectors **a** and **b**, namely $(N-1)^{-1}\mathbf{a}^t\mathbf{R}\mathbf{a}=(N-1)^{-1}\mathbf{b}^t\mathbf{C}\mathbf{b}=1$, the solutions $\{r_k; \mathbf{a}_k, \mathbf{b}_k\}$ arrived at in setting out to obtain scores which will distinguish maximally between either rows or columns are also one and the same. The two derivations can in fact be combined and the $\{r_k; \mathbf{a}_k, \mathbf{b}_k\}$ obtained *simultaneously* from the singular value decomposition of **G**. Let the singular value decomposition of **G** be

$$\underset{(r\times c)}{\mathbf{G}} = \underset{(r\times s')}{\mathbf{P}} \; \underset{(s'\times s')}{\mathbf{D}} \; \underset{(s'\times c)}{\mathbf{Q}^t} \tag{5.26}$$

where $\mathbf{P}^t\mathbf{P}=\mathbf{Q}^t\mathbf{Q}=\mathbf{I}_{s'}$, $\mathbf{D}=\text{diag}(d_0, \ldots, d_{s'})$ and $s'=\min(p+1, q+1)$. Here **P** consists of the orthonormalized eigenvectors associated with the s' largest eigenvalues of $\mathbf{G}\mathbf{G}^t$, **Q** consists of the corresponding orthonormalized eigenvectors of $\mathbf{G}^t\mathbf{G}$ and the d_k are squareroots of the eigenvalues of the symmetric, positive-semidefinite matrices $\mathbf{G}^t\mathbf{G}$ or $\mathbf{G}\mathbf{G}^t$. Then the scores for rows and columns are contained in the columns of **A** and **B** respectively, where

$$\underset{(r\times s')}{\mathbf{A}} = \underset{(1\times 1)}{(N-1)^{\frac{1}{2}}} \; \underset{(r\times r)}{\mathbf{R}^{-\frac{1}{2}}} \; \underset{(r\times s')}{\mathbf{P}} \; \underset{(s'\times s')}{\mathbf{D}^{\beta}}, \tag{5.27}$$

$$\underset{(c\times s')}{\mathbf{B}} = \underset{(1\times 1)}{(N-1)^{\frac{1}{2}}} \; \underset{(c\times c)}{\mathbf{C}^{-\frac{1}{2}}} \; \underset{(c\times s')}{\mathbf{Q}} \; \underset{(s'\times s')}{\mathbf{D}^{\beta}}, \tag{5.28}$$

and β is a scaling parameter open to choice. Singular value decomposition therefore provides a unified solution for dual scaling. Initially, we seemed to have two separate problems. Both solutions, however, are readily obtained from (5.26).

The conditions $\mathbf{b}_k^t\mathbf{c}=\mathbf{a}_k^t\mathbf{r}=0$ were introduced in order to simplify the algebra, and, as we have noted, are responsible for a solution $\{r_0^2=1;\ \mathbf{a}_0=\mathbf{1}\,(r\times 1),\ \mathbf{b}_0=\mathbf{1}\,(c\times 1)\}$ of no intrinsic interest. The unwanted solution can be eliminated by replacing **X** in (5.23) by the doubly-centered matrix $\mathbf{X}^*(r\times c)$, where

$$\mathbf{X}^*=\mathbf{X}-(N^{-1})\,\mathbf{r}\,\mathbf{c}^t. \tag{5.29}$$

With this adjustment to **X**, the matrix **G** in (5.26) has $s=\min(p, q)$ rather than $s'=\min(p+1, q+1)$ nonzero singular values d_k and corresponding singular vectors $\mathbf{p}_k$ and $\mathbf{q}_k$ $(k=1, \ldots, s)$.

Dual scaling was introduced above in two ways – as a means of maximizing the correlation for a 2-way table and as a means of maximizing the discrimination between the row *and* the column categories of a 2-way table. Both formulations yield the same results and are therefore equivalent. Nishisato (1980, pp. 59–65) has provided a proof of this assertion. Our purpose in this section has been to show that dual scaling is formally equivalent to a canonical correlation analysis in which all the variables are binary-valued dummy variables, and, in certain respects also to a *double* canonical variate or discriminant analysis. Where a solution of rank 2 or rank 3 is judged acceptable, the scores may be used to display the rows and columns of the original table simultaneously as points in a low-dimensional vector space.

5.3 Further aspects of dual scaling

Having derived dual scaling we are in a position to examine properties of the solution and to consider the interpretation of results. We shall see that a feature of singular interest about dual scaling is the symmetry which exists between the roles of the rows and columns of the original table in the analysis. We begin, however, by examining assumptions underlying use of the method.

Assumptions. Dual scaling has considerable appeal as a distribution-free procedure for the exploratory analysis of categorical data. Without distributional assumptions, dual scaling can be justified on the grounds that it maximizes association and discrimination between the rows and columns of a doubly-classified array. Nevertheless, for sensible interpretation it is necessary for the data to conform at least approximately to certain specifications. The joint distribution of the row and column attribute-states and the appropriateness of certain distance functions implicit in the method, in particular, merit attention.

The rows and columns of **X** can each be considered to define a set of points in higher-dimensional space. We shall have more to say about such geometric aspects of dual scaling in discussing duality later in this section. For present purposes it will suffice to confine our remarks to some rather general points concerning the *shape* of the configurations and the definition of *distance* in each space, about which assumptions are implicitly made in applications. While dual scaling is essentially distribution-free, it is advisable to check before embarking on analysis that the sample is homogeneous and free from distributional peculiarities associated with disjunct samples or subsets of samples. Such peculiarities can have profound, deleterious consequences on the outcome of analysis. Further, if a useful reduction of dimensionality is to be achieved, one would like the variation represented by the error sums of squares and cross-products matrices $(\mathbf{R}-\mathbf{X}\mathbf{C}^{-1}\mathbf{X}^t)$ and $(\mathbf{C}-\mathbf{X}^t\mathbf{R}^{-1}\mathbf{X})$ to be approximately elliptically symmetric (Nash, 1983). Graphical methods for shedding light on these various characteristics of a sample have been described by Gnanadesikan (1977, Chap. 6). Such preliminary inspection guards against gross violation of distributional norms or standards. In ecology, where dual scaling is often used in connection with vegetation surveys, it seems to me a considerable assumption that elliptically symmetric error distributions are likely where the vegetation is heterogeneous to any considerable extent. The influence of disjunctions in the data, where present, may be ameliorated by procedures described by Greenacre and Degos (1977) and Gabriel and Zamir (1979).

The distance between points in either of the two spaces referred to above is defined as a *weighted* sum of squared differences between the corresponding row or column entries of the table, the weights being simple functions of the column or row totals, respectively. These particular choices of distances are integral parts of dual scaling, one of their effects being to equalize the contribution of rare and abundant attribute-states to the metric structure of each space. Now, the prescription of a metric in multivariate analysis generally, is, or ought to be, guided by substantive considerations. In practice, however, as Gnanadesikan (1977, p. 90) has remarked, selection is often more a matter of trial and error, intuition and good luck. Nevertheless the pertinency of the distance function in dual scaling deserves to be pondered before committing oneself to the analysis. Reference may be made to Gower (1972) for a discussion of certain distance functions of the kind used in dual scaling.

Dual scaling has been presented here as a descriptive tool with little reference to possible inferential uses. Formal tests of statistical hypotheses are nevertheless possible, though their use involves additional assumptions. These additional assumptions are that the available data comprise a sample of independent observations from either a generalized multiple hypergeometric distribution (Lehmann, 1975, p. 382) or a multinomial distribution. Hypothesis testing is taken up in Sect. 5.4 below.

Low-rank matrix approximation. Dual scaling is sometimes regarded as a procedure for the *visual* interpretation of patterns in data summarized as a 2-way table. It may therefore be worthwhile to sketch the derivation of dual scaling as a solution to the problem of obtaining an optimal low-rank approximation of $\mathbf{X}$. Our task will be simplified without materially affecting the outcome if we work with the scaled version of $\mathbf{X}$ given in (5.23), that is with $\mathbf{G} = \mathbf{R}^{-\frac{1}{2}}\mathbf{X}\mathbf{C}^{-\frac{1}{2}}$, rather than with $\mathbf{X}$ itself. Suppose that the rank of $\mathbf{G}$ is $s' = \min(r, c)$. Let us find the least squares rank t approximation $\hat{\mathbf{G}}_{(t)}$ of $\mathbf{G}$ for which $1 \leq t < s'$. The appeal of such an approximation is that it consists of only $t(r+c)$ independent quantities, rather than the $r \times c$ of $\mathbf{G}$ itself, and is therefore more amenable to graphical display and interpretation. Writing $\hat{g}_{ij}$ for the ijth element of $\hat{\mathbf{G}}_{(t)}$, the least squares criterion requires the sum of squared discrepancies $\Sigma_i \Sigma_j (g_{ij} - \hat{g}_{ij})^2$ to be minimal for all rank t matrices in the class of $r \times c$ matrices. In practice, $\hat{\mathbf{G}}_{(t)}$ is conveniently found by minimizing the corresponding matrix norm $\|\mathbf{G} - \hat{\mathbf{G}}_{(t)}\|^2 = \mathbf{E}^2$. A key step in the derivation results from the observation (Eckart & Young, 1936; Corsten, 1976a) that the error matrix $\mathbf{E}$ will be minimal if all its eigenvalues are simultaneously minimal. This condition will be satisfied if $\hat{\mathbf{G}}_{(t)}$ is constructed from the singular value decomposition of $\mathbf{G}$ in (5.26) by suppressing the last $s' - t$ terms:

$$\begin{aligned} \underset{(r \times c)}{\hat{\mathbf{G}}_{(t)}} &= \underset{(r \times t)}{\mathbf{P}} \; \underset{(t \times t)}{\mathbf{D}} \; \underset{(t \times c)}{\mathbf{Q}^t} \\ &= d_1 \mathbf{p}_1 \mathbf{q}_1^t + d_2 \mathbf{p}_2 \mathbf{q}_2^t + \ldots + d_t \mathbf{p}_t \mathbf{q}_t^t. \end{aligned} \tag{5.30}$$

In other words, the least squares rank t approximation of $\mathbf{G}$, $\hat{\mathbf{G}}_{(t)}$, minimizing $\mathbf{E}$, is given by the t leading terms in the singular value decomposition of $\mathbf{G}$. Usually, the rank of $\hat{\mathbf{G}}_{(t)}$ is chosen to be much smaller than the rank of $\mathbf{G}$. The respective row and column spaces of $\hat{\mathbf{G}}_{(t)}$ are both t-dimensional. Graphical exploration and interpretation will clearly be easier in fewer dimensions, the most useful representation resulting when $t = 2$. The adequacy of the approximation for any value of t can be assessed by means of the goodness of fit criterion (5.7), replacing r_k^2 by d_k^2 and $\mathbf{K}$ by $\hat{\mathbf{G}}_{(t)}$. However, the approximation (5.30) is of little direct use in plotting the rows and columns of $\hat{\mathbf{G}}_{(t)}$ because these entities are expressed with respect to coordinate systems representing the original variables. The rows and columns of $\hat{\mathbf{G}}_{(t)}$ each define a set of points in t-space and it is the coordinates of these row and column points in relation to *principal axes* of the respective scatters that are required. These coordinates age given by Eqs. (5.27) and (5.28) on replacing $\mathbf{P}(r \times s')$, $\mathbf{Q}(c \times s')$ and $\mathbf{D}(s' \times s')$ by the corresponding reduced-rank matrix of (5.30).

The derivation of dual scaling as the solution of a graphical problem therefore leads us once again to singular value decomposition. This time it is the least squares properties of the decomposition which are exploited. The distribution-free basis of the least squares principle and its computational convenience as well as the facility it affords in matrix approximation all contribute to the appeal of dual scaling.

Duality. The adjective 'dual' in dual scaling refers to the *symmetry* with which the rows and columns of $\mathbf{X}$ are treated in the analysis. Duality has many aspects, of which we shall consider just a few. One of the clearest illustrations is provided by the observation that the analyses of $\mathbf{X}$ and $\mathbf{X}^t$ provide equivalent results.

At this point it will be helpful to regard both the rows and columns of $\mathbf{X}$ as specifying galaxies of points in higher-dimensional space. The rows, after being scaled by their respective totals, $r_i^{-1}\mathbf{x}_i = [x_{i1}/r_i, x_{i2}/r_i, \ldots, x_{ic}/r_i]^t$ for $i = 1, \ldots, p+1$, serve as the coordinates of r points in c-space. The scaling by r_i^{-1} has the effect of equalizing the contributions of rows with comparatively small and large totals, respectively, to the analysis. Similarly, the columns after scaling by their respective totals, $c_j^{-1}\mathbf{x}_j = [x_{1j}/c_j, x_{2j}/c_j, \ldots, x_{rj}/c_j]^t$ for $j = 1, \ldots, q+1$, determine a set of c points in r-space. Dual scaling enables the dispersions of row and column points to be analyzed simultaneously and in a completely symmetric way. Let us consider first the centroids of the respective scatters and the distance between pairs of points in the two spaces.

The centroid of the row points can be shown (Greenacre & Degos, 1977) to be specified by the $c \times 1$ vector of column sums $\mathbf{c}$, and, dually, the centroid of the column points by the $r \times 1$ vector of row sums, $\mathbf{r}$. The distance between any two row points in c-space is a function of the weighted sum of squares of differences in the scaled rows, the weights being the reciprocals of the column totals. More specifically, the squared distance d_{hi}^2 between the row points h and i is given by

$$d_{hi}^2 = (r_h^{-1}\mathbf{x}_h - r_i^{-1}\mathbf{x}_i)^t \mathbf{C}^{-1} (r_h^{-1}\mathbf{x}_h - r_i^{-1}\mathbf{x}_i) \tag{5.31}$$

which is shown by the presence of $\mathbf{C}^{-1}$ to be a *generalized* Euclidean distance. Dually, the squared distance between any two column points j and k in r-space is a weighted sum of squared differences in the scaled columns, the weights being reciprocals of the row totals:

$$d_{jk}^2 = (c_j^{-1}\mathbf{x}_j - c_k^{-1}\mathbf{x}_k)^t \mathbf{R}^{-1} (c_j^{-1}\mathbf{x}_j - c_k^{-1}\mathbf{x}_k). \tag{5.32}$$

The dual roles of the row and column totals is clearly revealed by Eqs. (5.31) and (5.32) – row and column totals each *weight* the points in one space while defining the *metric* in the other.

The symmetry in dual scaling is very much part of the singular value decomposition which largely defines the analysis. We have seen, for example, that the eigenvalues of $\mathbf{G}^t\mathbf{G}$ of (5.24), which leads to scores $\mathbf{a}_k$ for the rows, are the same as the eigenvalues of $\mathbf{G}\mathbf{G}^t$ of (5.25) which yields scores $\mathbf{b}_k$ for the columns. Further, Eq. (5.27) indicates that the row scores are functions of the left singular vectors of $\mathbf{G}$, while, dually, the column scores are shown by (5.28) to be functions of the right singular vectors of $\mathbf{G}$. Together, Eqs. (5.27) and (5.28) demonstrate that discrimination between the rows and columns of $\mathbf{X}$ is *simultaneously* maximized by dual scaling. As a consequence of these remarks it may be appreciated that the weighted dispersions of points in r-space and c-space are one and the same, and, moreover, that the dispersions can be decomposed identically into components which are orthogonal in the metric of the respective spaces.

Symmetry is exemplified also by the *transition formulae*, (5.17) and (5.22), which are repeated here as Eqs. (5.33) and (5.34):

$$\mathbf{a}_k = \mathrm{r}_k^{-1}\mathbf{R}^{-1}\mathbf{X}\,\mathbf{b}_k, \tag{5.33}$$

$$\mathbf{b}_k = \mathrm{r}_k^{-1}\mathbf{C}^{-1}\mathbf{X}^t\,\mathbf{a}_k. \tag{5.34}$$

Equation (5.33) shows the row scores $\mathbf{a}_k$ to be linear functions of the column scores, $\mathbf{b}_k$. More precisely, the rows scores are in fact weighted, rescaled averages of the column scores. Symmetrically, (5.34) shows the column scores to be linear functions of the row scores.

Interpreting results. The elements of the linear composites $\mathbf{u}_k(\mathrm{N}\times 1)$ and $\mathbf{v}_k$ $(\mathrm{N}\times 1)$ are the canonical scores of the N doubly-classified *individuals* in the original 2-way table. All r_i individuals in the ith $(i=1, \ldots, p+1)$ row of $\mathbf{X}$ have the same score on u_k. Consequently, $\mathbf{u}_k$ has only $p+1$ distinct elements. These distinct values are the scores or coordinates of the row *attribute-states* on u_k. In a similar way, the N-vector $\mathbf{v}_k$ contains precisely $q+1$ distinct values, which are the scores or coordinates of the $q+1$ column *attribute-states* on v_k. The u_k and v_k can be used to summarize the results of analysis graphically, as described in Sect. 2.4.3. In dual scaling graphical displays of two kinds are often worthwhile:

(a) separate displays of u_k against u_m and of v_k against v_m $(k, m=1, \ldots, t;\ k\neq m)$;
(b) joint displays of u_k and v_k against u_m and v_m $(k, m=1, \ldots, t;\ k\neq m)$.

The resulting configurations are often amenable to substantive interpretation. As we shall shortly see, however, not only are the two kinds of display open to interpretation in rather different terms, but, more fundamentally, the scaling adopted for the canonical variates has important consequences for interpretation.

Separate row or column displays. The concept of *distance* underlies the interpretation of row attribute-states displayed in u-space and of column attribute-states displayed in v-space. We consider first the case in which the canonical variates are scaled such that the variance of the kth pair of variates is equal to the kth root, r_k^2. We then have

$$\mathrm{var}(u_k) = \mathbf{a}_k^t\mathbf{S}_{11}\,\mathbf{a}_k = (\mathrm{N}-1)^{-1}\,\mathbf{a}_k^t\,\mathbf{R}\,\mathbf{a}_k = \mathrm{r}_k^2, \tag{5.35}$$

$$\mathrm{var}(v_k) = \mathbf{b}_k^t\mathbf{S}_{22}\,\mathbf{a}_k = (\mathrm{N}-1)^{-1}\,\mathbf{b}_k^t\,\mathbf{C}\,\mathbf{b}_k = \mathrm{r}_k^2, \tag{5.36}$$

($k \neq 0$). This is the standardization which results on setting $\beta = 1$ in (5.5) and (5.6), or, equivalently, in (5.27) and (5.28). Where the leading roots are dominant, displays based on the canonical variates corresponding to these roots summarize the salient relationships between row or column attribute-states with considerable efficiency. A further property of displays in which the canonical variates are standardized as in (5.35) and (5.36) merits attention. We have seen that the squared distances between the rows and between the columns of **X** defined by (5.31) and (5.32), respectively, are actually *generalized* distances. Yet when expressed in terms of the canonical variates, the squared distances between points corresponding to row or column attribute-states are Euclidean approximations of the generalized distances d_{hi}^2 or d_{kj}^2 (Gower, 1984, Sect. 17.5). This is a valuable property of the displays as it enables assessment to proceed as in the intuitive interpretation of physical space. On the other hand, in this case the displays do not represent least squares approximations of the scaled data matrix. Nevertheless, features of the resulting configurations, such as proximities, distances, clusters or trends are readily detected by the eye. Attention then turns to possible substantive implications of these features, which often prove more amenable to interpretation than the canonical variates themselves.

Let us now consider separate row or column displays in which the canonical variates are scaled in such a way that the residual error variance after removing the effect due to regression of the kth pair of variates, u_k and v_k, is set to unity. This is the scheme advocated by Nash (1983). We then have

$$\left.\begin{aligned} \operatorname{var}_e(u_k) &= \left\{\frac{(\mathrm{N}-1)}{(\mathrm{N}-c)}\right\}(1-\mathrm{r}_k^2)\,\mathbf{a}_k^{\mathrm{t}}\,\mathbf{S}_{11}\,\mathbf{a}_k = (\mathrm{N}-c)^{-1}(1-\mathrm{r}_k^2)\,\mathbf{a}_k^{\mathrm{t}}\,\mathbf{R}\,\mathbf{a}_k = 1, \\ \operatorname{var}_e(v_k) &= \left\{\frac{(\mathrm{N}-1)}{(\mathrm{N}-r)}\right\}(1-\mathrm{r}_k^2)\,\mathbf{b}_k^{\mathrm{t}}\,\mathbf{S}_{22}\,\mathbf{b}_k = (\mathrm{N}-r)^{-1}(1-\mathrm{r}_k^2)\,\mathbf{b}_k^{\mathrm{t}}\,\mathbf{C}\,\mathbf{b}_k = 1, \end{aligned}\right\} \quad (k \neq 0).$$

With this scaling, the Mahalanobis squared distance between any two rows or columns of **X** becomes an ordinary squared Euclidean distance in terms of the canonical variates u_k or v_k. Thus, it is easy to judge by eye how far apart two plotted row or column points are in terms of Mahalanobis distances. Such displays result in a wider spread of the canonical variates, and so to sharper discrimination, than is provided by alternative forms of standardization.

Simultaneous row and column displays. Separate row or column displays are mappings of the sample into subspaces of different vectorspaces. Yet, provided certain precautions are observed, such displays may be superimposed to provide a joint display of row and column points in a single subspace with respect to the same coordinate frame. In contrast to single row or column displays, for which distances between pairs of points were formally defined above, no corresponding formal specification for the distance between a row point and a column point in a joint display can be given. Nevertheless, provided due care and attention is paid to the scaling of the canonical variates, useful interpretations for the spatial relationships between attribute-states of different kinds can be given.

The scaling of canonical variates is entirely at the discretion of the user. Consequently, there is nothing to prevent different normalizations being used in separate and joint displays resulting from a single analysis. Having obtained separate row and column displays, there is much to be said for *renormalizing* the canonical variates in order to arrive at a simultaneous display for which the spatial proximities or angular relationships between pairs of row and column points are meaningful. Where the original 2-way array **X** is actually a contingency table, such a representation will be arrived at on setting the scale parameter β in (5.5) and (5.6), or, equivalently, in (5.27) and (5.28) to $\beta=\frac{1}{2}$. Then it can be shown (Gower & Digby, 1981; Heiser & Meulman, 1983a) that the cosine of the angle subtending a row point and a column point at the origin is proportional to the departure from expectation under the assumption of row and column independence of the corresponding cell of the contingency table. That is to say, the cosine of the angle is proportional to $(x_{ij}-r_i c_j/N)/(r_i c_j/N)^{\frac{1}{2}}$. As the cosine tends to unity the more the departure is likely to exceed the expectation for the cell; conversely, as the cosine approaches zero, the more likely the departure to fall short of expectation. With this scaling, therefore, the joint display is a representation of the deviations between the data and their expectations under independence, rather than of the data themselves.

Where **X** is not a contingency table the foregoing justification for a simultaneous plot cannot be sustained. The simultaneous display in these cases is merely a convenient device with no compelling theoretical basis. Nevertheless, by appropriately normalizing the canonical variates, proximities rather than angular relationships between row and column points are amenable to interpretation. The appropriate scaling results on setting β to unity in Eqs. (5.5) and (5.6) or in Eqs. (5.27) and (5.28). In order to justify the construction of a joint display in this case we appeal to three properties of the analysis – (a) the set of row points in c-space and the set of column points in r-space each occupy a subspace of dimension s with their origin at the centroid of each galaxy, s being equal to the rank of **X** minus 1; (b) the identity of the distribution of the eigenvalues in c-space and r-space; and (c) the duality of the transition formulae (5.33) and (5.34). The second property indicates that the shape and relative importance of the directions of spread of the weighted row and column configurations are one and the same, while the third property shows the row and column scores to be linearly related, so that each can be obtained from the other. The transition formulae are helpful in clarifying the nature of the spatial proximities of row and column points. By expanding (5.33) we discover how the score a_{ik} of the ith row category on the kth canonical variate arises:

$$a_{ik}=r_k^{-1}\sum_{j=1}^{q+1}\frac{x_{ij}}{r_i}b_{jk},\qquad i=1,\ldots,p+1. \tag{5.37}$$

Equation (5.37) shows that up to a constant r_k^{-1}, a_{ik} is a weighted mean of the column scores b_{jk} $(j=1,\ldots,q+1)$, where the weights are (x_{ij}/r_i). In other words, *all* $q+1$ column scores contribute to the score or coordinate of the ith row point. Consequently, a row point in a simultaneous display will lie in the vicinity of the column points whose corresponding attribute-states are best represented in the row in question. Essentially the same remarks are applicable to the location of column points based on the expansion of (5.34). For these reasons it is not permissible to comment on the proximity of a single row point to a single column point in a joint display, no such distance having been defined or intended.

What the joint display achieves is a concise representation of either row and column dependence or of the original table itself which conveys as many useful relations as possible and which is easily and quickly absorbed. In return for conciseness it seems that something has to be forfeited in terms of ease of interpretation. It may therefore be worthwhile to supplement assessment of the point configuration itself by considering possible substantive implications of the common reference frame provided by the canonical variates u_k and v_k. Correlation coefficients calculated between the canonical variates and the variables on which they are defined may be informative in this regard.

5.4 Hypothesis testing

Where the data for analysis form a contingency table, formal tests of hypotheses may be of interest in dual scaling. Provided the sample size N is large, procedures based on the normal theory described in Chap. 3 can be adapted to yield asymptotic tests. In addition, other tests are available whose application is not confined to contingency tables. We shall consider first a test for independence in contingency tables before proceeding to a test of dimensionality. The test for independence may be regarded as a test of linear association between the categorical variables defining the table, or, alternatively, as a test for between-row and between-column discrimination. Yet another way of regarding the test is as a means of assessing the independence of two partitions of the same N sample-units.

5.4.1 Independence

We wish to assess the independence of $\mathbf{z}_1$ and $\mathbf{z}_2$. Although neither $\mathbf{z}_1$ nor $\mathbf{z}_2$ is multinormal, tests based on normal theory may be applied by an appeal to asymptotic normality. An appropriate null hypothesis is that all s canonical correlation coefficients are jointly null:

$$H_0\colon \rho_1=\rho_2=\ldots=\rho_s=0, \tag{5.38}$$

the alternate being

$$H_1\colon \rho_k \neq 0 \quad \text{for at least one } k.$$

Likelihood ratio and union-intersection tests of hypothesis (5.38) which are applicable where at least one of the two sets of variables is multinormally distributed were described in Sect. 3.4.1. These procedures can also be used as asymptotic tests in dual scaling provided N and the x_{ij} are large, even though the distribution of neither $\mathbf{z}_1$ nor $\mathbf{z}_2$ is multinormal. The quantity

$$\chi^2(pq) = -\{(\mathrm{N}-1)-\tfrac{1}{2}(r+c-1)\} \log_e \Lambda, \tag{5.39}$$

where $\Lambda = \Pi_{k=1}^{s}(1-\mathrm{r}_k^2)$, is distributed asymptotically as a chi-squared variate on pq degrees of freedom ($p=r-1$ and $q=c-1$), may be used to test (5.38). Notice that the argument for degrees of freedom and the value of the multiplier $\{\cdot\}$ in (5.39) differ from those of (3.13). The adjustments follow by extension from the practice adopted in canonical variate analysis (Sect. 4.5.1), where the degrees of freedom were $p(g-1)$, and from the observation that in certain respects dual scaling may be regarded as a double canonical variate analysis.

The union-intersection test of Sect. 3.4.1 can also be used to test (5.38). The parameter values for $\theta(s, m, n)$ of that section are as follows:

$$s = \min(p, q), \qquad m = (|p-q|-1)/2, \qquad n = (\mathrm{N}-p-q-2)/2. \tag{5.40}$$

Rejection of (5.38) in a test at some suitable level of α would lead to the conclusion that the data are consistent with the association of A and B.

Two other tests of hypothesis (5.38) are available. Corsten (1976a) and O'Neill (1978a) have shown that under H_0 the asymptotic distribution of the quantity Nr_1^2 is given by

$$\mathrm{Nr}_1^2 \sim c_{\max}(p, q),$$

where $c_{\max}(p, q)$ denotes the largest root of a standard Wishart matrix with parameters $\max(p, q)$ and $\min(p, q)$. The distribution of $c_{\max}$ has been tabulated by Hanumara and Thompson (1968). Secondly, the statistic $\Sigma\, \mathrm{r}_k^2$ ($k=1, \ldots, s$) is distributed asymptotically under the null hypothesis as Pillai's trace criterion, V (Kshirsagar, 1972, p. 383), percentage points of the distribution of which are reported by Timm (1975, p. 650).

5.4.2 Dimensionality

Where hypothesis (5.38) can be rejected it is worthwhile to consider the dimensionality of the association. Dimensionality is equivalent to the number of non-zero canonical roots or to the rank of $\mathbf{X}$. The sequential tests of the joint nullity of the smallest $s-j$ roots, after first removing those roots assumed to be significant, of Sect. 3.4.2 are useful for this purpose.

The chi-squared approximation of the likelihood ratio criterion for testing hypothesis (3.18) is

$$\chi_j^2(f) = -\{(\mathrm{N}-1)-\tfrac{1}{2}(r+c-1)\} \sum_{k=j+1}^{s} \log_e(1-\mathrm{r}_k^2), \qquad j=1, \ldots, s-1. \tag{5.41}$$

For large N, this quantity is distributed approximately as a chi-squared variate on $f=(p-j)(q-j)$ degrees of freedom for the jth step. These tests of hypothesis (3.18) are conveniently summarized as shown in Table 3.2, with due regard for the degrees of freedom. It is the last χ^2 value in Table 3.2 which is the appropriate value for testing the jth hypothesis. A union-intersection procedure for establishing the number of non-zero roots is described in Sect. 3.4.2.

The likelihood ratio procedures outlined here are believed to yield tests which are broadly correct even though the distribution of neither set of variables is multivariate normal (E.J. Williams, 1952; Bartlett, 1965; Kshirsagar, 1972, p. 382). The significance levels provided, however, are not to be interpreted too precisely. Two further tests with application in dual scaling deserve mention. Lebart (1976) and Lebart, Morineau and Tabard (1977) have investigated the quantity C_t defined in (5.7), which expresses the proportion of total predictable variance accounted for by t retained canonical roots. In particular, the proportion of their sum which the successive roots can be expected to be has been evaluated, and graphs showing the upper .05 points of such proportions of total variability for the first five roots for contingency tables of different sizes provided. The distributional assumption underlying the graphs is the normal approximation to the multinomial distribution. Benzécri (1973) has derived confidence regions for individual points in graphical displays which are useful where the data analyzed are multinomially distributed.

5.5 Affinities with other methods

Insights into the nature of dual scaling are provided by both canonical correlation analysis and canonical variate analysis. Dual scaling is also related to several other statistical methods. We shall review some of these relationships with the aim of throwing further light on the method. In particular, we shall be concerned with affinities between dual scaling, on the one hand, and the analysis of contingency tables, correspondence analysis and principal component analysis, on the other.

5.5.1 Dual scaling and the analysis of contingency tables

Consider a 2-way contingency table $\mathbf{X}(r \times c)$ in which individuals are classified with respect to two characteristics. The attributes on each characteristic are represented by the r row categories and c column categories, respectively. We wish to examine the relationship between the two characteristics defining the table. The problem can be addressed in several ways. Here we employ a procedure described by E.J. Williams (1952) because of its similarity to dual scaling. Williams' method provides tests which are more sensitive than the usual overall χ^2 test of independence for a 2-way table and also yields scores characterizing the row and column categories of the table. The scores are optimal in the sense of maximizing the correlation between the two categorized variables. Our notation for the 2-way table follows that of Sect. 5.2.

The null hypothesis of independence may be expressed

$$\mathrm{H}_0\colon\ p_{ij}=p_{i\cdot}\,p_{\cdot j}, \qquad i=1,\ldots,r, \quad j=1,\ldots,c, \tag{5.42}$$

where p_{ij} is the probability of an individual belonging to the ith row category and the jth column category, $p_{i\cdot} = \Sigma_j p_{ij}$ is the marginal probability for the ith row and $p_{\cdot j} = \Sigma_i p_{ij}$ is the marginal probability for the jth column. The alternative hypothesis is that the row and column characteristics are not independent, H_1: $p_{ij} \neq p_{i\cdot}\, p_{\cdot j}$. The quantities in (5.42) are readily estimated from sample data, and, as is well known, the null hypothesis can be tested by determining the extent of departures from expectation under H_0 in the observed frequencies. The familiar test statistic is

$$\chi_o^2 = \sum_i \sum_j \frac{(x_{ij} - r_i c_j/N)^2}{r_i c_j/N}, \tag{5.43}$$

the subscript o indicating that chi-squared represents an overall or aggregate value for the table. Under the null hypothesis, χ_o^2 is distributed approximately as a chi-squared variate on $(r-1)(c-1)$ degrees of freedom. Where hypothesis (5.42) proves untenable, analysis can be carried further. In such cases χ_o^2 can be partitioned into several components. Various partitioning schemes exist. In the method to be outlined, the decomposition is provided by the latent roots of a matrix derived from the contingency table.

The rationale, briefly, is as follows. The association in the table is examined in terms of correlations between linear combinations of the row and of the column categories. A key concept is that of the *rank* of the association, which is equivalent to the number of pairs of linear combinations which can be fitted. When the characteristics defining the 2-way table are not associated, Williams (1952) has pointed out that the expected values corresponding to the frequencies in the table will form a matrix of unit rank. The usual null hypothesis (5.42) can therefore be equivalently stated by declaring that the expected values in the table are assumed to form a matrix of rank 1. Now the observed frequencies $\{x_{ij}\}$ will depart from expectation, if only because of sampling effects, to comprise a matrix the rank of which is at most $s' = \min(r, c)$. Since, when there is no association the expectations are of rank 1, an association of rank t will correspond to expected values of rank $t+1$. The object of the usual test of significance is to decide whether the departures are consistent with the assumption that the expectations are of unit rank.

In order to sketch Williams' (1952) procedure, we begin by re-expressing (5.43) in the alternative form

$$\chi_o^2 = N\left[\sum_i \sum_j \frac{x_{ij}^2}{r_i c_j} - 1\right]. \tag{5.44}$$

It is easy to verify that the first term within the brackets in (5.44) is the trace of the $r \times r$ symmetric matrix

$$\mathbf{T} = \mathbf{G}\mathbf{G}^t, \tag{5.45}$$

where $\mathbf{G} = [g_{ij}]$ is the $r \times c$ matrix obtained from the elements of $\mathbf{X}$ by setting

$$g_{ij} = x_{ij}/(r_i c_j)^{\frac{1}{2}}. \tag{5.46}$$

Equation (5.44) can now be written

$$\chi_o^2 = N[\mathrm{tr}(\mathbf{T}) - 1]. \tag{5.47}$$

It is not difficult to show that the term '−1' in (5.47) refers to a latent root of **T** which arises solely from the expected values of the observations in the contingency table. To see this, assume that the characteristics are independent and sampling effects absent. The observed and expected frequencies will then be identical and χ_o^2 therefore zero. Then, from (5.47), we find

$$\mathrm{tr}(\mathbf{T}) = 1. \tag{5.48}$$

Now the rank of **T** in the absence of both association and sampling errors is shown by Maxwell (1973) to be one. Hence **T** has only a single latent root, and, since by (5.48) its trace is unity, its single latent root must also be unity. That this root is responsible for the term -1 in (5.44) can be seen on comparing (5.43) and (5.44). This comparison shows the -1 to represent the effect of subtracting the expected from the observed frequencies when calculating χ_o^2 by (5.43). The single latent root of unity of **T** is therefore attributable to the matrix of expectations. When association is present, it follows that a non-zero value of χ_o^2 will depend on latent roots of **T** other than the root corresponding to the expectations.

Assume, now, that rows and columns are not independent. The rank of **T** will then be $1 < R(\mathbf{T}) \leq s' - 1$ and its latent roots therefore more than one in number. It can be shown that the largest root in such cases is always unity and is not relevant to the problem of association (E.J. Williams, 1952, 1967; Kendall & Stuart, 1973, p. 591). Denoting the latent roots of **T** after the largest (namely $\mathrm{r}_0^2 = 1$) by r_k^2 $(k = 1, \ldots, s)$ and recalling that the sum of the latent roots of a matrix is equal to its trace, we have

$$\sum_{k=1}^{s} \mathrm{r}_k^2 = \mathrm{tr}(\mathbf{T}) - 1. \tag{5.49}$$

where $s = \min(r, c) - 1$. Substituting for $\mathrm{tr}(\mathbf{T}) - 1$ in (5.47) we obtain

$$\chi_o^2 = N\left[\sum_{k=1}^{s} \mathrm{r}_k^2\right]. \tag{5.50}$$

This equation shows the sum of the latent roots r_k^2 $(k = 1, \ldots, s)$ to be a simple function of the familiar chi-squared criterion (5.43) for testing (5.42). Further, (5.50) indicates that the r_k^2 yield an additive *partition* of χ_o^2:

$$\chi_o^2 = N[\mathrm{r}_1^2 + \mathrm{r}_2^2 + \ldots + \mathrm{r}_s^2]. \tag{5.51}$$

The $(r-1)(c-1)$ degrees of freedom associated with χ_o^2 may be partitioned in a parallel fashion, $r + c - (2j + 1)$ degrees of freedom being associated with the jth term $(j = 1, \ldots, s)$ of the decomposition. It might therefore be supposed that the significance of individual terms in (5.51) could be tested by reference to percentage points of the appropriate chi-squared distribution. Lancaster (1963), however, has shown that this is not so. Tests of the *residual* association which remains after components presumed to be significant have been successively removed are nevertheless still feasible. With each root r_k^2 ($k=1, \ldots, s$) of $\mathbf{T}$ is also associated a pair of latent vectors, $\mathbf{a}_k(r \times 1)$ and $\mathbf{b}_k(c \times 1)$, satisfying the relations

$$\mathbf{G}\mathbf{G}^t\mathbf{a}_k = r_k^2\,\mathbf{a}_k, \tag{5.52}$$

$$\mathbf{G}^t\mathbf{G}\,\mathbf{b}_k = r_k^2\,\mathbf{b}_k. \tag{5.53}$$

The elements of $\mathbf{a}_k$ are the required row weights and the elements of $\mathbf{b}_k$ the required column weights for the kth pair of linear combinations of rows and columns referred to at the outset. The correlation between the kth pair of resulting composite variables is given by r_k ($k=1, \ldots, s$). The elements of $\mathbf{a}_k$ and $\mathbf{b}_k$ also provide the optimal scores for the rows and columns, respectively, of the contingency table. These scores are used in interpreting the individual components of χ_o^2 represented by the $N r_k^2$ in (5.51).

The original 2-way table can be reconstructed from the results by means of the relation:

$$x_{ij} = \frac{r_i c_j}{N}\left[1 + \sum_{k=1}^{s} r_k^{-1} a_{ik} b_{jk}\right],$$

for $i=1, \ldots, r$ and $j=1, \ldots, c$. In this expression r_i and c_j are the ith and jth diagonal elements of $\mathbf{R}$ and $\mathbf{C}$, respectively, and a_{ik} and b_{jk} are the ith and jth elements of the vectors $\mathbf{a}_k$ and $\mathbf{b}_k$ scaled in accordance with setting $\beta=1$ in (5.27) and (5.28). The analysis can therefore be seen to provide a structural model for the data.

Having sketched the use of scores for the analysis of association in contingency tables we are now in a position to examine the relationship between E.J. Williams' method and dual scaling. It will be profitable to first compare the methods in terms of the overall test of independence before proceeding to examine other characteristics.

Equivalent forms of the null hypothesis of the independence of row and column categories are expressed by (5.38) and (5.42). In (5.38) the hypothesis is formulated in terms of squared canonical correlation coefficients and in (5.42) in terms of cell and marginal probabilities. We have seen that (5.38) is conveniently tested using the quantity

$$\chi^2 = -\{(N-1) - \tfrac{1}{2}(r+c-1)\} \log_e \Lambda, \tag{5.54}$$

where $\Lambda = \Pi_{k=1}^{s}(1-r_k^2)$, while a suitable test criterion for (5.42) is provided by

$$\chi_o^2 = N\left[\sum_{k=1}^{s} r_k^2\right]. \tag{5.55}$$

The two criteria, though different, are closely related. Under independence, (5.54) and (5.55) each have limiting chi-squared distributions on $(r-1)(c-1)$ degrees of freedom. Provided N is large and the r_k^2 in (5.54) small, the criteria are equivalent. Nevertheless, E.J. Williams (1967) has remarked that the χ^2-distribution is a poor approximation of the distribution of $\chi_o^2 = N[\Sigma\, r_k^2]$ if the marginal totals of the 2-way table differ to any appreciable extent.

Where the hypothesis of independence is rejected, (5.54) and (5.55) can each be partitioned to yield further tests. With respect to (5.54), partition is in terms of the latent roots r_k^2 of $\mathbf{KK}^t$ while (5.55) is similarly partitioned by means of the latent roots r_k^2 of $\mathbf{GG}^t$. Now $\mathbf{K}(p \times q)$ and $\mathbf{G}(r \times c)$ are closely related. The relationship is most readily appreciated after re-expressing $\mathbf{G} = [x_{ij}/(r_i c_j)^{\frac{1}{2}}]$ in the equivalent form

$$\mathbf{G} = \mathbf{R}^{-\frac{1}{2}} \mathbf{X} \mathbf{C}^{-\frac{1}{2}}, \tag{5.56}$$

previously encountered in (5.23). Comparison of (5.56) and (5.3) shows $\mathbf{G}$ and $\mathbf{K}$ to be similarly constructed. It is readily verified from Table 5.2b that $\mathbf{G}$ is a scaled cross-products matrix between incidence matrices corresponding to the rows and columns of the initial 2-way table in *non-centered* form. Similarly, (5.2) and (5.3) show $\mathbf{K}$ to be the same scaled cross-products matrix for the leading p rows and q columns of the *centered* incidence matrices. The operation of centering in (5.2) prior to forming $\mathbf{K}$ has the effect of eliminating the extraneous solution $r_0^2 = 1$ of the contingency table analysis from the canonical correlation analysis of Sect. 5.2.1. If we neglect r_0^2, the canonical roots $r_1^2 \geq r_2^2 \geq \ldots \geq r_s^2$, where $s = \min(p, q)$, of $\mathbf{KK}^t$ are identical to be latent roots $r_1^2 \geq r_2^2 \geq \ldots \geq r_s^2$ where $s = \min(r, c) - 1$, of $\mathbf{GG}^t$. The equivalence of the respective partitions will now be apparent. Further, the latent vectors associated with the latent roots r_k^2 of $\mathbf{KK}^t$ and $\mathbf{GG}^t$ $(k \neq 0)$ are also the same, at least up to an arbitrary scale factor which may be absorbed by the vectors. Therefore the row and column scores provided by the two analyses are also equivalent.

In view of the relationships described it may be appreciated that the contingency table analysis of E.J. Williams (1952, 1967) is numerically exactly equivalent to dual scaling.

5.5.2 Dual scaling, correspondence analysis and principal component analysis

Correspondence analysis is a simultaneous scaling method applicable to two-way arrays, the resulting row scores being weighted averages of the column scores and, reciprocally, the column scores weighted averages of the row scores. The analysis is applicable to a wide class of arrays of non-negative numbers, including contingency tables and tables of presence-absences. In practice, correspondence analysis proceeds iteratively from the assignment of an initial, arbitrary vector of scores to rows or columns. Suppose that such an arbitrary vector of scores, $\mathbf{a}_0$, is assigned to the rows of the array. Using these values as weights, the weighted average for each of the columns is calculated. The resulting column scores, $\mathbf{b}_1$, are then used as weights to obtain the weighted average for each of the rows, which in turn are used to calculate new weighted scores for the columns. The sequence of operations can be summarized as follows:

$$\begin{aligned} \mathbf{b}_1 &= \mathbf{C}^{-1} \mathbf{X}^t \mathbf{a}_0, \\ \mathbf{a}_1 &= \mathbf{R}^{-1} \mathbf{X} \mathbf{b}_1, \\ &\cdots\cdots\cdots \\ \mathbf{a}_m &= \mathbf{R}^{-1} \mathbf{X} \mathbf{b}_m, \end{aligned} \tag{5.57}$$

in which new scores $\mathbf{b}_1, \mathbf{a}_1, \ldots$ are successively derived from an initial set $\mathbf{a}_0$. This reciprocal procedure is continued until the scores converge to stable values, $\mathbf{a}_m$ and $\mathbf{b}_m$, say, for the rows and columns, respectively. Nishisato (1980, p. 60) has shown that the procedure converges identically to the optimal vectors $\mathbf{a}$ and $\mathbf{b}$ of dual scaling. For reasons which will now be plain, correspondence analysis is sometimes referred to as *reciprocal averaging* (e.g., see Horst, 1935; Hill, 1973). The scaling of the vectors $\mathbf{a}$ and $\mathbf{b}$ in (5.57) is a matter of convenience. Often, the quantity 1/r, where r is the correlation between $\mathbf{a}$ and $\mathbf{b}$ with respect to the matrix $\mathbf{X}$, is adopted as the scale unit. In such cases, we may write

$$\begin{aligned} \mathbf{a} &= \mathrm{r}^{-1}\mathbf{R}^{-1}\mathbf{X}\mathbf{b}, \\ \mathbf{b} &= \mathrm{r}^{-1}\mathbf{C}^{-1}\mathbf{X}^{\mathrm{t}}\mathbf{a}. \end{aligned} \tag{5.58}$$

Correspondence analysis is also readily shown to be equivalent to a singular value decomposition. To demonstrate the equivalence we proceed as follows. First, notice that after a little manipulation, Eqs. (5.58) can be re-expressed as:

$$\begin{aligned} \mathrm{r}(\mathbf{R}^{\frac{1}{2}}\mathbf{a}) &= (\mathbf{R}^{-\frac{1}{2}}\mathbf{X}\mathbf{C}^{-\frac{1}{2}})\mathbf{C}^{\frac{1}{2}}\mathbf{b}, \\ \mathrm{r}(\mathbf{C}^{\frac{1}{2}}\mathbf{b}) &= (\mathbf{R}^{-\frac{1}{2}}\mathbf{X}\mathbf{C}^{-\frac{1}{2}})^{\mathrm{t}}\mathbf{R}^{\frac{1}{2}}\mathbf{a}. \end{aligned}$$

On substituting for $\mathbf{C}^{\frac{1}{2}}\mathbf{b}$ in the first of these equations we arrive at

$$\mathrm{r}^2(\mathbf{R}^{\frac{1}{2}}\mathbf{a}) = (\mathbf{R}^{-\frac{1}{2}}\mathbf{X}\mathbf{C}^{-\frac{1}{2}})(\mathbf{R}^{-\frac{1}{2}}\mathbf{X}\mathbf{C}^{-\frac{1}{2}})^{\mathrm{t}}\mathbf{R}^{\frac{1}{2}}\mathbf{a}.$$

Finally, letting $\mathbf{G} = \mathbf{R}^{-\frac{1}{2}}\mathbf{X}\mathbf{C}^{-\frac{1}{2}}$ we have

$$\mathrm{r}^2(\mathbf{R}^{\frac{1}{2}}\mathbf{a}) = \mathbf{G}\mathbf{G}^{\mathrm{t}}\mathbf{R}^{\frac{1}{2}}\mathbf{a}. \tag{5.59}$$

Similarly,

$$\mathrm{r}^2(\mathbf{C}^{\frac{1}{2}}\mathbf{b}) = \mathbf{G}^{\mathrm{t}}\mathbf{G}\mathbf{C}^{\frac{1}{2}}\mathbf{b}. \tag{5.60}$$

Equations (5.59) and (5.60) together show correspondence analysis to be equivalent to the singular value decomposition of $\mathbf{G}$, and, incidentally by virtue of (5.23)–(5.28), also to dual scaling.

Let us now examine the relationship between dual scaling and principal component analysis. Equations (5.59) and (5.60) will be helpful for this purpose. The matrices $\mathbf{G}\mathbf{G}^{\mathrm{t}}$ and $\mathbf{G}^{\mathrm{t}}\mathbf{G}$ are each symmetric, positive semidefinite. Equation (5.59) shows $\mathbf{a}^* \equiv \mathbf{R}^{\frac{1}{2}}\mathbf{a}$ to be an eigenvector of $\mathbf{R}^{-\frac{1}{2}}\mathbf{X}\mathbf{C}^{-1}\mathbf{X}^{\mathrm{t}}\mathbf{R}^{-\frac{1}{2}}$ corresponding to the eigenvalue r^2; similarly, (5.60) shows $\mathbf{b}^* \equiv \mathbf{C}^{\frac{1}{2}}\mathbf{b}$ to be an eigenvector of $\mathbf{C}^{-\frac{1}{2}}\mathbf{X}^{\mathrm{t}}\mathbf{R}^{-1}\mathbf{X}\mathbf{C}^{-\frac{1}{2}}$ which also corresponds to the eigenvalue r^2. Premultiplying (5.59) by $\mathbf{R}^{-\frac{1}{2}}$ and (5.60) by $\mathbf{C}^{-\frac{1}{2}}$ shows after a little manipulation that the vectors of row scores, $\mathbf{a} = \mathbf{R}^{-\frac{1}{2}}\mathbf{a}^*$, and of column scores, $\mathbf{b} = \mathbf{C}^{-\frac{1}{2}}\mathbf{b}^*$, are eigenvectors of the matrices $\mathbf{R}^{-1}\mathbf{X}\mathbf{C}^{-1}\mathbf{X}^{\mathrm{t}}$ and $\mathbf{C}^{-1}\mathbf{X}^{\mathrm{t}}\mathbf{R}^{-1}\mathbf{X}$, respectively, corresponding to the common eigenvalue r^2. Thus, $\mathbf{a}$ specifies the principal axes of the weighted scatter of row points corresponding to $\mathbf{R}^{-1}\mathbf{X}$ in *c*-space in the metric $\mathbf{C}^{-1}$, while $\mathbf{b}$ specifies the principal axes of the weighted scatter of column points corresponding to $\mathbf{C}^{-1}\mathbf{X}^{\mathrm{t}}$ in *r*-space in the metric $\mathbf{R}^{-1}$. The variances of the respective principal components are the eigenvalues, r^2, which numerically are also the canonical roots. Usually, the eigenvectors **a** and **b** of both spaces are normalized in the same way to yield components whose variance is identically r^2:

$$(N-1)^{-1}\mathbf{a}^t\mathbf{R}\,\mathbf{a}=(N-1)^{-1}\mathbf{b}^t\mathbf{C}\,\mathbf{b}=r^2.$$

Dual scaling therefore amounts to two simultaneous principal component analyses of the weighted matrices $\mathbf{R}^{-1}\mathbf{X}(r\times c)$ and $\mathbf{C}^{-1}\mathbf{X}^t(c\times r)$, respectively.

It will be instructive to pursue the connection between dual scaling and principal component analysis from a geometric viewpoint a little further. In components analysis it is customary to regard a matrix $\mathbf{X}(N\times p)$ as specifying a cloud of N points in p-space. Components analysis itself is a procedure for describing the dispersion of these points with respect to a set of orthogonal axes – the principal components. The first few components often provide a reference frame of low dimension onto which the points can be projected in such a way that the true higher-dimensional structure in the data is substantially preserved. Dual scaling can be comprehended in broadly the same terms. There are, however, two important differences. First, we have seen that in dual scaling a *pair* of dual vector spaces are invariably associated with $\mathbf{X}$ $(r\times c)$; the rows of $\mathbf{X}$ specify N points in c-space while the columns specify N points in r-space. Second, in row (column) space it is the *weighted* dispersion of points which is decomposed along principal axes, the weights being the row (column) totals, and the projection of points being orthogonal with respect to the *metric* of row-space (column-space). Defined in this way, the decomposition along principal axes is identical in each space. This is why dual scaling yields a visual interpretation of the relative positions of *both* clouds in a common subspace of low dimension.

5.6 Relationships among statistical methods

Several connections between canonical correlation analysis and other multivariate procedures were described in Chap. 4 and in preceding sections of the present chapter. Specifically, canonical variate analysis and dual scaling were shown to be open to interpretation as particular cases of canonical correlation which arise under specialization of the variables. Further, close formal relationships were shown to exist between canonical variate analysis, multivariate analysis of variance and principal component analysis, on the one hand, and between dual scaling, the canonical scoring of contingency tables, correspondence analysis and principal component analysis on the other. The affinities of canonical analysis are evidently many and varied. Such relationships are pursued further in the present section. In particular, we shall see that many familiar methods of classical univariate, bivariate and multivariate statistics are open to interpretation as special cases of canonical correlation analysis. Bartlett (1947) seems to have been the first to have drawn attention to the connections between several important multivariate techniques.

Binary-valued dummy variables were introduced in Sect. 4.2 as a device for conveying information on the partition of a sample. Binary dummy variables are also useful for sketching something of the nature of the connections between canonical analysis and other methods. Hotelling (1936) introduced canonical analysis as a means of investigating the linear correlation structure between two sets of continuous variables. Specialization of either or both sets of variables was later shown to yield canonical variate analysis and dual scaling, respectively. Where the criterion variables are further specialized by reducing them to a single member ($q=1$), several familiar univariate and bivariate procedures result. Table 5.3 summarizes several cases of canonical analysis which arise under restrictions on either the kind or number of variables employed, or on both. The table is largely self-explanatory. The methods of the right-hand side of the body of the table are more general than those of the left-hand side in the sense that the number of criterion variables, q, is unrestricted provided that $q>1$. Similarly, the methods of the last row are more general than those above them in that under appropriate restrictions they embody all of the preceding methods. Kshirsagar (1972, pp. 206, 347, 354, 379) has discussed relationships between several of these methods in some detail. Table 5.3 is far from exhaustive, containing only a selection of methods from a considerably wider class. Thus, by restricting the variables of both sets to a single member ($p=q=1$), such familiar quantities or procedures as the t-statistic and simple correlation and regression analysis result. Moreover, principal component analysis does not appear despite its close formal connections with canonical analysis, noted earlier.

Evidently, canonical analysis can be used to relate and unify the methods of parametric data analysis. Some insight as to why this should be so is provided by two observations. First, that a dummy variable in canonical analysis functions as a device to account for the degrees of freedom associated with the

Table 5.3. Special cases of canonical correlation analysis which arise under restrictions on the kind and number of variables employed

Nature of variables		*Number of criterion variables, q*	
Predictor	*Criterion*	$q=1$	$q>1$
Binary	*Continuous*	Analysis of variance	Multivariate analysis of variance; T^2
Continuous	*Binary*	Discriminant analysis; T^2	Canonical variate analysis; multiple discriminant analysis
Mixed binary and continuous	*Continuous*	Analysis of covariance	Multivariate analysis of covariance
Binary	*Binary*	Binary regression; χ^2 $2\times c$ table	Dual scaling; canonical scoring of $r\times c$ tables
Continuous	*Continuous*	Multiple regression	Multivariate regression; canonical correlation analysis

partition of a sample into groups. Secondly, that differences between means and correlations between variables are analogous concepts – a difference between sample means on a single variable conveys essentially the same information as a correlation coefficient between that variable and a dummy variable of sample membership. These two notions together enable the class of techniques comprising analysis of variance and regression analysis to be subsumed under the multivariate general linear model (Morrison, 1976, p. 170). Put rather differently, the multivariate general linear model generates a family of methods of which canonical analysis is the most versatile. Several benefits follow from awareness of the unifying capacity of canonical analysis. Perhaps the most useful advantage is that comprehension and appreciation of a large number of statistical tools is facilitated. Secondly, theoretical advances made with respect to canonical analysis may be found to be applicable in a variety of other contexts. Lastly, it may be that sophisticated computer codes developed with canonical analysis in mind can be used for other purposes with only minor adaptation. At the same time, however, it is well to bear in mind that the derivation of methods which are embeded in canonical analysis is a response to problems whose formulation is generally most natural in terms other than that of the linear correlation between two sets of variables.

The affinities which unite parametric statistical methods can be approached from several viewpoints in addition to that adopted above. Rao (1960), for example, has shown that most such methods can be subsumed under the *general theory of regression*. Krzanowski (1971) has demonstrated that the *algebra* underlying many classical multivariate techniques is very similar. Such methods are algebraically related in that they involve the optimization of a criterion which can be expressed as a quadratic form or as a ratio of quadratic forms. Robert and Escoufier (1976) have pointed out that these methods are also unified by noting that they may be regarded as resulting from a search for *metrics* which will optimize an alternative criterion, namely Escoufier's *RV*-coefficient (Escoufier, 1973). Yet another unifying principle is provided by Knapp's (1978) demonstration that the *tests of significance* associated with simple correlation, the *t*-test in independent samples, multiple regression and correlation analysis, one-way analysis of variance, analysis of covariance, the *t*-test in correlated samples, discriminant analysis and the chi-squared test of independence are all special cases of the test of the null hypothesis of canonical correlation analysis. The interrelatedness of several univariate and multivariate techniques has been discussed by Pruzek (1971) and Baggaley (1981) from much the same standpoint. A further unifying concept is that of *low-rank matrix approximation*, as Greenacre and Underhill (1982) have shown. We observe that *more general* models than either canonical analysis or the multivariate general linear model exist. Foremost among these are Jöreskog's (1977) structural relations model and the family of nonlinear partial least squares models of Wold (1966a, b, 1980) and of Forrest Young and his co-workers (e.g., see Young, 1981). The extension of canonical analysis to situations in which the data are *continuous functions* of a variable such as time (Cailliez & Pages, 1976; Ramsay, 1982; Ramsay & de Leeuw, 1983) also merits mention at this point.

5.7 Concluding remarks

The formulation of the problem of obtaining scores for the analysis of contingency tables as a special case of canonical correlation analysis can be traced to Fisher (1938) and Maung (1941). In the canonical analysis of discrete variables they were however anticipated by the work of Hirschfeld (1935).

The applicability of dual scaling is not restricted to contingency tables and in fact embraces rectangular arrays of positive and negative numbers of several kinds. The chief requirement of the data in this regard is that the rows and columns of the array be commensurable. Accordingly, in addition to contingency tables, dual scaling is directly applicable to $\mathrm{N} \times p$ arrays of counts pertaining to p variables in N samples. As a special case, the variables may be binary, extending analysis to tables of presence/absences. Other types of categorical data can sometimes be re-expressed in forms suitable for analysis. Rank-ordered data, for example, are transformed as follows. Suppose $\mathbf{R}(\mathrm{N} \times p)$ represents a matrix of N observations on p ranked variables. For the ijth element r_{ij} of $\mathbf{R}$ define g_{ij}^* as

$$g_{ij}^* = p + 1 - 2r_{ij}, \tag{5.61}$$

where r_{ij} is the rank of the jth variable ($j = 1, \ldots, p$) for the ith sample and rank r is defined as the rth choice out of p. Once $\mathbf{G}^*(\mathrm{N} \times p)$ is formed by applying (5.61), analysis proceeds as described previously on replacing $\mathbf{G}$ in (5.26) by $\mathbf{G}^*$. As another example, consider an $\mathrm{N} \times p$ array for which the p variables are each open to some specified number k_i of possible outcomes. Then the binary *response pattern* matrix (Nishisato, 1980, p. 4), $\mathbf{F}^*(\mathrm{N} \times k)$, say, where $k = k_1 + k_2 + \ldots + k_p$ and k_i is the number of options for variable i, is used to replace $\mathbf{G}$ in (5.26). Reference may be made to Nishisato (1978; 1980, Chap. 6), Greenacre (1981) and Skinner and Sheu (1982) for numerical examples of the use of these and related transformations. Procedures for dealing with missing values in dual scaling have been developed by Mutombo (1973), Nora-Chouteau (1974), Nishisato (1980, p. 198) and by Gabriel and Zamir (1979).

The theory of the *canonical scoring of contingency tables* has been described or extended by several workers. Among these, mention is made of E.J. Williams (1952, 1967), Lancaster (1957, 1958, 1963, 1969), Good (1965), McKeon (1965), Srikantan (1970), Kshirsagar (1972, p. 379), Kendall and Stuart (1973, p. 588), de Leeuw (1973), Maxwell (1973), Hill (1974), Corsten (1976a), O'Neill (1978a, b), Kaiser and Cerny (1980), Nishisato (1980, p. 74), Goodman (1981) and Haberman (1981). *Correspondence analysis* in one guise or another has been discussed by Guttman (1941, 1946), Bartlett (1951, 1965), Benzécri (1969, 1973), David, Campiglio and Darling (1974), Hill (1974), Teil (1975), David, Dagbert and Beauchemin (1977), Lebart, Morineau and Tabard (1977), Greenacre and Degos (1977), Nishisato (1980, 1984), Greenacre (1981, 1984), Greenacre and Underhill (1982), Gordon (1981, p. 117; 1982), Gower and Digby (1981), Deville and Saporta (1983) and Lebart, Morineau and Warwick (1984). With respect to *low-rank matrix approximation*, the contributions of Corsten (1976a), Gabriel (1978a), Gabriel and Zamir (1979), Rao (1980) and Greenacre and Underhill (1982) are especially pertinent. In relation to the class of methods for quantifying qualitative data based on the principles of *optimal scaling and alternating least squares* reference may be made to de Leeuw (1973), Young, de Leeuw and Takane (1976), Perreault and Young (1980) and Young (1981). The *extension* of dual scaling to *m*-way arrays ($m > 2$) has been investigated by Hill (1974), Gower (1977), Lebart, Morineau and Tabard (1977), Nishisato (1980, p. 98) and Deville and Saporta (1983).

Several workers have addressed themselves to the *interrelations among statistical methods*. Relationships between canonical analysis and procedures such as multiple regression, multiple correlation, principal component analysis, multiple discriminant analysis, factor analysis and scaling theory have been described by Bartlett (1938, 1948), McKeon (1965), Rozeboom (1965) and Glahn (1968). McDonald (1968) has provided a unified treatment of the weighting problem as it arises in canonical analysis, principal component analysis, canonical factor analysis and optimal scaling, and has shown that in these contexts weighting leads to a search for the extreme values of a ratio of quadratic forms. Lyttkens (1972) has discussed regression aspects of canonical correlation, while Robinson (1973) has shown canonical correlation analysis to be equivalent to the problem of estimating a matrix of linear regression coefficients **B** of less than full rank. Relationships between correspondence analysis and methods such as canonical correlation, component analysis, simultaneous linear regression, bivariate correlation and classical multidimensional scaling have been investigated by Hill (1974), Greenacre and Degos (1977), Nishisato (1980, p. 59), Buyse (1983) and Heiser and Meulman (1983a, b). Akaike (1976) has discussed connections between canonical correlation analysis, singular value decomposition and the canonical representation of a linear stochastic system. Affinities between canonical correlation and interbattery factor analysis have been explored by Browne (1979, 1980) and by Huba, Newcomb and Bentler (1981).

Izenman (1980) has used a generalization of the multivariate linear regression model as a means of unifying the classical multivariate procedures of principal component analysis, canonical variate analysis and canonical correlation analysis. The value of the multivariate general linear model as a unifying structure for the analysis of statistical models has been mentioned by Pruzek (1971) and by Burdick (1982), while Muller (1981, 1982) has examined relationships between canonical analysis, principal component analysis, redundancy analysis and multivariate regression in terms of the general linear model. Relationships between canonical analysis, redundancy analysis and principal component analysis have also been studied by DeSarbo (1981), Tyler (1982) and Van de Geer (1984). Young, de Leeuw and Takane (1976), Perreault and Young (1980) and Young (1981) have discussed similarities between general linear model procedures (multivariate analysis of variance, multivariate regression, discriminant analysis, canonical correlation analysis, etc.) and a class of nonmetric alternating least squares models with optimal scaling features. Van de Geer (1981, 1984) has shown that ridge regression estimation, oblique Procrustes rotation, principal component analysis and various forms of canonical analysis are all open to interpretation in terms of a particular determinantal equation. Connections between singular value decomposition, Karhunen-Loève transformation and principal component analysis have been discussed by Gerbrands (1981). Keller and Wansbeek (1983) have provided a unified treatment of several multivariate techniques, including ordinary least squares, principal component analysis, orthogonal regression and canonical analysis, in terms of the error-in-variables model. Pertinent contributions to the clarification of relations among statistical methods have also been made by McNeill (1974), Scobey and Kabe (1980), de Leeuw (1982) and Huberty (1983).

Several treatments of canonical analysis as a special case of other yet *more general* models are available. Bagozzi, Fornell and Larcker (1981) and Fornell (1982) have discussed canonical analysis as a particular case of a linear structural equations (LISREL) model, while Wold (1966a, b), Young (1981) and Dijkstra (1983) have discussed canonical analysis in the context of partial least squares (PLS) models. For an account of canonical analysis as a special case of the linear errors-in-variables (EV) model, reference may be made to Keller and Wansbeek (1983).

Further reading. Dual scaling is widely used in ecology and related fields. Applications in which the model is carefully fitted and the results adequately reported, however, are by no means common. Of the numerous applications, reference is made to studies by Hatheway (1971), Lacoste and Roux (1971, 1972), Hill (1973), Hill and Smith (1976), Teil and Cheminee (1975), Blanc, Chardy, Laurec and Reys (1976), David, Dagbert and Beauchemin (1977), Greenacre and Degos (1977), Lebart, Morineau and Tabard (1977), Bonin and Roux (1978), Bachacou, Chessel and Croze (1979), Laurec, Chardy, de la Salle and Richaert (1979), Benzécri (1980), Greenacre (1981, 1984) and Gordon (1982).

Part II
Applications

General introduction

Seven applications of canonical analysis are described. The first two applications are in the nature of experiments. Their purpose is to enable an assessment of the ability of canonical analysis to recover known relationships between sets of variables of ecological interest to be made. The first experiment deals with relationships between the abundances of three plant species and the geographical position of the samples in which the species occurred. The second study concerns relationships between several soil properties and the representation of a number of plant species in a limestone grassland community. Both investigations are purposely characterized by the relative simplicity of the ecological relationships involved with the intent of making an appraisal of canonical analysis possible in fairly clear-cut terms. If the analyses fail to recover the simple relationships expected, then doubt would certainly be cast on the efficacy of canonical analysis in ecology. On the other hand, if successful, the analyses would demonstrate that the method can contribute usefully, at least where the ecological relationships involved are of a similar order of complexity to those encountered here; moreover, the studies would then provide a foundation from which further exploration of the opportunities offered by canonical analysis could confidently be based.

The remaining applications are more realistic in character. The purpose of these analyses is threefold:

(a) to illustrate the varied opportunities offered by canonical analysis in exploring relationships of ecological interest and in this way contribute to a better understanding of the role of the method in ecology;
(b) to show something of the range and flexibility of canonical analysis which results from various specializations on the nature of the variables employed; and
(c) to draw attention to connections between canonical analysis and other methods of data analysis widely used in ecological work.

The first two of this group of analyses involve rain forest vegetation. The first analysis deals with relationships between soil properties and the occurrence of different forest communities, and the second with relationships between the seedling and the mature tree composition of several different forest communities. In both analyses all the variables of interest are continuous.

These studies of rain forest vegetation are followed by two analyses which illustrate the use of canonical analysis to explore comparative relationships among a number of multivariate universes. The first analysis concerns relationships between three plant communities in terms of their overall species' composition; the second with relationships among eight grass species with respect to their responses to added treatment levels of nitrogen in a mineral nutrition experiment. In each of these analyses, *one* of the two measurement domains consists of binary-coded dummy variables. Certain connections between canonical analysis, multivariate analysis of variance and multiple discriminant analysis are examined in the context of the second study.

The final application concerns relationships between several study areas and the presence of seven species of large herbivore in East African rangeland. *Both* sets of variables in this study consist of binary-valued dummy variables. The formal equivalence of this form of canonical analysis to the analysis of association in $r \times c$ tables and to dual scaling is demonstrated in connection with this example.

Experimental studies. In view of doubts which have been expressed concerning the suitability and usefulness of canonical analysis in ecology, it seemed desirable to attempt to establish the soundness of the method in a variety of ecological contexts. With this purpose in mind, a number of experimental analyses were made. The analyses were based on ecological data for which the major relationships between the variables involved had already been established with some degree of confidence. It is therefore possible to assess the ability of canonical analysis to recover the *known* interrelationships among the variables. Two investigations of this kind are reported here. Both studies involve relatively few variables ($p+q$ equals 6 and 11, respectively). The results obtained therefore also provide a convenient means of showing how to go about interpreting a canonical analysis, as well as revealing the kind and extent of the information provided. Accordingly, the analyses are worked through in some detail.

6. Experiment 1: an investigation of spatial variation

6.1 Introduction

The experiment concerns the joint pattern of spatial variation common to three plant species belonging to a limestone grassland community in Anglesey, North Wales. The species were *Phleum bertolonii* DC., *Dactylis glomerata* L. and *Galium verum* L. From previous work, the species were known to share the same relatively simple generalized distribution pattern, which was centered on the southwestern extremity of the study area. Furthermore, the direction and extent of the departure of the distribution of each species from the common trend was also known. *P. bertolonii* was the most circumscribed of the species, being strictly confined to the southwestern corner. *G. verum*, on the other hand, was the most widespread species, occurring throughout the study area although its maximum representation occurred in the southwest. *D. glomerata* was less extreme with a distribution intermediate between those of the other species, but resembling that of *P. bertolonii* more closely than that of *G. verum*.

Estimates of the abundance of the species in a random sample of N=45, 10×10 m stands were available. These estimates provided observations on one set of variables for canonical analysis while the geographical coordinates of stands comprised observations on a second set. Stand position was determined in relation to an arbitrary (X, Y)-coordinate system whose origin was located close to the center of the area surveyed. The X-coordinate specified stand position in the east-west direction and the Y-coordinate in the north-south direction. In addition, the crossproduct term, XY, between X and Y was also calculated. In view of the known southwesterly character of the species' distributions it was anticipated that the inclusion of the cross-product term might improve the fit of the model. Thus, in all, three spatial measures were used. The purpose of the experiment was to determine to what extent canonical analysis could recover known facts about the spatial variability of the species examined. Briefly, these facts were:

(a) that the species share a common distribution pattern centered on the southwestern extremity of the area studied; and
(b) that there exist departures from the joint trend in the detailed distribution of the species, *P. bertolonii* being the most circumscribed and *G. verum* the most widespread.

For the analysis, we have N=45 random samples, $p=3$ ecological variables, $q=3$ spatial variables. The two measurement domains will be denoted by **x** and

y, respectively. Data used in the analysis are reported in Appendix A.2 (Table A-1) and the results of the analysis are summarized in Tables 6.1 and 6.2.

6.2 Results

6.2.1 The canonical correlation coefficients

The canonical correlation coefficients r_k $(k=1, \ldots, 3)$ are reported in Table 6.1a. The magnitude of r_k indicates the degree of linear correlation between u_k and v_k. The first canonical correlation coefficient (.79) is appreciably larger than the second (.46) while the third (.09) is very small.

The squared correlation coefficients r_k^2 (Table 6.1a) express the proportion of the variance of the kth canonical variate, u_k, say, that is explained by its conjugate, v_k, or *vice versa*. It can be seen that $r_1^2=.631$. Thus, 63.1% of the variation in the linear combination of species specified by $u_1=\mathbf{a}^t\mathbf{z}^{(x)}$ is attributable to variation in that particular linear combination of the spatial measures specified by $v_1=\mathbf{b}^t\mathbf{z}^{(y)}$. Similarly, for the second and third pairs of canonical variates we find that 21.4% and 0.8% of the variation in the species' composites u_2 and u_3 are attributable to v_2 and v_3, respectively. From the magnitude of r_1^2, it appears that the overall relationship between the two domains is reasonably strong. Moreover, that aspect of the overall relationship defined by u_1 and v_1 is three times stronger than that specified by u_2 and v_2, while that correspond-

Table 6.1. Limestone grassland, Anglesey. Canonical analysis. Spatial variation in the representation of three plant species. Canonical correlation coefficients and tests of the linear independence of **x** and **y**

(a) The canonical correlation coefficients r_k and union-intersection tests of independence and dimensionality. Approximate critical values of Roy's largest-root criterion $\theta_\alpha(s, m, n)$ are shown for $\alpha=.05$ and $\alpha=.01$ $(m=-\frac{1}{2};\ n=18\frac{1}{2})$.

k	s	r_k	r_k^2	$\theta_{.05}(s, m, n)$	$\theta_{.01}(s, m, n)$	p
1	3	.795	.631	.280	.350	$\ll .01$
2	2	.462	.214	.198	.268	$< .05$
3	1	.088	.008	.175[a]	.243[a]	$> .05$

[a] From the U-distribution: $\theta_\alpha(1, m, n)=1-U_\alpha(p, 1, n)$.

(b) Likelihood ratio tests of independence and dimensionality. Bartlett's approximate chi-squared criterion for assessing the joint nullity of the smallest $s-k$ canonical correlation coefficients.

k	Roots	Chi-squared	df	p
0	1, 2, 3	50.47	9	$< .0001$
1	2, 3	10.04	4	.04
2	3	.31	1	.58

ing to the third pair of canonical variates is negligible. When we come to consider redundancy in Sect. 6.2.4, we shall approach these various relationships from a rather different viewpoint and one which will cause us to modify these assessments somewhat.

The squared correlation coefficients are essential for the tests of independence and dimensionality which follow.

6.2.2 Independence

Union-intersection and likelihood ratio tests of hypothesis (3.9) of the linear independence of **x** and **y** are reported in the *first* line of Table 6.1a, b, respectively. Both procedures lead decisively to the rejection ($p \ll .01$) of hypothesis (3.9). There are grounds, therefore, for asserting that species' abundance varies systematically with geographical position. In order to pursue the nature of the dependency it is necessary to establish its *dimensionality* – that is, in other words, to determine the number of noteworthy linear relationships between domains.

6.2.3 Dimensionality

The sequential hypotheses (3.18) of Sect. 3.4.2 are of interest here. Union-intersection and likelihood ratio tests of hypothesis (3.18) are shown in Table 6.1a, b, respectively (lines 2 and 3 of the body of each table). It is apparent from the table that no significant relationship between **x** and **y** remains after the first two canonical correlation coefficients are eliminated. Thus, it appears that the dimensionality of the linear association between **x** and **y** is two. Accordingly, it seems likely that the *canonical variates* u_k and v_k $(k=1,2)$ corresponding to r_1^2 and r_2^2 will be required in order to fully comprehend and account for the relationship. It is to the canonical variates and their ecological implications that we now turn.

6.2.4 The canonical variates

In interpreting the canonical variates we rely principally on the correlations between the variates and the original variables, as well as on several indices calculated from them.

Intraset structure correlations. The intraset correlation coefficients appear in the left-hand side of Table 6.2 and are helpful in establishing the nature of the canonical variates defined on each set of variables. The square of each intraset correlation coefficient represents the proportion of the variance of a variable which is directly associated with a particular canonical variate.

The correlations of the spatial variables with v_1 differ in sign. These signs suggest that v_1 represents a *contrast* of east (X) and north (Y) against southwest/northeast (XY). XY shows the strongest correlation (−.70) with v_1 closely followed by Y (.62). The corresponding correlations of the species with u_1 are

Table 6.2. Limestone grassland, Anglesey. Canonical analysis. Spatial variation in the representation of three plant species: correlations between the original variables and the canonical variates

Canonical variate	u_1	u_2	u_3	h_w^2	v_1	v_2	v_3	h_b^2
Species								
G. verum	−.433	.126	−.893	1.000	−.344	−.058	−.078	.128
D. glomerata	−.877	−.479	−.027	.999	−.697	−.221	−.002	.535
P. bertolonii	−.982	.175	.077	1.000	−.780	.081	.077	.615
Variance extracted	.640	.092	.268	1.000	.404	.020	.002	.426
Redundancy	.404	.020	.022	.426	.404	.020	.002	.426

Canonical variate	v_1	v_2	v_3	h_w^2	u_1	u_2	u_3	h_b^2
Spatial coordinates								
X (east)	.483	.164	.860	1.000	.384	.076	.075	.159
Y (north)	.625	.378	−.684	1.000	.497	.175	−.060	.281
XY	−.697	.534	−.479	1.000	−.554	.247	−.042	.370
Variance extracted	.370	.152	.479	1.000	.234	.032	.004	.270
Redundancy	.233	.032	.004	.269	.234	.032	.004	.270

all alike in sign, indicating that u_1 expresses some feature or trait *common* to all three species with respect to the linear composite of spatial variables represented by v_1. Notice that the strength of the species' correlations varies appreciably, from that of *G. verum* (−.43) to that of *P. bertolonii* (−.98). v_1 accounts for 37% of the variance common to the spatial domain ($V_1^2=.370$), while u_1 accounts for 64% of the variance of the species' measures ($U_1^2=.640$).

Turning to the second pair of canonical variates we see that the correlations tend to be smaller than those with u_1 and v_1. The intraset correlations of the spatial variables with v_2 are alike in sign; evidently, v_2 expresses some characteristic common to the three spatial variables in relation to the linear composite of the species represented by u_2. The strongest correlation (.53) belongs to XY; v_2 therefore appears to represent in part some aspect of southwest/north-easterly-ness related to u_2 which is at the same time uncorrelated with that previously specified by v_1. From the intraset species' correlations with u_2 we see that the strongest relationship is with *D. glomerata* (−.48), *P. bertolonii* and *G. verum* being characterized by quite small positive correlations. Evidently, u_2 represents a subsidiary distinction among the species previously unified by u_1. In terms of u_2 at least, *D. glomerata* is seen to be equally distinct from the remaining species. Thus u_2 appears to express what is unique to the distribution of *D. glomerata*, namely its approximately equal relationship to the distribution of *both* of the other species. The second pair of canonical variates account for only small amount of the variability of their respective domains, the variances of v_2 and u_2 being $V_2^2=.152$ and $U_2^2=.092$, respectively; v_2 and u_2 are evidently appreciably weaker components of their respective domains than are the first pair of canonical variates.

The third canonical variate v_3 of the spatial variables is characterized by sizeable correlations which contrast the variable X (.86) against both Y (−.68) and XY (−.48). The corresponding variate u_3 of the species' domain very clearly distinguishes *G. verum* from the remaining species. Indeed, u_3 is in fact essentially a pure, though inverse, measure of *G. verum*. It is of considerable interest to find that v_3 is the *strongest* canonical variate of the spatial domain ($V_3^2 = .479$), accounting for 48% of the total variance, that is, in other words, of trace ($\mathbf{R}_{22}$); u_3, although somewhat weaker ($U_3^2 = .268$), is nevertheless a sharply defined and relatively strong component of the species' domain.

Before proceeding it may be worthwhile to summarize the salient characteristics of the canonical variates to have emerged so far. The first canonical variate of the species' domain, u_1, clearly expresses some aspect of distribution *common* to all three species which is exemplified above all by *P. bertolonii*; u_2 and u_3, by way of contrast, are relatively pure measures of *D. glomerata* and *G. verum* respectively, and therefore specify features *unique* to the distribution of each. The nature of the v_k of the spatial domain are less easily characterized in this way. However, their dominant features can very roughly be subsumed under the headings 'southwest/northeast', 'southwest/northeast' and 'east plus south', respectively. The question of the ecological significance of relationships *between* the two sets of canonical variates can now be taken up.

Interset structure correlations. The interset variable/canonical variate correlations appear in the right-hand side of Table 6.2. These quantities are helpful in clarifying the nature of interrelationships between the two sets of variables. The square of each interset correlation coefficient is analogous to the coefficient of determination, R^2, of multiple regression. Thus each squared correlation expresses the proportion of the variance of a variable which is predictable from a particular canonical variate of the 'other' measurement domain. In interpreting these correlations, it is necessary to bear in mind the interpretation of the canonical variates arrived at from consideration of the intraset correlations.

For completeness, two sets of interset correlations are reported in Table 6.2. In the present analysis, however, only one set could have any possible ecological significance – namely those between the linear composites v_k of the spatial measures and species' abundance; relationships in the other direction, that is of the 'effect' of species abundance on spatial position, are clearly devoid of ecological meaning. In many applications, however, the *directed* nature of this analysis is lacking, being supplanted by an interest which is symmetric with respect to variables of both sets. In such cases, both sets of correlations will require attention.

The species correlations with v_1 show that the species all respond in the same sense to the spatial contrast represented by v_1. From the negative signs of the correlations, and recalling the nature of v_1, it is apparent that species vary *directly* with XY, the spatial variable which best characterizes v_1. In other words, it seems that all three species tend to increase in abundance southwestwards and northeastwards from the center of the study area with increase in XY. Similarly, it is clear that from the signs of the correlations, the three species vary inversely with X (east) and Y (north). This is equivalent to saying that the species *increase* in abundance westwards and southwards from the center of the study area. The absolute values of the correlations with v_1 show that *P. bertolonii* is the species most strongly (.78) associated with the linear composite represented by v_1, closely followed by *D. glomerata* (.70).

At this point it will be helpful to clarify the meaning of the cross-product term XY. The area of vegetation examined has an irregular shape and the disposition of the community in relation to the coordinate system used is such that the northeast quadrant is virtually without samples. This simplifies interpretation of the XY term. In view of the virtual absence of samples NE from the origin, it emerges that the XY term can for practical purposes be regarded as accounting for variation *only* in a southwesterly direction. We can now see that jointly the canonical variates u_1 and v_1 give rise to a consistent and intuitively reasonable picture: the abundances of all three species increase westwards, southwards and above all southwestwards from the center of the area surveyed. Furthermore, Table 6.2 (upper right) reveals that a substantial part (40.4%) of the total variance of the species' domain is accounted for in this way ($V^2_{\mathbf{x}|v_1} = .404$).

The interset correlations between the species and the second canonical variate of the spatial measures, v_2, are all small in absolute terms ($\leq .22$). The strongest relationship is the weak ($-.22$) inverse correlation of *D. glomerata*, the correlations of the remaining species with v_2 being close to zero. Recalling that v_2 perhaps represents some aspect of southwesterly-ness, the inverse relationship with *D. glomerata* implies that the species increases in abundance, at least to some extent, in the opposite direction. This result might conceivably indicate that, while *Dactylis* is predominantly southwestern in distribution, as shown by its relationship with v_1, it is by no means strictly so, being 'centered' somewhere between the southwestern extremity and the center of the study area. *Phleum* and *G. verum* are alike in being virtually uncorrelated with v_2 and hence are totally lacking in the 'intermediate' type of distribution which characterizes *Dactylis*.

We shall see shortly that the explanatory power of the third canonical variate of the spatial measures, v_3, in the species' domain is very small ($V^2_{\mathbf{x}|v_3} = .002$). This result indicates that little would be gained by efforts to interpret the third canonical relationship, and, for this reason, we may safely disregard the interset correlations of the species with v_3.

We are now in a position to summarize the joint relationships between the canonical variates of the ecological and spatial variables. These are as follows:

(a) the canonical variates u_1 and v_1 indicate that the representations of *P. bertolonii*, *D. glomerata* and *G. verum* all increase in a generally southwesterly direction from the origin; yet there are differences among the species in the degree of their response, that of *P. bertolonii* being the strongest, and that of *G. verum* the weakest;

(b) the second pair of canonical variates u_2 and v_2 appear to correspond to a weak though largely uncontaminated expression of *D. glomerata*, specifying what is unique to the distribution of this species, namely, that, while it is substantially southwesterly in character, it is not strictly so.

We turn to the redundancy in the species' domain to obtain an indication of the relative weight or importance of the relationships identified.

Redundancy. Redundancy expresses the explanatory power of a canonical variate of one domain with respect to the observed variables of the other.

It will be instructive to proceed by first examining the variance attributable to u_1 and v_1 in relation to the domain on which each is defined and also with respect to the 'other' domain. Reference to Table 6.2 shows that while 64% of the total variance of the species comprising **x**, on which u_1 is defined, is accounted for by this canonical variate, only 40% of the variance in **x** is accounted for by the corresponding canonical variate v_1 of the *spatial* domain. Similarly, 37% of the total variance of the spatial variables comprising **y** is accounted for by the first canonical variate v_1 of this domain, while the corresponding canonical variate u_1 of the *species* domain accounts for only 23% of the variance in **y**. Evidently, a canonical variate of a specified domain is a stronger explanatory construct with respect to that domain than is its conjugate of the other. Furthermore, we find that the percentage of the species' variance attributable to the linear combination of spatial measures represented by v_1 is appreciably smaller (40.4) than that given by the squared canonical correlation coefficient considered earlier, r_1^2 (63.1). Bearing in mind that the proportion of the multivariate variance of a measurement domain attributable to a canonical variate of the 'other' is, indeed, redundancy, it emerges that redundancy and a squared canonical correlation coefficient provide different indications of the strength of a particular relationship, that given by redundancy invariably being the more conservative. We shall now approach redundancy from a different viewpoint in order to further clarify its nature and its relationship to a squared canonical correlation coefficient.

It is easy to verify from Tables 6.1a and 6.2 that redundancy can be calculated as the within-set variance (U_k^2 or V_k^2) of a canonical variate attenuated by the between-set variance (r_k^2) of the variate in question. For u_1 and v_1, for example, we find:

$$\begin{aligned} U^2_{\mathbf{y}|u_1} &= V^2_{1(\mathrm{y})} r_1^2 = .370 \times .631 = .233, \\ V^2_{\mathbf{x}|v_1} &= U^2_{1(\mathrm{x})} r_1^2 = .640 \times .631 = .404. \end{aligned} \tag{6.1}$$

These quantities are the redundancies of the spatial and species domains, respectively; the redundancy in **y**, given the availability of u_1 of the species domain, is .233 while the redundancy in **x**, given the availability of v_1 of the spatial domain, is .404. From (6.1) it is plain that redundancy does in fact provide a more comprehensive index of the explanatory power of a canonical variate across domains than a squared canonical correlation coefficient. The desirability of taking account of the variance absorbed by a canonical variate with respect to the variables on which it is defined in assessing the explanatory power of the variate is intuitively appealing. The origin of the seeming discrepancy between a squared canonical correlation coefficient and redundancy as an index of explanatory power noted above is also apparent from (6.1).

Turning to the second pair of canonical variates we find that u_2 accounts for 9% of the variance of the species domain on which it is defined, and for 3% of the variance of the spatial domain (Table 6.2). Similarly, v_2 accounts for 15% of the variance of the spatial domain of which it is a linear composite, and for 2% of the variance of the species' domain. Although moderately correlated ($r_2 = .46$), u_2 and v_2 are evidently rather weak components of their respective domains. Both features are reflected in the redundancies associated with these canonical variates:

$$U^2_{\mathbf{y}|u_2} = V^2_{2(y)} r^2_2 = .152 \times .214 = .032,$$
$$V^2_{\mathbf{x}|v_2} = U^2_{2(x)} r^2_2 = .092 \times .214 = .020.$$

The explanatory power of both canonical variates across domains is very small. A similar result is found in connection with the third pair of canonical variates, although for different reasons. In contrast to the second pair of variates, u_3 and v_3 are relatively strong components of their respective domains but are themselves only weakly correlated ($r_3 = .09$); hence the shared variance common to u_3 and v_3 is very small ($r^2_3 = .008$). The redundancies associated with u_3 and v_3 are, respectively:

$$U^2_{\mathbf{y}|u_3} = V^2_{3(y)} r^2_3 = .479 \times .008 = .004,$$
$$V^2_{\mathbf{x}|v_3} = U^2_{3(x)} r^2_3 = .268 \times .008 = .002.$$

The effect of the virtual absence of covariation between u_3 and v_3 ($r^2_3 = .008$) on the explanatory power of these canonical variates across domains is obvious.

Redundancy leads to rather a different perception of the strengths of not only the overall dependence of **x** and **y** but also of each of its component parts from that arrived at in considering the r^2_k. The overall relationship is decidedly weaker in terms of redundancy than of squared canonical correlations; moreover, the first canonical relationship proves to be *twenty* times stronger than the second while the third is entirely negligible.

The dominance of the first canonical relationship enables us to assert that the linear relationship between **x** and **y** is, for all practical purposes, subsumed by u_1 and v_1 alone. Accordingly, it may be permissible to disregard the second and third pairs of canonical variates. This is equivalent to declaring that a model of rank $r = 1$ is considered to provide the most acceptable fit to the data. Such a position differs from that arrived at earlier on the basis of the hypothesis tests of Sect. 6.2. Bearing in mind the comprehensiveness of redundancy, however, we prefer to place greater emphasis on this index rather than on the canonical roots, r^2_k. In other words, it may sometimes be justifiable to allow an assessment of rank based on redundancy alone to override an assessment based on squared canonical correlations, irrespective of any statistical significance which may attach to the latter. We therefore reaffirm our choice of a rank 1 model.

Total redundancy expresses the proportion of the total variance of one measurement domain which is predictable from all the canonical variates of the other. We see from Table 6.2 that 42.6% of the variance of the species' domain is predictable from the spatial measures, while 27% of the variance of the latter is predictable by the species. However, recalling the acceptance of a model of rank 1, we ought properly to adjust the total redundancy estimates to take account of this. Acceptance of a rank 1 model leads to revised total redundancies which in this case are the redundancies generated by u_1 and v_1 alone. Thus

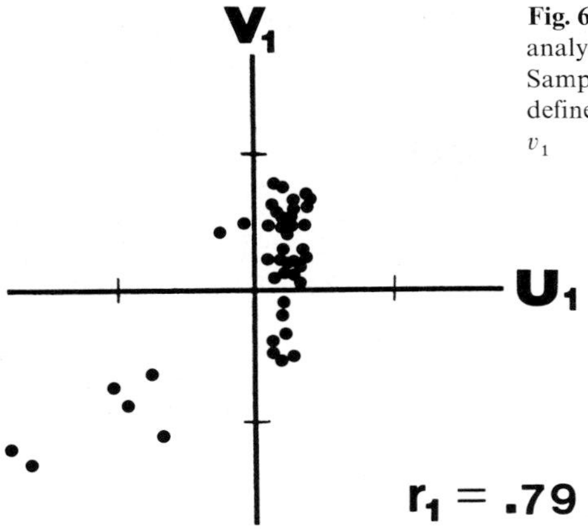

Fig. 6.1. Limestone grassland, Anglesey. Canonical analysis of spatial variation in three plant species. Sample of N=45 stands mapped into the subspace defined by the first pair of canonical variates, u_1 and v_1

the revised quantities are 40% and 23% for the **x**- and **y**-domains respectively. We conclude that 40% of the observed species' variability is attributable to variation in the spatial measures examined.

As a final step in interpreting the canonical variates graphical representation proved helpful. Figure 6.1 shows the N=45 sample-points mapped into the subspace defined by the first pair of canonical variates, u_1 and v_1. The correlation between u_1 and v_1 is $r_1 = .79$ and is expressed in the figure by a tendency for points to fall roughly on the diagonal extending from upper right to lower left. There are two distinctive features to the graph:

(a) the very noticeable scatter of points extending along the diagonal at the lower left;
(b) the concentration of points and the truncation of the scatter on the positive side of both axes close to the origin.

The first feature reflects the positive correlation between species' abundance and stands with a large XY coordinate, *i.e.* stands located at the southwestern extremity of the area. The second characteristic arises both from the scarcity of stands in the northeastern quarter of the survey area and from the absence of two of the species and the comparative rarity of the third from virtually all stands except those in the southwest. Consequently, potential sample points in the first (NE) quadrant of the sample space of Fig. 6.1 are unrealized in the sample.

Figure 6.1 very effectively summarizes the known facts about the joint distribution of the species provided the meaning of the canonical variates is kept in mind.

An alternative representation of the same phenomenon is provided by mapping the field distribution of canonical variate u_1 of the species' domain. In Fig. 6.2 the canonical variate scores of samples on u_1, the first variate of the species set, were plotted on a map of the survey area at the position of the samples. Contours have been inserted to show the form of the corresponding *canonical trend-surface*. The principal feature of interest is the clear 'centering' of the surface on the southwestern extremity of the map. Its level nature over

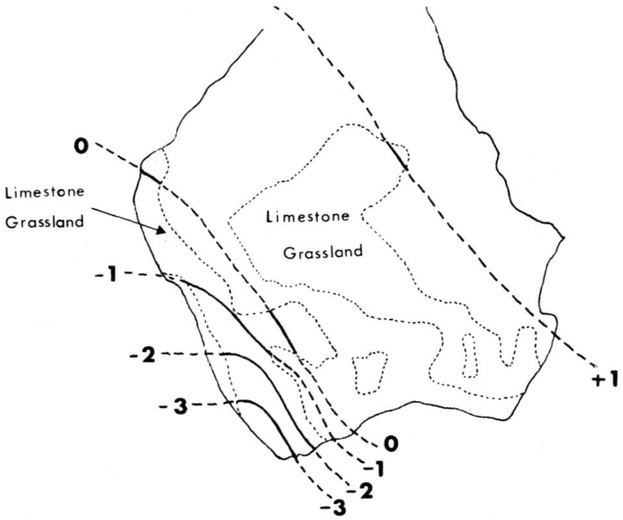

Fig. 6.2. Limestone grassland, Anglesey. Canonical trend-surface analysis. Spatial distribution of the first canonical variate, u_1, of the species' domain

much of the remainder of the area reflects the absence of two of the species and the comparative rarity of the third over a large part of the region. The canonical trend-surface summarizes that part of the spatial distribution of the species which is common to all three species. In other words the surface represents the generalized joint distribution pattern of the species. From the direction of the species correlations with u_1 and from the shape of the surface, it is apparent that species' abundance increases towards the southwest. Notice that the trend-surface does not show the absolute abundance of the species, the trend being expressed in terms of arbitrary units. The question of *why* the canonical trend should have the form that it does is a question of considerable ecological interest. Pursuit of questions of this kind would take us too far, however, and fall outside the scope of the present review.

Only one feature of the analysis remains to be considered. This is the extent to which the model accounts for the variances of the variables comprising each measurement domain.

6.2.5 Variable communalities

The *intraset* communalities (Table 6.2, h_w^2) of the variables of both sets are all essentially unity. Thus, the variances of the species and of the spatial variables are fully accounted for by the canonical variates. This is a consequence of the facts that in this analysis there are an equal number of variables in each set ($p=q=3$) and that Table 6.2 reports the full rank solution ($r=3$). Appreciable differences are nevertheless apparent between variables with respect to the extent to which the variance of each is accounted for by the rank $r=1$ solution.

The *interset* species' communalities (Table 6.2, $h_{\bar{b}}^2$) show that sizeable proportions of the variance of *P. bertolonii* (.61) and *D. glomerata* (.53) are accounted for by the spatial measures. On the other hand, the variance of *G. verum* is less well explained (.13). This is not altogether surprising as *G. verum* is more widely distributed than either of the other species. It is to be expected that the *joint* distribution pattern common to all three species will be strongly influenced by the close similarity of the distributions of *P. bertolonii* and *D. glomerata*. Consequently, it is not altogether surprising to find that the somewhat disparate distribution of *G. verum* is less adequately accounted for than those of the remaining species.

Notice that the interset communalities, $h_{\bar{b}}^2$ of Table 6.2 refer to the full rank 3 model. In the present analysis, however, the communalities of the rank 1 model do not differ to any appreciable extent from those of the rank 3 model.

6.3 Conclusions

The first canonical relationship recovers both:

(a) the joint southwesterly trend in spatial variation common to all three species; and
(b) departures in the distribution of particular species from the joint trend, *G. verum* being the species least closely associated with the overall trend and *P. bertolonii* the species most closely associated with it.

Recall that these are precisely the characteristics which at the outset were considered to comprise the salient ecological features of the data. The relationship accounts for 40% of the species' variation. This may be regarded as a satisfactory result, especially if the somewhat singular spatial distribution of *G. verum* is borne in mind.

The second and third canonical relationships also appear to distinguish meaningful ecological aspects of the data. However, in comparison with the first pair of canonical variates the contributions of the subsequent variates are altogether subordinate. The second and third canonical relationships apparently focus on rather fine points of detail, singling out features unique to the distributions of *D. glomerata* and *G. verum*, respectively.

In some circumstances it might be considered desirable to try to improve the fit of the model. Two steps could be taken in efforts to achieve this:

(a) addition of further polynomial terms of the spatial variables, X and Y;
(b) deletion of variables from the 'response' set – *G. verum* would be an obvious candidate for removal – as well as those antecedent variables (polynomial terms in X and Y) found not to contribute usefully to the overall fit.

7. Experiment 2: soil-species relationships in a limestone grassland community

7.1 Introduction

This experiment concerns relationships between the abundances of several plant species and associated soil characteristics in a limestone grassland community in Anglesey, North Wales. Previous work (Gittins, 1969) suggested that the factor-complexes of soil moisture and soil fertility are influential in determining the representation of many of the species present. Knowledge of the salient soil-species relationships provides a basis against which the ability of canonical analysis to recover such relationships can be assessed.

Estimates of the composition of the community and of the intensity of three soil properties based on a random sample of forty-five 10×10 m stands were available. Thirty-three species were encountered in all. The soil variables examined were depth (d), a surrogate for the more direct probable ecological control of soil moisture, extractable phosphate (P) and exchangeable potassium (K). For the purpose of the experiment, eight species were selected from the thirty-three for which data were available. Selection was guided primarily by a desire to use species whose response to variation in soil depth was as diverse as possible. The species selected, in approximate order of their response, were: *Helictotrichon pubescens* (Huds.) Pilger, *Trifolium pratense* L., *Poterium sanguisorba* L., *Phleum bertolonii* DC., *Rhytidiadelphus squarrosus* (Hedw.) Warnst., *Hieracium pilosella* L., *Briza media* L., and *Thymus drucei* Ronn. Species at the *H. pubescens* end of the sequence tend to be most abundant on the deeper soils, while those at the *T. drucei* end tend to be most abundant on the shallower soils; species towards the middle are relatively independent of soil depth.

This choice of species was expected to give rise to a data matrix for analysis whose inherent structure was basically simple. The inclusion of *P. bertolonii* and *P. sanguisorba*, however, was purposely designed to enrich this fundamentally simple structure, the representation of both species being believed to be influenced by soil P in addition to depth. Furthermore, there was evidence that the two species respond in *opposite* senses to variation in P, *P. bertolonii* increasing sharply in abundance with increase in P over the range encountered in the field, while *P. sanguisorba* tends to be most abundant on soils deficient in P. The remaining species were all considered to be largely independent of variation in soil P over the range encountered. Soil potassium, on the other hand, unlike either depth or phosphate, was believed to have a negligible effect on the behavior of *all* the species examined. Accordingly, the matrix for analysis was expected to be characterized by two features:

(a) a major structural component corresponding to differential species' response to variation in soil depth; and
(b) a subsidiary component corresponding to the contrasting behavior of *P. bertolonii* and *P. sanguisorba* in relation to soil P.

In fact, the relationships involved were known to be somewhat less clear-cut than for reasons of simplicity they are stated here. Complications arise because in the field soil depth and phosphate tend to be correlated to some extent, the deeper soils in general also being those of highest phosphate status. Thus species might conceivably respond more to the interaction of depth × phosphate than to either variable acting alone. Such considerations indicated that it might be worthwhile to include cross-product terms between all three soil variables in the analysis. Consequently, it was decided to make use of the variables d × P, d × K and P × K as well as the three observed soil variables, giving a total of six soil properties in all.

For the canonical analysis, therefore, we have a random sample of N=45 stands, $p=8$ estimates of species' abundance and $q=6$ soil variables. The two measurement domains will be denoted **x** and **y**, respectively. Interest centers on the nature of the relationship between the representation of the species and the intensity of the soil properties examined. For the analysis to be judged successful, it would need to identify the connections between the soil properties and species described above, and, in so doing, draw attention to the difference in magnitude between the overall effects of soil depth and phosphate on the species as a whole. Moreover, the totally subordinate role of soil potassium would also need to be exposed. The data on which the analysis is based appear in Appendix A.2 (Table A-2) while the results obtained are summarized in Tables 7.1 and 7.2.

7.2 Results

7.2.1 The canonical correlation coefficients

The canonical correlation coefficients r_k $(k=1, \ldots, 6)$ are reported in Table 7.1a. Three of the correlations exceed 0.5, the values of the first two, in particular, being $r_1=.92$ and $r_2=.70$. Certainly, the magnitude of the largest of these is suggestive of the existence of a linear relationship between **x** and **y**.

From Table 7.1a we find that the canonical roots, r_k^2, for $k=1, \ldots, 3$ are $r_1^2=.851$, $r_2^2=.484$ and $r_3^2=.271$, respectively. Recall that a squared canonical correlation coefficient expresses the explanatory power of each of a pair of canonical variates with respect to its conjugate. It therefore follows from the values of the roots that 85% of the variation in the linear combination of species given by $u_1=\mathbf{a}_1^t\mathbf{z}^{(x)}$ is attributable to variation in that linear combination of soil properties defined by $v_1=\mathbf{b}_1^t\mathbf{z}^{(y)}$; similarly, 48% and 27% of the variation in the species' composites u_2 and u_3 is accounted for by variation in v_2 and v_3, respectively. Evidently, a sizeable part of the variation in u_1 is explained

Table 7.1. Limestone grassland, Anglesey. Canonical analysis. Relationships between six soil properties and eight plant species. Canonical correlation coefficients and tests of the linear independence of **x** and **y**

(a) The canonical correlation coefficients r_k and union-intersection tests of independence and dimensionality. Approximate critical values of Roy's largest-root criterion $\theta_\alpha(s, m, n)$ are shown for $\alpha = .05$ and $\alpha = .01$ ($m = \frac{1}{2}$; $n = 14\frac{1}{2}$).

k	s	r_k	r_k^2	$\theta_{.05}(s, m, n)$	$\theta_{.01}(s, m, n)$	p
1	6	.923	.851	.557	.616	$\ll .01$
2	5	.696	.484	.511	.575	$> .05$
3	4	.521	.271	.457	.525	$> .05$
4	3	.374	.140	.392	.466	$> .05$
5	2	.257	.066	.313	.391	$> .05$
6	1	.218	.048	.280[a]	.350[a]	$> .05$

[a] From the U-distribution: $\theta_\alpha(1, m, n) = 1 - U_\alpha(p, 1, n)$.

(b) Likelihood ratio tests of independence and dimensionality. Bartlett's approximate chi-squared criterion for assessing the joint nullity of the smallest $s-k$ canonical correlation coefficients.

k	Roots	Chi-squared	df	p
0	1, 2, 3, 4, 5, 6	115.06	48	$< .0001$
1	2, 3, 4, 5, 6	45.35	35	.11
2	3, 4, 5, 6	21.33	24	.62
3	4, 5, 6	9.77	15	.83
4	5, 6	4.28	8	.83
5	6	1.78	3	.62

by variation in v_1. Thereafter the explanatory power of the v_k ($k > 1$) declines sharply, though in a remarkably regular fashion.

To establish whether a firm basis for regarding **x** or **y** as dependent does indeed exist, the tests of Sect. 7.2.2 are useful.

7.2.2 Independence

Results of union-intersection and likelihood ratio tests of the overall hypothesis of independence, $H_0: \Sigma_{12} = \mathbf{0}$, are shown in Table 7.1. For the union-intersection test we find (Table 7.1a) that $\theta_s = .851 > \theta_\alpha(6, \frac{1}{2}, 14\frac{1}{2}) = .616$ for $\alpha = .01$, where s denotes the sample value of the test statistic θ. Similarly, for the likelihood ratio test we see (Table 7.1 b) that $\chi_s^2 = 115.06 > \chi_\alpha^2(48) = 93.22$ for $\alpha = .0001$. The tests point rather decisively towards the null hypothesis being untenable. We therefore reject H_0 and conclude that **x** and **y** are likely to be linearly related. Interest now centers on the dimensionality of the relationship.

7.2.3 Dimensionality

Results of tests of the sequential hypotheses (3.18) are reported in Table 7.1a and b (lines 2 to 6 of the body of each table). Union-intersection and likelihood ratio tests are consistent in indicating that, after the largest root, r_1^2, is eliminated, no significant relationship remains. In other words, the data support the hypothesis $H_0(1)$: $\rho_1 \neq 0$, $\rho_2 = \ldots = \rho_6 = 0$. Acceptance of this hypothesis is equivalent to declaring that the weight of the evidence suggests that the dimensionality of the linear relationship between **x** and **y** is one.

We turn now to examine the ecological implications, if any, of the canonical variates corresponding to the largest root, r_1^2. We shall see also that it will be prudent to examine the canonical variates corresponding to at least some of the statistically non-significant roots.

7.2.4 The canonical variates

Correlation coefficients between the original variables and the canonical variates u_k and v_k ($k=1, \ldots, 3$), as well as several other interpretive indices calculated from the correlations, are shown in Table 7.2.

Intraset structure correlations. We consider first the correlations between the soil variables and the canonical variates $v_k = \mathbf{b}_k^t \mathbf{z}^{(y)}$ defined on them.

The soil variables all contribute in the same direction to v_1 and, with one exception (K), all have sizeable correlations with this variate. v_1 is however

Table 7.2. Limestone grassland, Anglesey. Canonical analysis. Relationships between six soil characteristics and eight plant species: correlations between the original variables and canonical variates

Canonical variate	u_1	u_2	u_3	h_w^2	v_1	v_2	v_3	h_b^2
Species								
H. pubescens	.968	−.018	−.008	.937	.893	−.010	−.004	.798
P. bertolonii	.499	−.666	.346	.812	.460	−.464	.180	.459
T. pratense	.492	−.050	−.490	.485	.454	−.035	−.255	.272
P. sanguisorba	.381	.449	−.277	.423	.352	.313	−.144	.243
R. squarrosus	.015	−.316	−.386	.249	.014	−.220	−.201	.089
H. pilosella	−.296	.433	−.037	.276	−.273	.301	−.019	.165
B. media	−.759	.200	.150	.639	−.700	.139	.078	.515
T. drucei	−.895	−.013	−.197	.840	−.826	−.009	−.103	.693
Variance extracted	.380	.122	.081	.583	.323	.059	.022	.404
Redundancy	.323	.059	.022	.404	.323	.059	.022	.404

Canonical variate	v_1	v_2	v_3	h_w^2	u_1	u_2	u_3	h_b^2
Soil property								
d	.834	.291	−.348	.901	.769	.203	−.181	.655
P	.639	−.466	.553	.931	.590	−.324	.288	.536
K	.160	.031	.952	.933	.148	.022	.496	.268
d×P	.910	−.275	.234	.958	.840	−.191	.122	.757
d×K	.636	.186	.552	.737	.587	.117	.288	.441
P×K	.414	−.346	.821	.965	.382	.246	.428	.390
Variance extracted	.422	.088	.394	.904	.360	.042	.107	.509
Redundancy	.360	.042	.197	.509	.360	.042	.107	.509

characterized particularly by d × P (.91) and d (.83). Above all, therefore, v_1 seems to be an expression of soil depth together with the interaction between depth and phosphate. The correlations of the soil variables with v_2 are noticeably weaker than their correlations with v_1; the strongest correlations are those of P (−.47), P × K (−.35) and d × P (−.27). Accordingly, v_2 may be regarded as essentially an expression of P. In a similar way we see that v_3 is predominantly an expression of K, the strongest relationships being those of K (.95), P × K (.82) and d × K (.55). The first three canonical variates of the soil domain can therefore very roughly be taken to be expressions of d together with d × P, P and K, respectively. From the variances of the v_k we see that v_1 and v_3 each account for some 40% of the total variance of soil variables, while v_2 accounts for only 9% ($V_2^2 = .088$). Together, the three canonical variates absorb 90.4% of the variance of the soil domain; the v_k ($k = 1, \ldots, 3$) therefore efficiently summarize the total variance of the soil domain on which they are defined.

We next consider the corresponding canonical variates $u_k = \mathbf{a}_k^t \mathbf{z}^{(x)}$ of the species' domain. The opposing signs of the species correlations with u_1 indicate that u_1 expresses some characteristic of the species which distinguishes among them with respect to the linear composite of soil variables represented by v_1. A majority of the species have sizeable correlations with u_1, the strongest being those of *H. pubescens* (.97) and *T. drucei* (−.89), though in opposite senses. Only *R. squarrosus* (.02) is uncorrelated with u_1. On arranging the species in terms of the magnitude of their correlations with u_1 we see that the arrangement closely approximates the order of their response to soil depth arrived at from prior ecological considerations described in Sect. 7.1. The second component u_2 is characterized largely by a contrast between *P. bertolonii* (−.67) and *P. sanguisorba* (.45). Three species (*H. pubescens, T. pratense* and *T. drucei*) make virtually no contribution to this canonical variate. The last component, u_3, is basically a weakly defined contrast of *T. pratense* (−.49) and *R. squarrosus* (−.39) against *P. bertolonii* (.35). Of the three canonical variates, u_1 is much the strongest, accounting for 38% of the total variance of the species' domain; u_2 and u_3 account for some 12% and 8% of the species' variance, respectively. Collectively, the three canonical variates absorb 58.3% of the variance of the species' domain. They are therefore somewhat less representative of their domain than are the corresponding variates v_k of the soil damain.

Interset structure correlations. In the present study ecological interest is confined strictly to the possible effects of the soil properties examined on species' abundance. In other words, the analysis is *directed* in nature, and, for this reason, we shall confine our attention to the structure correlations between the species and the canonical variates v_k of the soil domain. These correlations appear in the upper right-hand section of Table 7.2.

From the *direction* of the species' correlations with v_1 it is clear that considerable differences exist between the species. Several, notably *H. pubescens* (.89) and *P. bertolonii* (.46), vary directly with v_1 while other species, notably *T. drucei* (−.83) and *B. media* (−.70) are inversely related to v_1. There are also considerable differences between species in the *strength* of their relationship. The correlations of *H. pubescens* and *T. drucei*, for example, are strong while that of

R. squarrosus (.01) is decidedly weak. Moreover, the *order* of species in terms of the size of their correlations with v_1 is substantially that of the species' responses to variation in soil depth mentioned previously (Sect. 7.1). Recalling that v_1 may broadly be interpreted as an expression of soil depth together with d × P, it is clear that the first canonical relationship largely recovers the relationship between the species and soil depth (and d × P) described at the outset.

Phleum bertolonii (−.46), *P. sanguisorba* (.31) and *H. pilosella* (.30) are the species most closely related to v_2. Recalling that v_2 appears to correspond largely to soil P, and taking note of the directions of the correlations, we see that *P. bertolonii* varies directly with soil P while *P. sanguisorba* and *H. pilosella* vary inversely with P. Thus the second canonical relationship retrieves the subsidiary relationship between *P. bertolonii* and *P. sanguisorba* described earlier (Sect. 7.1). Furthermore, from the magnitude of the correlations it is clear that the relationships involved are considerably weaker than those of several species with v_1. Notice that the similarity between *H. pilosella* and *P. sanguisorba* was not anticipated. Reference back to the original data confirmed that *H. pilosella* does indeed exhibit a tendency towards greater abundance on P deficient soils. Thus the analysis properly directs attention to this previously overlooked though comparatively minor point.

The majority of the species' correlations with v_3 do not depart appreciably from zero. The strongest correlations are those of *T. pratense* (−.25) and *R. squarrosus* (−.20). Evidently none of the species has much in common with the linear composite of soil variables specified by v_3, which was regarded as corresponding essentially to K.

Redundancy. The redundancy in the species' domain generated by each of the canonical variates v_k of the soil domain is $V^2_{\mathbf{x}|v_1} = .323$, $V^2_{\mathbf{x}|v_2} = .059$ and $V^2_{\mathbf{x}|v_3} = .022$, respectively (Table 7.2). It is clear that the explanatory power of the v_k falls systematically across the variates. This is rather a different situation from that obtaining within the soil domain. There we saw that v_2 accounts for less than a quarter of the variance accounted for by either v_1 or v_3.

From the redundancies it is clear that v_1 (which represents essentially d and d × P), accounts for much the greater part (32.3%) of the total variance of the species' domain. In contrast, v_2 (roughly soil P) and v_3 (roughly soil K) account for only some 6% and 2% of the total species' variance, respectively. Thus, v_2 and v_3 are manifestly weaker explanatory constructs than v_1. More specifically, it is clear that the apparent effect of soil depth together with d × P on the representation of the species examined is roughly five times that of soil P and its concomitants and some fifteen times that of soil K and its concomitants. It is also worth noticing that v_2 accounts for considerably more (6%) of the species' variance than v_3 (2%), despite the strength of v_3 in the soil domain.

The redundancies associated with the v_k are broadly consistent with the earlier finding (Sect. 7.2.3) that all linear relationship between **x** and **y** is concentrated in a single dimension. However, we should not overlook that the canonical variates corresponding to r_2 also proved to have considerable ecological interest.

Bearing this in mind, it is clear that there may be grounds for overriding the earlier assessment of the number of trustworthy relationships between the variables. The explanatory power of v_2 across domains ($V^2_{x|v_2}=.06$) together with the ecological interest attaching to the relationship suggest that a model of rank 2 would indeed provide a more acceptable combination of fit, insight and parsimony than the original model ($r=1$). For this reason we disregard the original finding and opt now for a model of rank $r=2$.

The total redundancy of the species' domain given the first three canonical variates of the soil domain is 40.4% (Table 7.1). For the rank 2 model the revised total redundancy is 38.2%. In Sect. 7.2.5 we shall examine the interset species' communalities in order to see how the variance accounted for by the rank 2 model is distributed among the species.

At this point it will be useful to summarize the salient points of ecological interest to have emerged so far. These are as follows:

(a) that variation in d and d × P appears to have a profound effect on species' abundance, explaining 32% of the variation in the representation of the species studied;
(b) that quite small differences in P appear to have relatively large effects on the representation of certain species and account for 6% of the total species' variance; and
(c) that relatively large differences in soil K appear to have a negligible effect on the species studied, explaining just 2% of the total variation in species' abundance.

Figure 7.1 summarizes the essential aspects of the dependency between the two sets of variables graphically. The figure shows the N=45 sample-points mapped into the four potentially most informative two-dimensional spaces of the analysis. To facilitate discussion, certain stands are identified by numerical labels. We mention also that the direction of u_1 and v_1 have been reflected in all four graphs in order to allow comparison of the results with those of a quite different analysis of the same data (see Gittins, 1965). Reflection, however, does not have a crucial bearing on interpretation and its effects are ignored in the following account.

Figures 7.1a and b show the sample mapped into subspaces specified by the canonical variates corresponding to r_1 and r_2, respectively. The graphs show the linear correlation between the species and the soil factor-complexes which, for convenience, may be subsumed under the headings of soil depth and soil phosphate status, respectively. Reference to the original data reveals that sites 1, 3 and 43 (Fig. 7.1a) are characterized by shallow, P deficient soils in which *T. drucei* and *B. media* tend to achieve their maximum representation. Sites such as 5, 13 and 35, on the other hand, tend to have comparatively deep, P-rich soils in which *H. pubescens*, *P. bertolonii* and *T. pratense* are all well represented. A distinctive feature of the plot is the essentially continuous character of the point scatter. Ecologically, continuity implies the existence of a trend in the sample between the extremes of soil conditions and species' abundancies noted above. We remark also that the configuration is open to precisely the

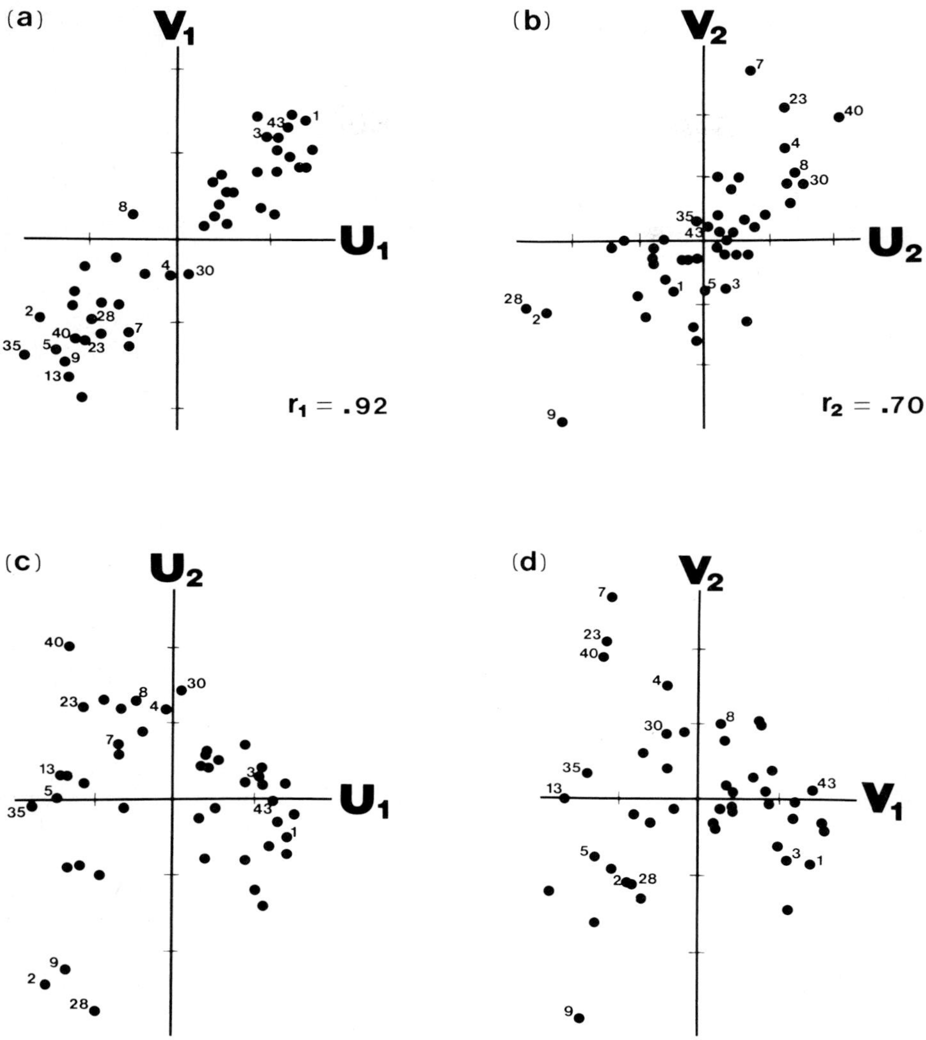

Fig. 7.1 a–d. Limestone grassland, Anglesey. Canonical analysis. Relationships between six soil properties and eight plant species. N = 45 stands mapped into subspaces defined by canonical variates **(a)** u_1 and v_1; **(b)** u_2 and v_2; **(c)** u_1 and u_2 of the species' domain; **(d)** v_1 and v_2 of the soil domain. (Direction of u_1 and v_1 reflected throughout)

interpretation arrived at here by direct appeal to the meanings of the canonical variates u_1 and v_1 themselves.

The interpretation of Fig. 7.1b may be approached in a similar manner. Sites 7, 23 and 40 are shown by reference to the original data to be P deficient and to support *P. sanguisorba* and *H. pilosella* in comparative abundance. On the other hand, sites 2, 9 and 28 are characterized by soils rich in P which support *P. bertolonii* in profusion.

Figure 7.1c is a mapping of the sample into a subspace (u_1, u_2) of species' space which is inherently likely to possess *ecological* significance. This follows because the projection is influenced not only by the variables comprising the *soil* domain but also by the relationship between the species and soil variables studied. We have seen that the linear combinations of species represented by u_1 and u_2 are open to interpretation in terms of the linear combinations of soil variables given by $v_1 = \mathbf{b}_1^t \mathbf{z}^{(y)}$ and $v_2 = \mathbf{b}_2^t \mathbf{z}^{(y)}$. Notice, however, that strictly speaking u_1 and u_2 are not orthogonal, although for convenience in the figure the axes representing them are drawn at right angles, as is customary. The normalized vectors of canonical weights, $\mathbf{a}_1$ and $\mathbf{a}_2$, show that u_1 and u_2 in fact intersect at an angle of 72° in species' space; nevertheless, the projections of the samples onto u_1 and u_2 are uncorrelated. The figure summarizes the structure of the simplified 'community' of eight selected species with respect to the ecological control or influence exerted by the factor-complexes of soil depth and P status.

Figure 7.1d is a mapping of the sample into a comparable subspace of the soil domain. In principle, it may be interpreted in exactly the same kind of way as the preceding graph. We do not pursue the details here. Two general points, however, are worth making based on a comparison of Figs. 7.1c and d. It is apparent from the graphs that:

(a) the point-scatter has considerably more coherence in species' space than in soil space; and that

(b) the second canonical relationship affects principally sites with negative scores on u_1 and v_1.

Ecologically, these features indicate that the sample is more variable in terms of the soil properties than of the species examined and that the canonical variates u_2 and v_2 describe a subsidiary relationship which, unlike u_1 and v_1, affects essentially only a subset of the $N = 45$ samples.

The four graphs comprising Fig. 7.1 together provide what may be considered to be the minimum sufficient graphical summary of the analysis. Figure 7.1a and b draw attention in a very direct and economical way to the salient characteristics of the dependency between $\mathbf{x}$ and $\mathbf{y}$; they are essentially visual expressions of the canonical correlation coefficients, r_1 and r_2. Figures 7.1c and d are alternative representations of the same phenomenon. They each display the sample in a two-dimensional subspace of just *one* of the two measurement domains, the mapping in each case being profoundly influenced by the variables comprising the 'other' domain. Although we have not relied above on the meanings given to the canonical variates themselves in interpretation, the canonical variates do provide a direct and effective alternative basis for interpretation.

7.2.5 Variable communalities

The intraset communalities (h_w^2) of the soil variables for the rank 3 model are reported in Table 7.2. With the exception of that for $d \times K$ (.74), the communalities all exceed .90. Thus the v_k $(k = 1, \ldots, 3)$ uniformly account for substantial proportions of the variances of each of the soil variables. The equivalent species' communalities are more variable. Nevertheless, they show that in general the u_k do account for sizeable proportions of the species' variances.

The interset species communalities, h_b^2 for the rank 3 model (Table 7.2, upper-right) vary considerably in size from those of *H. pubescens* (.80), *T. drucei* (.69) and *P. bertolonii* (.46) to that of *R. squarrosus* (.09). Thus the explanatory power of the model is unevenly distributed among the species. Species at either end of the sequence described at the outset (Sect. 7.1) are those which on *a priori* ecological grounds can be considered to be most susceptible to variation in the soil properties examined. It is clear from the h_b^2 that variation in the abundance of such species is at least moderately well accounted for. Species towards the center of the sequence, on the other hand, are those whose representation can be shown from general biological considerations to be least related to the explanatory variables used. This point is borne out by briefly considering *R. squarrosus*, whose variance is scarcely accounted for at all by the model ($h_b^2 = .09$). *R. squarrosus* is unique among the species studied in that it is a bryophyte. In its ecology, therefore, it is likely to be quite distinct from the remaining species. In particular, it seems likely that the representation of *R. squarrosus* will be influenced more by variables such as pH at the soil surface, humidity and competition, which were not studied, than by any of the soil properties actually used. For this reason, it is in no way disconcerting to find that the variation of this species remains virtually unaccounted for. Essentially the same kind of argument can be extended to other species whose variation is similarly poorly explained, notably *H. pilosella*.

The earlier acceptance of a rank 2 model strictly calls for revised estimates of both intraset and interset communalities. Fortunately, this does not require us to drastically alter the preceding remarks, at least as far as the interset species' communalities are concerned. Finally, we remark that, while on the whole the fitted model may be held to possess a reasonable degree of explanatory power, its lack of fit may itself be turned to good advantage. The lack of fit, as expressed by the interset communalities, directs attention to the most promising additional variables which might be used in efforts to improve explanatory power.

7.3 Conclusions

The canonical variates corresponding to the first two canonical correlation coefficients substantially retrieve the ecological features of interest believed to have been imparted to the data analyzed by the process of species selection. Specifically:

(a) the species' correlations with v_1 identify the anticipated differences between species with respect to variation in soil depth and depth × phosphate;
(b) the species' correlations with v_2 identify the anticipated responses of species to variation in soil P;

(c) the explanatory power of v_1 and v_2 in the species' domain is consistent with expectations based on prior knowledge of the relative importance of soil depth (and $d \times P$), and soil P, respectively, on the behavior of the species examined.

The variance of those species considered to be most influenced by the soil factors studied is adequately accounted for by the fitted model; conversely, the variance of species considered to be largely independent of the soil variables studied remains unaccounted for. The analysis confirms the anticipated negligible ecological influence of soil K and also draws attention to a point of some ecological interest concerning *H. pilosella*, previously overlooked. Thus there are grounds for concluding that a canonical model of rank 2 effectively identifies and communicates the salient ecological features of the data analyzed. Furthermore, it seems likely that with ingenuity the explanatory power of the model could be increased by the addition or deletion of appropriate variables.

Experimental analyses: appraisal and conclusions. Canonical analysis identified the expected relationships in both experiments. In each case analysis resulted in some 40% of the total variance of the variable set of interest being accounted for. Although the percentage of explained variance is not high, it is well to remember that in comparable procedures such as multiple regression, where explained variances (R^2) of perhaps 70% are not uncommon, the ratio of predictor variables to the criterion variable is often of the order of 5:1. In both experiments the ratio of predictor to criterion variables was nearer 1:1. In such circumstances it is perhaps only to be expected that in general the percentage of predictable variance will fall. Nevertheless, in my view the results obtained demonstrate that canonical analysis is well able to efficiently recover relationships of ecological interest between two sets of variables. Each analysis *completely* extracted the known ecological content of the data analyzed, notwithstanding the comparatively modest level of explained variance achieved. The second analysis also led to the recognition of a previously overlooked ecological feature of some interest. In ecological terms, therefore, there can be little doubt as to the success of the analyses and hence of the effectiveness of the method in these applications. The seeming discrepancy between the assessment of the analyses ecologically and in terms of explained variance is open to interpretation as follows. The variance *not* accounted for has two components – an error component and a component attributable to systematic departure from linearity. Figures 6.1 and 7.1a, b suggest that any inadequacy due to the linear form of the fitted models is negligible. Consequently, the comparatively large unexplained variances may well be largely attributable to the high unique variances which characterize ecosystems generally.

On the strength of the results obtained it seems justifiable to conclude that in small scale studies the use of canonical analysis in ecology may be rewarding. The results also indicate that a more penetrating ecological appraisal of the method would be justified. The experimental analyses described above, based as they are on selected data sets of limited complexity, can hardly be considered to be representative of real ecological investigations. Accordingly, it seems appropriate to proceed to examine the performance of canonical analysis in the context of a varied range of ecological problems of the kind encountered in practice. Five such applications are considered below (Chaps. 8–12). The quantities of interest in canonical analysis are almost all functions of the covariance matrix. For this reason it is important that the covariance matrix provide an adequate summary of the data for analysis. An assumption underlying all the analyses, therefore, was that the joint distribution of variables would in fact be well-characterized by their sample mean vector and sample covariance matrix. In practical terms this condition was equivalent to assuming in each case that the distribution was elliptically symmetric, not too long-tailed and uncontaminated by extraneous observations. Further assumptions arose in the context of particular applications. Where statistical tests were used it was necessary to assume that samples were selected independently from an appropriate multivariate normal universe. In analyses involving spatially distributed vector-variables this condition required that samples be spatially uncorrelated. In employing canonical variate analysis the assumption of a reasonably stable covariance structure across groups was made.

None of the assumptions mentioned was in fact systematically investigated prior to analysis. Accordingly, it would be wise to accept inferential and other aspects of the results presented with circumspection.

8. Soil-vegetation relationships in a lowland tropical rain forest

8.1 Introduction

The aim of the following application of canonical analysis is to clarify relationships between soils and vegetation in a 1 km^2 area of rain forest in the Bartica Triangle region of Guyana, South America. The available field data consist of estimates of the composition of twenty-five 100 m × 100 m stands of forest vegetation together with determinations of selected soil characteristics of the sites. Stands were arranged on a square lattice at intervals of 200 m, and, in recording their vegetation, attention was confined to the predominant, woody component of the vegetation. One hundred and seventy species of woody plant were encountered in all, the representation of these being estimated in terms of basal area. Ten soil samples were collected in each site and pooled to form a single, composite sample on which determinations of twenty physical and chemical properties were later made in the laboratory. A comprehensive account of the field sampling procedures and of methods of soil analysis has been given by Ogden (1966).

Before embarking on the canonical analysis it was necessary to re-express the data in a form better suited to analysis of this kind. The most pressing need in view of the small size of the sample (N=25) and large number of variables ($p+q=190$) was to reduce the number of variables in some way consistent with the overall objective of the study. It was also desirable to arrive at some more comprehensive description of site *vegetation* than that provided by a statement simply of species composition, given the stated aim of the analysis.

A reduction in species number was achieved in the following manner. Species were first ranked according to their independent contributions to the total species' sum of squares. After examining the order of species in the light of field knowledge of their behavior (Ogden, 1977a) and in relation to a stress index calculated from the data, all but the top ranking thirty-three species were eliminated. In a separate analysis, the soil variables were similarly reduced from twenty to nine.

The overall vegetational composition of stands was next determined in relation to the thirty-three retained species. Non-centered principal component analysis of site-normalized data followed by varimax rotation of the components was used for this purpose. The choice of method here was guided by evidence that the data structure determined by the field observations was not uniformly continuous (Gittins & Ogden, 1977) and by the declared interest in soil-vegetation relationships (see Noy-Meir, 1973; Noy-Meir, Walker & Williams, 1975).

Six non-centered varimax components were found to account for 97% of the total stand sum of squares, each component corresponding to a recognizable and reasonably distinct forest community. These components were considered to adequately characterize the vegetation and accordingly were used as one set of variables in the canonical analysis. The forest community defined by each component was named after a representative species; the communities identified in this way, together with an indication of the 'importance' of each, as expressed by the percentage of the total sum of squares associated with the defining varimax component were: Greenheart (*Ocotea rodiaei* (Schomb.) Mez.), 44%, Wallaba (*Eperua falcata* Aubl.), 22%, Morabukea (*Mora gonggrijpii* (Kleinh.) Sandw.), 15%, *Mora* (*Mora excelsa* Bth.), 6%, *Pentaclethra* (*Pentaclethra macroloba* (Willd.) Kze.), 8%, and *Eschweilera* (*Eschweilera sagotiana* Miers.), 3%.

The nine retained soil variables comprised the second set of variables in the canonical analysis. These variables were: pH, base saturation (%); phosphorus (ppm P); sand (% particles .05–.20 mm); silt (% particles .002–.05 mm); active acidity (KCl acid); potassium (m-equiv. K/100 g air dry soil); potential cation exchange capacity (CeCg); and moisture retaining capacity (% H_2O in air dry soil).

The canonical analysis thus came to be based on a systematic sample of $N=25$ stands of rain forest vegetation, $p=6$ generalized vegetation variables (**x**) and $q=9$ soil properties (**y**). The ecological objective was to achieve a preliminary understanding of how forest differentiation in the area surveyed might be related to the soil properties examined. The data underlying the analysis are reported in Table A-3 of Appendix A.2 and the results of analysis are summarized in Tables 8.1 and 8.2.

Before turning to the results themselves it is worth remarking that the field observations on which the analysis was based were collected during the course of an expedition which was in the field for a period of six weeks. The expedition's vegetation research program had several objectives and the design of the vegetation survey was influenced by a need to satisfy the sometimes conflicting sampling requirements of different objectives. For this reason the data obtained are not ideally suited to canonical analysis, even following the modifications described above. In particular, the small size of the sample is a source of special concern, while the lack of independence among samples might also be questioned. Provided these points are borne in mind in interpreting the results, however, the analysis may nevertheless prove useful in illustrating the kind of insight which canonical analysis can yield in conjunction with exploratory ecological surveys of the kind involved.

8.2 Results

8.2.1 The canonical correlation coefficients

The canonical correlation coefficients r_k ($k=1, \ldots, s$) are reported in Table 8.1. The correlation between u_1 and v_1 is evidently strong, since $r_1=.93$. Thus,

Table 8.1. Lowland tropical rain forest, Guyana. Canonical analysis. Relationships between nine soil properties and six forest communities. Canonical correlation coefficients, r_k, canonical roots, r_k^2 and percentage $\mathrm{tr}(\mathbf{R}_{22}^{-1}\mathbf{R}_{21}\mathbf{R}_{11}^{-1}\mathbf{R}_{12})$ absorbed by the roots individually (%tr) and cumulatively (C%)

k	r_k	r_k^2	%tr	C%
1	.926	.857	29.08	29.08
2	.904	.817	27.72	56.80
3	.745	.556	18.87	75.67
4	.653	.427	14.49	90.16
5	.468	.220	7.47	97.63
6	.265	.070	2.38	100.01
Total	–	2.947	100.00	–

at first sight at least, r_1 is certainly suggestive of the existence of a linear relationship between **x** and **y**. Thereafter, the r_k $(k>1)$ decline in a rather systematic way until the smallest, $r_6=.26$. The strength of several of these correlations appears to indicate the occurrence of more than one linear relationship between domains.

The squared canonical correlation coefficients, r_1^2 and r_2^2 (Table 8.1) show that the canonical variates v_1 and v_2 of the soil domain each account for some 80% of the variance of the species' composites represented by u_1 and u_2, respectively. Thus, the roots r_1^2 and r_2^2 appear to reflect the existence of well-defined relationships between the soil properties and forest communities of interest. From the table it can also be seen that the sum of all s roots, $\Sigma\, r_k^2$, is equal to 2.95. This quantity is identically the sum of the diagonal elements or trace of the matrix product $\mathbf{R}_{22}^{-1}\mathbf{R}_{21}\mathbf{R}_{11}^{-1}\mathbf{R}_{12}$, which is a measure of the predictable variance of the system. We shall find $\Sigma\, r_k^2$ useful in connection with estimating dimensionality (Sect. 8.2.3).

8.2.2 Independence

In the absence of independent samples we are unable to rely on formal test procedures to provide guidance in assessing the independence of **x** and **y**. Moreover, in attempting to arrive at an informal assessment based simply on the magnitude of r_1, it would be prudent to regard r_1 as an upwardly biased estimate of the parameter of interest, ρ_1, having regard for the large number of variables $(n=15)$ relative to sample size $(N=25)$ – cf. Sect. 2.4.1. Even so, the size of r_1 is such as to provide grounds for cautiously accepting that **x** and **y** may indeed be linearly related. Accordingly, we shall tacitly assume this to be the case. The question of the dimensionality of the relationship, therefore, has now to be addressed.

8.2.3 Dimensionality

Lacking formal tests, we shall rely on the percentage $\mathrm{tr}(\mathbf{R}_{22}^{-1}\mathbf{R}_{21}\mathbf{R}_{11}^{-1}\mathbf{R}_{12})$ attributable to the kth root, r_k^2, as a heuristic to throw light on dimensionality.

Table 8.2. Lowland tropical rain forest, Guyana. Canonical analysis. Relationships between nine soil properties and six forest communities: correlations between the original variables and canonical variates

Canonical variate	u_1	u_2	u_3	u_5	h_w^2	v_1	v_2	v_3	v_5	h_b^2
Community										
Greenheart	−.218	−.697	−.398	.441	.886	−.203	−.630	−.297	.207	.569
Wallaba	−.516	.798	.023	−.169	.932	−.481	.721	.017	−.079	.758
Morabukea	.168	−.412	.671	−.335	.760	.157	−.372	.500	−.157	.438
Mora	.798	.172	−.368	−.225	.852	.744	.152	−.274	−.105	.664
Pentaclethra	.470	−.065	−.090	.662	.671	.438	−.059	−.067	.310	.296
Eschweilera	.419	−.006	.379	.478	.548	.391	−.005	.283	.224	.283
Variance extracted	.229	.221	.149	.175	.775	.198	.181	.083	.039	.501
Redundancy	.197	.181	.083	.039	.500	.198	.181	.083	.039	.501
Canonical variate	v_1	v_2	v_3	v_5	h_w^2	u_1	u_2	u_3	u_5	h_b^2
Soil factor										
pH	.347	−.076	−.451	.420	.506	.323	−.069	−.336	.197	.261
Base saturation	−.092	.370	−.004	−.732	.681	−.086	.335	−.003	−.343	.237
Phosphorus	.888	.238	−.157	.285	.951	.828	.215	−.117	.134	.763
Sand	−.311	.769	−.202	−.001	.729	−.290	.695	−.151	−.000	.590
Silt	.427	−.249	.048	−.351	.370	.398	−.225	.036	−.164	.237
Active acidity	.321	−.212	−.175	.676	.636	.299	−.192	−.130	.317	.244
Potassium	.204	−.514	.001	−.208	.349	.190	−.465	.001	−.097	.262
Cation exchange	.378	−.766	−.351	.178	.885	.352	−.693	−.262	.083	.680
Moisture factor	.119	−.507	−.110	.304	.376	.111	−.458	−.082	.142	.249
Variance extracted	.167	.223	.049	.171	.609	.144	.182	.027	.038	.391
Redundancy	.143	.182	.027	.038	.390	.144	.182	.027	.038	.391

From Table 8.1 it can be seen that together the roots r_k^2 $(k=1, \ldots, 3)$ account for 75% of the predictable variance, while inclusion of the fourth root increases this percentage to 90. The remaining roots, r_5^2 and r_6^2, are by comparison, of negligible importance. These results suggest that for practical purposes three or four linear relationships may be considered to substantially exhaust the presumed dependency between **x** and **y**.

Further light on dimensionality is shed by the redundancy in the forest domain generated by the canonical variates v_k of the soil domain. The redundancies attributable to the v_k $(k=1, 2, 3, 5)$ of a particular rank 4 model are reported in Table 8.2 (upper-right; in addition we mention that the redundancies attributable to v_4 and v_6 are .03 and .01, respectively). The explanatory power of the v_k $(k>3)$ in the forest domain is clearly small. This result is not inconsistent with that given by the canonical roots. Yet a preliminary ecological assessment of the canonical variate v_5 was instrumental in the selection of the rank 4 model given in Table 8.2. The strength of v_5 in the soil domain $(V_5^2=.171)$ together with its not entirely negligible explanatory power in the forest domain $(V_{\mathbf{x}|v_5}^2=.039)$ provides some justification for declaring that the model defined by the roots r_k^2 $(k=1, 2, 3, 5)$ provides the most acceptable combination of fit, insight and parsimony. In other words, for practical purposes we choose to regard the dimensionality of the relationship between **x** and **y** as four.

Before turning to consider possible ecological implications of the canonical variates of the $r=4$ model, we return briefly to Table 8.1. Recalling from Sect. 8.2.1 that $\mathrm{tr}(\mathbf{R}_{22}^{-1}\mathbf{R}_{21}\mathbf{R}_{11}^{-1}\mathbf{R}_{12})$ expresses the predictable variance of the system and that $\Sigma\, r_k^2 = \mathrm{tr}(\mathbf{R}_{22}^{-1}\mathbf{R}_{21}\mathbf{R}_{11}^{-1}\mathbf{R}_{12})$, it may be appreciated from the table that the predictable variance ($\Sigma\, r_k^2 = 2.95$) is neither increased nor decreased in transforming the data to canonical variates. What canonical analysis does is to systematically *reallocate* the predictable variance among the canonical variates in such a way that the best prediction can be made in terms of the fewest variates (see Table 8.1, column four, %tr).

8.2.4 The canonical variates

In interpreting the canonical variates we shall rely principally on the structure correlations given in Table 8.2. Although, for completeness, the table reports the structure correlations pertaining to both sets of canonical variates, u_k and v_k, we shall confine our attention to the intraset and interset correlations of the v_k alone.

Intraset structure correlations. From the correlations of the soil variables with the v_k it can be seen (Table 8.2, lower-left) that v_1 is characterized principally by phosphorus (.89), although silt (.43) and potential cation exchange capacity (.38) are to some extent also related in a direct sense to this canonical variate; we note also the inverse correlation between sand (−.31) and v_1. The second canonical variate, v_2, is essentially a contrast of percentage sand (.77) against cation exchange capacity (−.77), potassium (−.51), and moisture retaining capacity (−.51). In a similar way we see that for simplicity v_3 may roughly be characterized by pH (−.45), while v_5 can be equated approximately with base saturation (−.73) and active acidity (.68).

Of the four canonical variates v_2 is the strongest, absorbing 22% of the total variance of the soil domain ($V_2^2 = .223$), followed by v_1 and v_5, each of which accounts for some 17%; v_3 is a much weaker composite accounting for only 5% of the variance of the soil variables. Collectively, the four canonical variates account for 61% of the total variance of the soil domain.

Interset structure correlations. The *magnitude* of the correlations of the forest communities with v_1 show that for the most part the communities are all related to this canonical variate, at least to some extent (Table 8.2, upper-right); the *signs* of the correlations, however, indicate that the communities respond to variation in v_1 in different ways. *Mora* (.74), *Pentaclethra* (.44) and *Eschweilera* (.39) forests all respond in a direct sense, while Wallaba (−.48) and Greenheart (−.20) vary inversely with v_1; only Morabukea (.16) appears to be substantially independent of this canonical variate. Recalling that for simplicity v_1 may roughly be equated with soil P and silt, it appears that *Mora* and to some extent also *Pentaclethra* and *Eschweilera* forests tend to be associated with silty soils of relatively high P status, while, in contrast, Wallaba and Greenheart tend to occur on comparatively sandy soils deficient in P.

The forest communities are also sharply differentiated by their correlations with v_2. Wallaba (.72) and Greenheart (−.63) clearly respond in opposite senses to variation in v_2, while, with the exception of Morabukea (−.37) forest, the remaining communities are largely invariant with respect to v_2. Thus v_2 represents essentially a contrast between the forest communities of P deficient sites. Recalling the earlier interpretation of v_2 and observing the signs of the structure correlations, we see that Wallaba forest tends to be associated with sandy, infertile soils of low water retaining capacity, while Greenheart is inversely related to soils of this kind, *i.e.* Greenheart tends to be found on comparatively base rich, clayey soils of relatively high water retaining capacity. Furthermore, we see that Morabukea forest resembles Greenheart in its association with such soils, at least to some extent.

The third canonical variate, v_3, is not well defined in the vegetation domain, its strongest correlation being +0.5 with Morabukea. As v_3 can crudely be equated with pH and potential cation exchange capacity, it appears that Morabukea forest shows some tendency to develop on acid soils of low potential cation exchange capacity. In addition, Greenheart (−.30) and *Mora* (−.27) forests seem to respond to the same soil factor-complex, though in an inverse sense.

It is clear from the magnitude of the interset structure correlations with v_5 that this canonical variate is an appreciably weaker explanatory construct in the vegetation domain ($V^2_{\mathbf{x}|v_5}=.039$) than in the soil domain on which it is defined ($V^2_{5\,(y)}=.171$). The strongest interset correlations are with *Pentaclethra* (.31) and *Eschweilera* (.22) forests. Bearing in mind the earlier interpretation of v_5, the correlations suggest that the differentiation of these communities may be determined, at least in part, by the occurrence of soils of low base saturation and high active acidity.

In conclusion, it is worth remarking that, in seeking to comprehend the relationships between soils and vegetation it is often instructive to examine the interset structure correlations of Table 8.2 from two further points of view – namely, by columns and by rows. The correlations by:

(a) columns express the apparent effect of the soil factor-complex subsumed by the kth canonical variate v_k on the vegetation as a whole; while by
(b) rows they represent the apparent response of a particular community to the soil factor-complexes corresponding to the v_k ($k=1, \ldots, r$) collectively.

With respect to the first point we have seen, for example, that v_1 distinguishes between communities of P rich and P depleted soils, respectively, while v_2 operates within the vegetation of P deficient soils to differentiate communities in relation to the particle size composition etc. of their associated soils. This process of specifying the apparent effect of the different soil factor-complexes is readily extended to the remaining soil canonical variates, v_k. The second point suggests that the *profile* of a community over the canonical variates may be helpful in identifying the edaphic controls of the community, at least in a provisional way. Thus, the profile of Wallaba forest, for example, suggests that P deficient, sandy soils of low potential cation exchange and water retaining capacities may well be among the edaphic controls of this community.

Redundancy. From the redundancy in the vegetation domain attributable to the canonical variates v_k of the soil domain we see (Table 8.2, upper-right) that the explanatory power of the v_k is largely concentrated in v_1 and v_2. Together these canonical variates account for 38% of the total variation in the forest communities. Thus, variation in the soil factor-complexes corresponding to v_1 and v_2, which for convenience may be characterized respectively by P and by particle size, appears to have pronounced effects on forest composition. On the other hand, v_3 and v_5 are weaker explanatory constructs, accounting respectively for 8% and 4% of the total variance of the vegetation domain. Evidently, variation in the factor-complexes represented by v_3 (roughly pH) and v_5 (base saturation) is associated with comparatively small changes in forest composition. It is interesting to observe in this connection, however, that small differences in soil pH appear to have a proportionately *greater* influence on the development and composition of the vegetation than appreciably larger differences in the factor-complex characterized by base saturation. This is apparent from the table on comparing the variances associated with v_3 and v_5 in the soil and vegetation domains, respectively; v_3 proves to account for only .29 times the variance attributable to v_5 in the soil domain while in the vegetation domain the explanatory power of v_3 is 2.13 times that of v_5.

Collectively, the v_k account for 50% of the total variance of the vegetation domain. The fit of the rank 4 model of Table 8.2 may therefore be considered to be reasonably satisfactory.

Figure 8.1 shows the sample (N=25) projected into four two-dimensional subspaces, associated with the analysis. The affiliation of samples with respect to the principal forest communities is indicated on each graph. Figure 8.1a and b display the sample in the subspaces defined by the canonical variates corresponding to r_1^2 and r_2^2, respectively. The scattergrams illustrate the linear correlation between the forest communities and the soil factor-complexes subsumed under the headings of phosphorus (v_1) and particle size (v_2), respectively. The differentiation of the vegetation apparently in response to variation in these factor-complexes is clearly evident; *Mora* forest in particular stands out as a very distinctive community while the *trends* in forest composition associated with changes in v_1 and v_2 are other notable features of the plots.

Figures 8.1c and d display the sample mapped into subspaces defined by the canonical variates (u_1, u_2) and (v_1, v_2) of the vegetation and soil domains, respectively. The extent of the overall similarity of the two representations is a direct expression of the relatedness of the two measurement domains. While a considerable degree of similarity is evident, the projection into (v_1, v_2) is manifestly less coherent than that into (u_1, u_2); indeed, a discontinuity separating sites bearing Wallaba forest from the remainder is a distinctive feature of the mapping into (v_1, v_2) which is completely lacking in the other. We do not pursue the ecological implications of such features here; our purpose is simply to emphasize that the opportunity exists to map the sample into a variety of subspaces associated with the analysis, each of which is at least potentially able to contribute towards comprehension of the ecological relationships of interest. The interpretation of the scattergrams in terms of the meanings ascribed to the canonical variates above is left as an exercise for the interested reader.

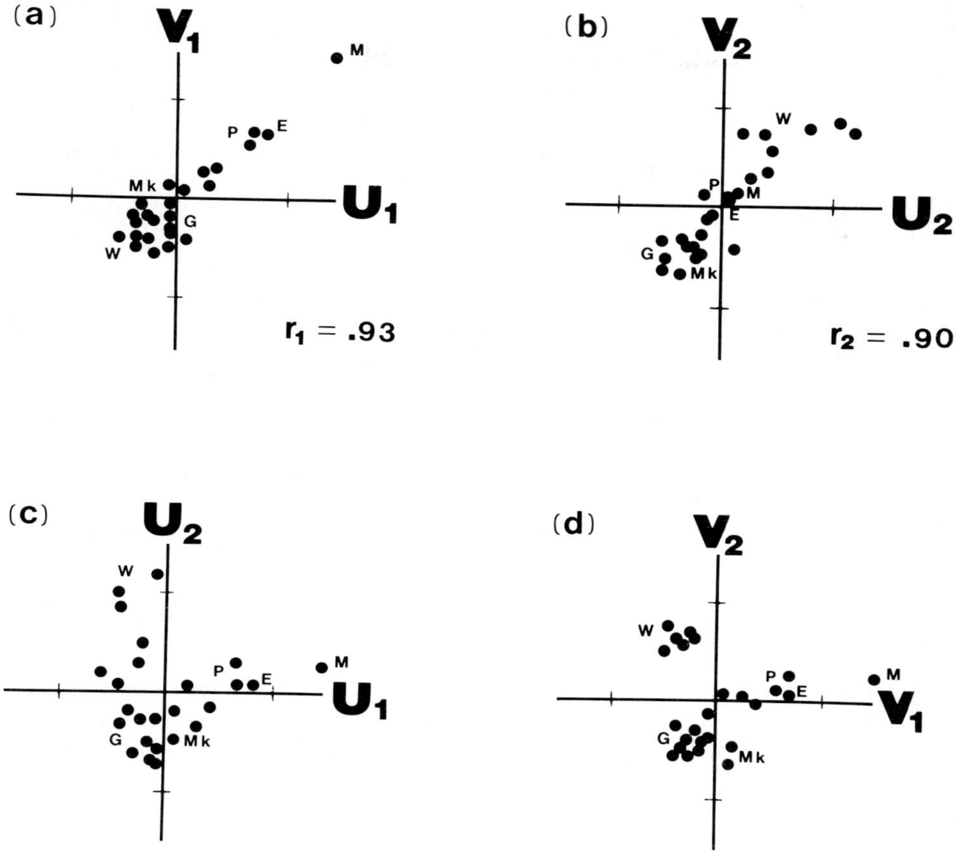

Fig. 8.1 a–d. Lowland tropical rain forest, Guyana. Canonical analysis of relationships between nine soil properties and six forest communities. N = 25 stands mapped into subspaces defined by **(a)** u_1 and v_1; **(b)** u_2 and v_2; **(c)** u_1 and u_2 of the forest domain; **(d)** v_1 and v_2 of the soil domain. The canonical variates normalized to unity throughout. E: *Eschweilera* forest; G: Greenheart; M: *Mora;* Mk: Morabukea; P: *Pentaclethra*; W: Wallaba

Figure 8.2 portrays the sample mapped into the subspace specified by the first three canonical variates u_k ($k = 1, 2, 3$) of the forest domain. The u_k in this instance have been normalized to be proportional to the explained variance of the respective v_k ($k = 1, 2, 3$) of the soil measures in the *vegetation* domain. The stereogram enables the differentiation of the forest communities in apparent response to variation in all *three* major soil factor-complexes to be viewed simultaneously. Specifically, the figure shows the differentiation of:

(a) *Mora, Pentaclethra* and *Eschweilera* forests in response to variation in the factor-complex corresponding to v_1;

(b) Wallaba, Greenheart and Morabukea forests within the vegetation of P impoverished sites, in response to variation in v_2; and of

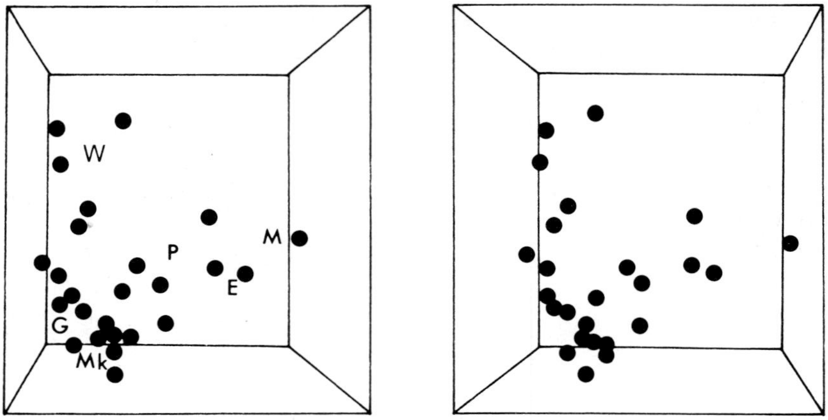

Fig. 8.2. Lowland tropical rain forest, Guyana. Canonical analysis of relationships between nine soil properties and six forest communities. N = 25 stands mapped into the subspace defined by the canonical variates u_k $(k = 1, 2, 3)$ of the forest domain. Key as in Fig. 8.1; u_1 is the horizontal axis, u_2 the vertical axis and u_3 the axis at right angles to the page

(c) Morabukea forest, and to some extent also, Greenheart, *Eschweilera* and *Mora,* in response to variation in v_3.

The stereogram enables the relative importance of the major components of the overall relationship between soils and vegetation to be appreciated, as well as the mutual affinities of the forest communities themselves with respect to the soil properties examined.

8.2.5 Variable communalities

The intraset variable communalities h_w^2 of the soil variables (Table 8.2, lower-left) range between those of potassium (.35) and phosphorus (.95). On the whole they indicate that the variance of a majority of the variables is at least moderately well explained by the canonical variates of the rank 4 model. The interset communalities h_b^2 of the forest communities (Table 8.2, upper-right) are somewhat weaker. They show that the explanatory power of the model is rather unevenly distributed among the six communities. Wallaba (.76), *Mora* (.66), Greenheart (.57), and, to a lesser extent, Morabukea (.44) are the communities whose variation is best accounted for.

8.3 Ecological assessment of the results

We must first examine the effects of the size of the sample and of the sampling design on the results obtained.

Sample size. The number of samples used in the analysis is very small (N=25). Nevertheless, the stands investigated comprise 25% of the study area or *areal domain* of interest. It seems, therefore, that any unbiased, replicate sample of the same size from the same 1 km^2 locality might reasonably be expected to yield substantially the results described above. Hence the adequacy of the sample might appear to be assured. Such an argument, however, neglects the crucial issue of the size of the sample in relation to the *subject domain* of interest, particularly the number of variables involved. The ratio of variables to samples ($^{15}/_{25}=.60$) is very high. The effects of a ratio of this kind on any analysis are twofold:

(a) to exaggerate the strength of the canonical correlation coefficients, structure correlations and interpretive devices derived from them (cf. Sect. 2.4.1);
(b) to limit the applicability of conclusions drawn from the results to the sample actually examined.

The second point follows because the maximization procedure on which canonical analysis is based is known to be particularly sensitive to sample-specific variation, especially in the case of small samples.

Sampling design. The lack of independence among samples is not regarded as detrimental to the analysis. Indeed, it has been argued convincingly by W.T. Williams (1971) that systematic samples are the preferred kind in most vegetation surveys. As a consequence of the systematic design, the results obtained are expected to reflect chiefly the *overall* trends and relationships which are the focus of interest in the present survey.

It appears therefore that the main consequences of the sampling frame which need to be borne in mind in assessing the results are its effects in *exaggerating* the strength of the relationships involved and in rendering the results *sample-specific*. For these reasons we regard the strength of the relationships described with some scepticism, and, in considering the wider applicability of the results, will lean on evidence external to the analysis itself. We are now in a position to take up the question of the worth of the analysis in ecological terms.

The soil-vegetation relationships established can themselves be related to more fundamental processes. The area surveyed consists physiographically of a dissected plateau with a relief some 25 m in amplitude. The topographic gradient from plateau top to creek bottom has been recognized by Ogden (1966) as a major determinant of the soil-vegetation relationships of the area. Ogden showed that the topographic gradient was associated with a catenary soil series, which in turn appeared to influence the composition of the vegetation at different positions on the gradient. The canonical variates of the rank $r=4$ model can all be related to features of the topographic and catenary soil-vegetation system. The interset structure correlations of v_1 reflect the soil-vegetation changes associated with the *global* topographic gradient from plateau surface to valley floor, while the structure correlations of the remaining v_k ($k>1$) correspond to *local* differentiation within the overall system at different positions on the gradient. Furthermore, the results obtained are broadly consistent with those of quite different analyses based on the same data by Ogden (1966) and with the results of independent investigations by Davis and Richards (1934) and Schultz (1960) in nearby areas. This similarity gives confidence in the ability of canonical analysis to identify relationships of ecological significance. The results obtained, however, do differ from those of previous studies in specifying the covariance structure of interest with greater precision and clarity than has hitherto been achieved. Further work in the field would be necessary, however, to substantiate the additional information provided. Nevertheless, on the basis of the evidence available, Odgen (1977a) has expressed the view that the canonical analysis represents a useful contribution to the description and comprehension of soil-vegetation relationships in the area surveyed.

8.4 Conclusions

The essential features of ecological interest may be summarized as follows:

(a) the soil factor-complex represented by phosphorus is associated with major changes in forest composition;
(b) the factor-complex represented by particle size is associated with major changes in the composition of the vegetation of phosphorus depleted sites;
(c) soil pH, base saturation and active acidity appear to be associated with the occurrence of the smaller forest communities.

The analysis also enabled some of the edaphic influences or controls of particular communities to be recognized, at least in a provisional way.

The results obtained are broadly compatible with those of several independent investigations. Such consistency goes some way towards validating the results of canonical analysis in this application and hence also of the method itself. Nevertheless, there are grounds for supposing that the strength of the relationships described is exaggerated, though to an unknown extent. Furthermore, the analysis in this case provides no basis for supposing that the conclusions drawn are applicable beyond the sample actually examined. Yet, the external substantiation of salient features of the results does indicate that the conclusions reached may conceivably possess greater generality.

It is not claimed on the basis of the results obtained that analysis has succeeded in identifying soil properties which are themselves necessarily among the immediate ecological or physiological controls of particular communities. We need only consider that, in the initial selection of soil variables or in reducing their number prior to analysis, we may well have overlooked or eliminated variables of greater physiological significance than any of those actually used to appreciate this point. What the analysis has achieved is a reasonably concise specification of the covariance structure of the soil properties and forest communities examined. The ecological interpretation of this structure may prove useful in directing attention to those soil-vegetation relationships among the variables examined which seem most likely to repay attention in subsequent endeavors to explain the occurrence and distribution of the communities studied.

9. Dynamic status of a lowland tropical rain forest

9.1 Introduction

The stability of vegetation over time is frequently of interest to plant ecologists. Stability, however, is a property of vegetation which is not readily investigated because its definitive study calls for sequential observation over time. This is especially true of forest vegetation where the time-scales involved are often too long to permit direct observation. One means of circumventing the difficulty in the case of forest communities is to regard size class measurements *as if* they constituted observations at different times. Differences in size correspond to differences in age which, cautiously interpreted, may be substituted for observations at particular points in time or of particular seral stages. In this way it may be possible to gain insight into the dynamic status of forest vegetation over the short term, at least. The present chapter concerns a study of this kind.

Field observations consisting of estimates of the composition of a systematic sample of $N=25$, 100 m × 100 m stands of lowland rain forest in Guyana, South America were available. In all, $p=170$ species of woody plant were encountered during the sampling and density estimates of these for $m=2$ size-classes obtained ('trees' > 5 cm dbh; 'seedlings' < 30 cm in height). Ecological interest centered on obtaining a preliminary assessment of the dynamic status of the forest. More specifically, the objective was to discover whether the field observations were consistent with the view that the forest was likely to be replaced by vegetation of substantially the same composition, as the trees composing it aged and died and were progressively replaced by the present generation of seedlings. Thus, attention focused on tree/seedling relationships; if composition was changing with time we might expect these relationships to be weaker than would be the case were the vegetation dynamically stable. In operational terms, therefore, we are led to an interest in the *correlation structure* of the tree/seedling density estimates. Canonical analysis is readily identified as an appropriate algebraic model for exploring the correlation structure between any m sets of variables ($m \geq 2$).

In order to express the field observations in a form better suited to canonical analysis, the data were modified using the procedures referred to in Sect. 8.1. Let the $N \times p$ matrix of tree densities be denoted by $\mathbf{A}_0^*$ (25 × 170) and the corresponding matrix of seedling densities by $\mathbf{A}_1^*$ (25 × 170). The species were first reduced in number to thirty-three and the remaining observations then standardized by site-norm. The matrices of reduced, site-normalized data are denoted by $\mathbf{A}_0$ (25 × 33) and $\mathbf{A}_1$ (25 × 33), respectively. To obtain composite indices or descriptors of site *vegetation*, separate non-centered principal component analyses of $\mathbf{A}_0$ and $\mathbf{A}_1$ were performed followed by varimax rotation and postnormalization of the components. In each case, six components were found to adequately characterize the tree and seedling counterparts of the vegetation, accounting for 93% and 89% of the total sum of squares of the cross-products matrices $\mathbf{A}_0\mathbf{A}_0^t$ and $\mathbf{A}_1\mathbf{A}_1^t$, respectively. Each rotated component could be equated with and was named after a more or less clearly defined forest community. The names of the communities together with an indication of the importance of each, as expressed by the percentage $\mathrm{tr}(\mathbf{A}_i\mathbf{A}_i^t)$ associated with its defining component, appear in Table 9.1. The rotated components were collected to yield the matrices $\mathbf{X}_0$ (25 × 6) and $\mathbf{X}_1$ (25 × 6) suitable for canonical analysis. The formal objective of analysis was to examine the relationship between the two sets of components, that is, in ecological terms between tree and seedling counterparts of the forest communities.

Table 9.1. Lowland tropical rain forest, Guyana. Communities identified following non-centered components analyses of tree and seedling matrices $\mathbf{A}_0$ (25 × 33) and $\mathbf{A}_1$ (25 × 33), respectively. The relative importance of communities is indicated by the percentage trace $(\mathbf{A}_i\mathbf{A}_i^t)$ associated with the defining component

Community[a]	% $\mathrm{tr}(\mathbf{A}_0\mathbf{A}_0^t)$	% $\mathrm{tr}(\mathbf{A}_1\mathbf{A}_1^t)$
Greenheart	23	23
Wallaba	20	17
Pentaclethra	13	9
Morabukea	13	14
Mora	{17	6
Eschweilera		20
Jessenia	7	–
Total	93	89

[a] More specifically, the species after which the communities were named are: Greenheart, *Ocotea rodiaei* (Schomb.) Mez.; Wallaba, *Eperua falcata* Aubl.; *Pentaclethra, Pentaclethra macroloba* (Willd.) Kze.; Morabukea, *Mora gonggrijpii* (Kleinh.) Sandw.; *Mora, Mora excelsa* Bth.; *Eschweilera, Eschweilera sagotiana* Miers.; *Jessenia, Jessenia bataua* (Mart.) Burret.

Thus the canonical analysis came to be based on a systematic sample of N = 25 100 m × 100 m stands of rain forest vegetation, each characterized by $p=6$ and $q=6$ generalized variables expressing the contribution of tree and seedling communities, respectively, to the total vegetation. The ecological objective was to arrive at a preliminary assessment of the dynamic status of the vegetation. The data employed are given in Table A-4 of Appendix A.2 and the results of the analysis are summarized in Tables 9.2 and 9.3 and in Figs. 9.1–9.3.

9.2 Results

9.1 The canonical correlation coefficients

The canonical correlation coefficients are reported in Table 9.2. It can be seen that the first four correlations exceed .85 and so are unquestionably large. The fifth canonical correlation, $r_5 = .65$, also departs appreciably from zero and is hardly negligible. Only $r_6 = .06$ is sufficiently close to zero to suggest that it may be neglected with impunity. It is important to observe, however, that, in attempting to assess the worth of the canonical correlation coefficients, the ratio of variables to samples in the analysis is high ($^{12}/_{25} = .48$). The magnitude of this ratio serves as a clear warning that the larger correlations, at least, are almost certainly positively biased. The squared canonical correlations, r_k^2 $(k = 1, \ldots, 4)$, show that the linear composites u_k and v_k corresponding to the kth root each account for more than 75% of the variance in their conjugate. Thus, even after allowing for bias, it seems that the canonical variates may be capable of predicting an appreciable part of the variance in their corresponding composite.

Table 9.2. Lowland tropical rain forest, Guyana. Canonical analysis of tree-seedling relationships. Canonical correlation coefficients, r_k, canonical roots, r_k^2, and percentage $\mathrm{tr}(\mathbf{R}_{22}^{-1}\mathbf{R}_{21}\mathbf{R}_{11}^{-1}\mathbf{R}_{12})$ absorbed by the r_k^2 individually (% tr) and cumulatively (C %)

k	r_k	r_k^2	% tr	C %
1	.988	.977	24.60	24.60
2	.956	.914	23.01	47.61
3	.946	.894	22.51	70.12
4	.871	.760	19.13	89.25
5	.651	.424	10.67	99.92
6	.056	.003	.08	100.00
Total	–	3.972	100.00	–

9.2.2 Independence

In the absence of independent samples and hence of significance tests, we shall address the question of the independence of the two sets of variables informally. Despite the likely bias of the larger correlations, the departure of r_1 from zero is nevertheless such as to be rather convincing in rendering an hypothesis of independence untenable. Accordingly, it appears that there are grounds for regarding the data as consistent with the view that tree and seedling counterparts of the vegetation are associated. Interest therefore turns towards establishing the dimensionality of the association.

9.2.3 Dimensionality

The magnitudes of the r_k $(k = 1, \ldots, s)$ are certainly suggestive of the existence of several linear relationships between the two measurement domains. Further, from Table 9.2 it can be seen that the first three canonical roots, r_k^2, each absorbs some 23–25% of the predictable variance represented by $\operatorname{tr}(\mathbf{R}_{22}^{-1}\mathbf{R}_{21}\mathbf{R}_{11}^{-1}\mathbf{R}_{12})$, while r_4^2, which accounts for 19%, is only slightly less efficient. The fifth root, r_5^2, is somewhat weaker, accounting for 11% of the predictable variance.

Together, these results point towards a dimensionality of four or five. After a preliminary assessment of the interpretability of the corresponding canonical variates, I have opted for five as the effective dimensionality of the linear association between the tree and seedling communities.

We now turn to the interpretation of the relationships associated with the roots, r_k^2, $k=1, \dots, 5$. For this purpose the canonical variates u_k and v_k will be required.

9.2.4 The canonical variates

The correlations between the canonical variates and the original variables are shown in Table 9.3. As it is reasonable to regard the existing tree vegetation as largely determining the composition of the seedling vegetation, rather than the converse, for simplicity we shall confine our attention to just one set of canonical variates, namely those of the tree domain, v_k.

Table 9.3. Lowland tropical rain forest, Guyana. Canonical analysis of tree-seedling relationships. Correlations between the original variables and the canonical variates

Canonical variate	u_1	u_2	u_3	u_4	u_5	h_w^2	v_1	v_2	v_3	v_4	v_5	h_b^2
Seedling Community												
Greenheart	.038	−.542	.763	.286	.118	.971	.038	−.519	.721	.249	.072	.858
Wallaba	−.952	.047	−.256	.153	.002	.997	−.941	.045	−.242	.133	.001	.964
Pentaclethra	.198	.847	.326	−.205	.300	.995	.196	.810	.308	−.179	.195	.859
Morabukea	.639	−.179	−.624	.405	.054	.997	.631	−.171	−.590	.353	.035	.901
Mora	.026	.040	−.010	−.662	.277	.518	.026	.038	−.009	−.577	.180	.367
Eschweilera	.411	−.469	.250	−.567	.058	.776	.406	−.449	.236	−.494	.038	.667
Variance extracted	.254	.211	.201	.179	.031	.876	.248	.193	.180	.136	.013	.770
Redundancy	.248	.193	.180	.136	.013	.770	.248	.193	.180	.136	.013	.770
Canonical variate	v_1	v_2	v_3	v_4	v_5	h_w^2	u_1	u_2	u_3	u_4	u_5	h_b^2
Tree Community												
Greenheart	.292	−.560	.730	.211	−.115	.989	.289	−.536	.690	.184	−.075	.887
Wallaba	−.936	.000	−.264	.206	.088	.996	−.925	.000	−.250	.180	.057	.954
Pentaclethra	.438	.742	.285	−.300	.277	.991	.433	.710	.269	−.261	.180	.864
Morabukea	.754	−.173	−.504	.372	−.067	.995	.745	−.165	−.477	.324	−.044	.917
Mora/Eschweilera	.548	−.297	.078	−.661	−.100	.842	.542	−.284	.074	−.576	−.065	.716
Jessenia	.410	−.241	.000	−.355	.699	.841	.405	−.230	.000	−.309	.455	.519
Variance extracted	.365	.173	.157	.146	.100	.942	.357	.159	.141	.111	.042	.810
Redundancy	.356	.159	.141	.111	.042	.809	.357	.159	.141	.111	.042	.810

Intraset structure correlations. The first canonical variate, v_1, of the tree domain (Table 9.3, lower-left) is dominated by a sharp contrast between Wallaba (−.94) and Morabukea (.75) forests. The remaining communities resemble Morabukea in having sizeable positive correlations with v_1 and hence also participate in this relationship to a greater or lesser extent. Thus v_1 expresses:

(a) a sharp compositional distinction between Wallaba and the other communities, particularly Morabukea;
(b) some degree of compositional similarity among the non-Wallaba communities, at least with respect to Wallaba forest.

The second canonical variate, v_2, operates entirely within the non-Wallaba vegetation. There, v_2 differentiates between *Pentaclethra* (.74) and Greenheart (−.56) forest, and at the same time suggests some degree of compositional affinity between Greenheart, *Mora/Eschweilera* (−.30) and *Jessenia* (−.24). v_3 and v_4 are open to interpretation in broadly the same terms. Thus compositional differences are suggested between Greenheart and Morabukea and between *Mora/Eschweilera* and Morabukea, respectively, while other communities appear to be related at least to the extent of the direction of their involvement with a particular variate. The final canonical variate, v_5 is largely an uncontaminated expression of *Jessenia* (.70).

The overwhelming impression is of a sharp contrast in composition between Wallaba and Morabukea forests in relation to the vegetation as a whole, while within the non-Wallaba vegetation the distinctiveness of Morabukea, *Pentaclethra* and Greenheart forests is striking. 36% of the total variance associated with the tree communities is absorbed by v_1. This canonical variate is twice as strong as each of the remaining variates, which themselves are of roughly equal strength; evidently the distinction between Wallaba and the non-Wallaba communities constitutes a major compositional feature of the vegetation examined. Collectively, the v_k ($k=1, \ldots, 5$) account for 94% of the total variance of the tree domain and hence effectively summarize this variation.

Interset structure correlations. We find from the interset correlations (Table 9.3, upper-right) that the first canonical variate, v_1, of the tree domain is characterized in the seedling domain above all by the contrast in sign between Wallaba (−.94), Morabukea (.63), and, to a lesser extent, *Eschweilera* (.41) forests; the correlations of the three remaining seedling communities with v_1 hardly depart appreciably from zero. Nevertheless, to a considerably degree, v_1 bears the same relationship to the seedling communities as to the tree communities on which it is defined. Glancing at the structure correlations of the seedling communities with the remaining v_k ($k=2, \ldots, 5$) and comparing these with the *intraset* correlations of corresponding tree communities, reveals that covariation between particular seedling and tree communities and the v_k is generally of the same sign and magnitude. Indeed, the fidelity of the intraset and interset correlations of the v_k within and across domains in both direction and strength is a remarkable feature of the results. Interpretation is facilitated by noting that by column the interset correlations may be thought of as expressing the apparent effect of a particular v_k of the tree domain on the seedling vegetation, while by row the correlations express the apparent response of a particular seedling community to the different components of the tree vegetation represented by the v_k collectively. Recall that the v_k themselves correspond to compositional contrasts or affinities among the tree communities. Bearing these points in mind, it is clear that:

(a) each v_k has substantially the same apparent effect on the seedling vegetation as it does in relation to the tree vegetation on which it is defined;
(b) the profiles or 'responses' of corresponding tree and seedling communities in the space of the v_k are substantially the same.

Evidently, tree and seedling counterparts of particular communities vary in a direct sense with one another, while the mutual relationships among the seedling communities are very much the same as those among the tree communities. Nevertheless, the relationship between tree and seedling communities falls short of an exact one-to-one correspondence. For example, we find that v_5, which above all is an expression of *Jessenia* forest, is related to two seedling communities, namely *Pentaclethra* (.19) and *Mora* (.18).

Redundancy. The redundancy in the seedling domain attributable to the canonical variates v_k of the tree domain falls systematically across the v_k, from $V^2_{\mathbf{x}|v_1} =$ 248 to $V^2_{\mathbf{x}|v_5} = .013$ (Table 9.3, upper-right). With the exception of v_5, it seems from the redundancies that each canonical variate contributes usefully to our understanding of tree/seedling relationships. Collectively, the v_k ($k=1, \ldots, 5$) account for 77% of the total variance of the seedling communities, so that the fit of the rank 5 model is reasonably satisfactory.

In Figs. 9.1–9.3 the canonical variates are used to summarize selected aspects of the covariance structure of the data graphically. Figures 9.1 a and b show the sample (N=25) plotted in the two-dimensional subspaces associated with the first two canonical correlation coefficients, r_1 and r_2. The most striking feature of the figure is the tendency for sample-points to fall in each case close to a diagonal line through the origin. The configurations very clearly express the respective canonical correlations ($r_1 = .99$; $r_2 = .96$). It is easy to appreciate from the figure that analysis does in fact lead to mappings which expose the correlation structure of the data. Figure 9.1 a displays the sample projected onto the first pair of canonical variates, u_1 and v_1. It is plain that Morabukea and Wallaba forests contribute most to the relationship shown, though in opposite senses. Provided that the ecological significance of the canonical variates is borne in mind (e.g., see the structure correlations of u_1 and v_1 in Table 9.3), it can be appreciated that stands with high positive scores on v_1 are characterized above all by Morabukea tree vegetation *and* that it is precisely these stands whose seedling vegetation is also characterized chiefly by Morabukea; similarly, it can be seen that stands with large negative scores on v_1 consist predominantly of Wallaba forest *and* that such stands are precisely those whose seedling vegetation is also composed largely of Wallaba. As a corollary, it is abundantly clear that Morabukea and Wallaba forests, in particular, are most unlikely to be in any way dynamically related.

Figure 9.1 b is a mapping onto the (u_2, v_2)-plane and reveals an entirely different aspect of the sample. It is readily seen that *Pentaclethra* and Greenheart

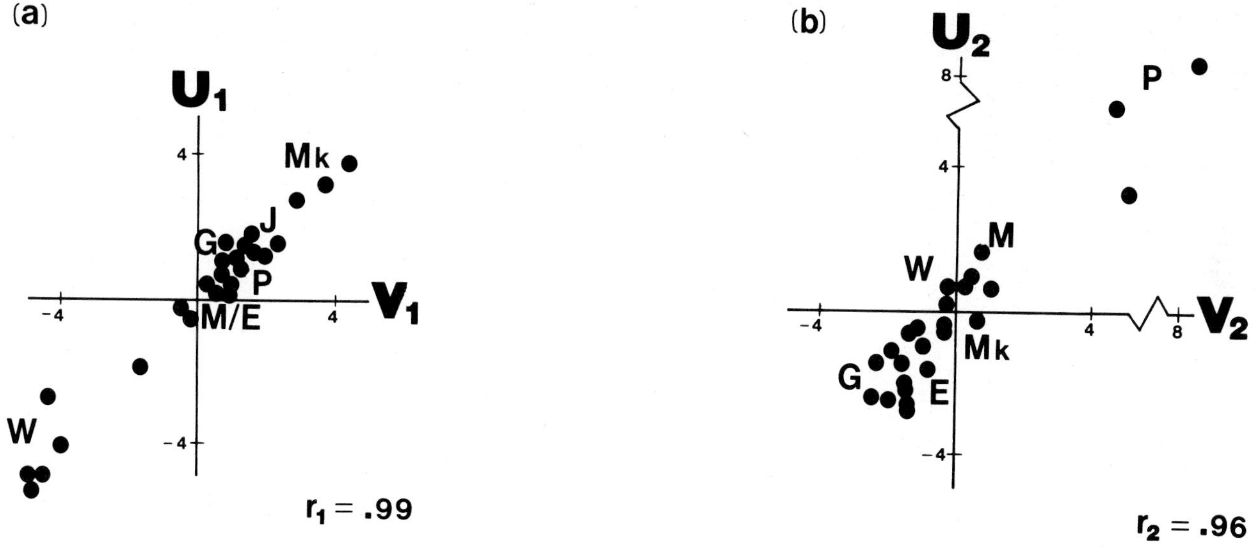

Fig. 9.1 a, b. Lowland tropical rain forest, Guyana. Canonical analysis of tree-seedling relationships. N = 25 stands mapped into subspaces defined by canonical variates (**a**) u_1 and v_1; (**b**) u_2 and v_2. E: *Eschweilera* forest; G: Greenheart; J: *Jessenia*; M: *Mora*; M/E: *Mora/Eschweilera*; Mk: Morabukea; P: *Pentaclethra*; W: Wallaba

forests contribute most to the linear relationship corresponding to r_2. From the structure correlations of Table 9.3 it may be appreciated that stands with high positive scores on u_2 and v_2 consist chiefly of *Pentaclethra;* furthermore, stands with sizeable negative scores are similarly characterized in terms of Greenheart.

Continuing in the same way, the sample could be mapped into a variety of other subspaces associated with the non-zero r_k. However, the picture of tree/seedling relationships which would emerge would necessarily be fragmented. For this reason, a *single,* comprehensive representation of the sample which simultaneously provided insight into the disposition of the sample in the *full* five-dimensional tree *and* seedling spaces of the analysis would obviously be advantageous. It is to such a representation that we shall turn shortly. Before doing so, however, it will be useful to summarize the ecological implications of the results to have emerged so far.

From the overwhelming similarities between the tree and seedling species' assemblages of Morabukea, Wallaba, *Pentaclethra* and Greenheart forests it seems clear that each of these communities has the potential at least to be succeeded in the short term by forest of substantially the same composition, as the current seedling generations mature. Moreover, from the strength of the tree/seedling relationships in each case, it is abundantly clear that the communities in question are most unlikely to be in any way dynamically related among themselves. The tree/seedling relationships defined by the canonical variates corresponding to the smaller roots, r_k^2 $(k>2)$, although involving other forest communities, are open to interpretation in substantially the same terms. The structure correlations of Table 9.3 summarize the relationships in question.

A canonical representation. A unified representation of the sample in the full five-dimensional tree and seedling spaces of the analysis is readily constructed using the procedure introduced by Andrews (1972). In this way, as described in Sect. 2.4.3, each stand in u_k- or v_k-space comes to be represented by a curve in 2-space (see Fig. 9.2). To avoid overburdening the function-plot, however, we have elected below to work with the *centroids* of particular communities, rather than with the individual samples composing the communities themselves. This step reduces the total number of curves for consideration from fifty to twelve. The results are shown in Figs. 9.2 and 9.3. Figure 9.2 shows the sine-cosine curves corresponding to the centroids of the tree communities and Fig. 9.3 the curves of the tree *and* seedling community centroids plotted simultaneously.

The most striking feature of Fig. 9.2 is the wide separation and distinctiveness of the curves. Compare, for example, the curves corresponding to Morabukea (Mk), Wallaba (W), *Pentaclethra* (P) and Greenheart (G) forests. Their distinctiveness indicates clearly that in v_k-space $(k=1, \ldots, 5)$, the mean separation between communities is appreciable. It seems clear, therefore, that in ecological terms the forest as a whole is made up of a number of compositionally *distinct* communties.

Turning to the joint plot of the sample in both tree and seedling spaces of the canonical analysis (Fig. 9.3), attention is drawn to three points:

(a) the well-separated curves corresponding to tree and seedling counterparts of *different* communities;

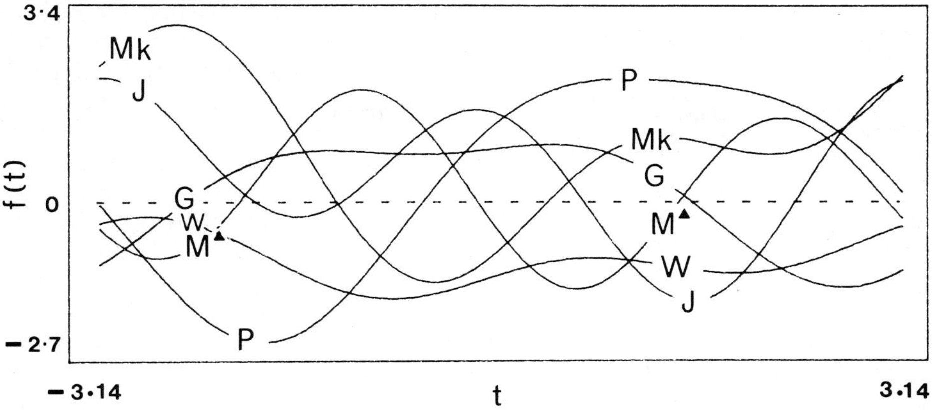

Fig. 9.2. Lowland tropical rain forest, Guyana. Canonical analysis of tree-seedling relationships. High-dimensional plot of the canonical variates. Centroids of tree communities of v_k-space ($k = 1, \ldots, 5$) mapped into an arbitrary trigonometric function. G: Greenheart forest; J: *Jessenia*; M*: *Mora/Eschweilera*; Mk: Morabukea; P: *Pentaclethra*; W: Wallaba

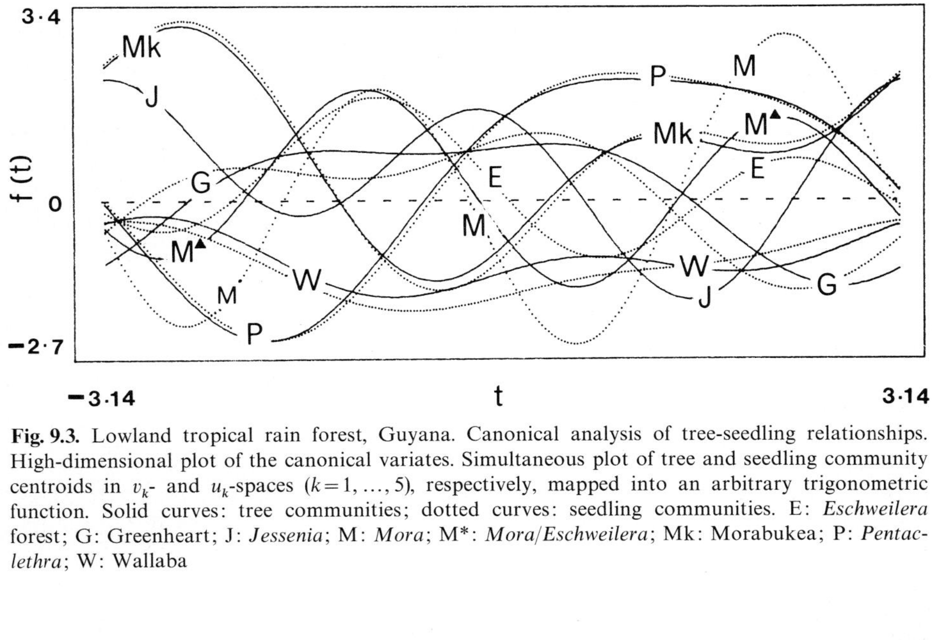

Fig. 9.3. Lowland tropical rain forest, Guyana. Canonical analysis of tree-seedling relationships. High-dimensional plot of the canonical variates. Simultaneous plot of tree and seedling community centroids in v_k- and u_k-spaces ($k = 1, \ldots, 5$), respectively, mapped into an arbitrary trigonometric function. Solid curves: tree communities; dotted curves: seedling communities. E: *Eschweilera* forest; G: Greenheart; J: *Jessenia*; M: *Mora*; M*: *Mora/Eschweilera*; Mk: Morabukea; P: *Pentaclethra*; W: Wallaba

(b) the proximity and overall similarity of curves corresponding to tree and seedling counterparts of the *same* community;
(c) the singular nature of the curve for *Jessenia* (J) which alone is comparatively unrelated to the curve of any other community.

Thus, the separation of the curves for different communities noted in connection with the tree component of the vegetation is equally well defined in terms of the seedling component. On the other hand, the remarkable fidelity of the relationship between curves for tree and seedling counterparts of particular communities as t varies over its entire range, stands in marked contrast to the highly distinctive curves of different communities. The affinity between tree and seedling curves is especially strong in the case of Morabukea and *Pentaclethra* forests, rather less so in the case of Wallaba and Greenheart and somewhat weaker still in the case of *Mora/Eschweilera* forest and its seedling affiliates, *Mora* and *Eschweilera*. Finally, in the case of *Jessenia,* which is represented by forest vegetation only, the relationship vanishes entirely. In connection with *Mora/Eschweilera* forest, it is of interest to observe that the curve for this community maintains a position mid-way between and symmetrically related to the comparatively well-separated *Mora* and *Eschweilera* seedling curves.

Before turning to consider the ecological implications of these results we shall briefly examine the variable communalities in order to ascertain how the explanatory power of the model is distributed among the tree and seedling communities.

9.2.5 Variable communalities

The intraset communalities of the tree communities, h_w^2 (Table 9.3, lower-left), vary from that of *Jessenia* (.84) to that of Wallaba (1.00). Further, we see that four communities have communalities in excess of .99. The v_k also account for a large part (>.80) of the variance, h_w^2, of a majority of the seedling communities (Table 9.3, upper-right). Only the variance of *Mora* (.37) is relatively poorly explained. It is therefore abundantly clear that for the most part the v_k ($k=1, \ldots, 5$) do effectively account for the variances of the tree communities on which they are defined as well as of the seedling communities of the 'other' domain.

9.3 Ecological assessment of the results

Having regard for the small size of the sample (N=25), it is necessary to make allowance for the effects of this aspect of the analysis on the results obtained. For precisely the reasons advanced in connection with the analysis of the previous chapter (see Sect. 8.3), it is clear that it would be prudent to adopt a sceptical attitude towards the *strength* of the relationships described, and, in addition, to seek for evidence external to the analysis itself if attempts to extend the *applicability* of the results beyond the sample actually examined are to be made.

The salient point to have emerged is that the vegetation examined appears to consist of communities which tend to be compositionally distinct but between the tree and seedling counterparts of which there exists for the most part a considerable degree of floristic and vegetational similarity. Broadly speaking, therefore, each community appears potentially able at least to be replaced in due course by a community of substantially the same composition. In a word, therefore, the dynamic status of the vegetation seems to be essentially stable. The status of *Mora/Eschweilera* and its seedling affiliates and that of *Jessenia* forest, however, are open to interpretation in rather different terms. From Fig. 9.3 it is conceivable that more or less distinctive *Mora* and *Eschweilera* seedling communities may be in the process of segregating from *Mora/Eschweilera* forest. It emerges also that *Jessenia* forest has little affinity with *any* seedling community. Thus, it may be that *Jessenia* represents a forest community which is not maintaining itself and which therefore may shortly come to be replaced by a community of a different kind. Thus, of the communities recognized, *Jessenia* and *Mora/Eschweilera* are those for which the likelihood of dynamic changes in composition seem most plausible.

Several of the most striking points to have emerged from the present analysis have been commented on by previous workers. In particular, Ogden (1966, p. 33), in an analysis of the same data by quite different methods, arrived at essentially the same view of the close compositional similarity between tree and seedling counterparts of the various communities as expressed here. In addition, several workers have commented on the remarkable compositional similarity of tree and seedling species' assemblages of Morabukea and *Pentaclethra* forests, which are notable features of the present analysis (Davis & Richards, 1933; Fanshawe, 1952; Ogden, 1966, 1977b). Ogden (1966, p. 33) has further remarked that the seedling population of Wallaba forest, which is itself a highly distinctive forest community, contains elements from adjacent forest communities which, although germinating, fail to survive to maturity. This feature may well account for the small but nevertheless discernible separation between the Wallaba tree and seedling curves apparent on Fig. 9.3. Such independent substantiation of certain aspects of the results goes some way towards validating the efficacy of canonical analysis in this application. Furthermore, the results of the canonical analysis go beyond those of previous studies in the penetrating nature of the picture of tree/seedling relationships which emerges. Further field work would be necessary, however, to establish the worth of the additional insights provided.

9.4 Conclusions

The results obtained yield insight into the dynamic status of the vegetation studied. They enable a distinction to be made between communities which appear to be in stable equilibrium from communities which seem less likely to be so, and, in the case of the latter, provide an indication of the changes which might be in progress. More specifically, it seems justifiable to conclude that:

(a) in the short term at least, the dynamic status of the vegetation appears to be essentially stable; and that

(b) *Jessenia* and *Mora/Eschweilera* are the communities which above all may be most susceptible to any departure from stable equilibrium.

Certain features of the results are substantiated by the results of independent investigations. Such external support goes some way towards validating the results presented here, and, moreover, attests to the general usefulness and appeal of canonical analysis. In addition, it suggests that the conclusions based on the results may possess some degree of generality beyond the sample of forest actually examined.

It is important to note that the results throw little if any light on any medium or long term changes which may be taking place. The limitation arises because the time interval represented by the difference in size between the trees and seedlings examined is short relative to any possible longer-term changes in the vegetation. Moreover, the use of size-classes as a means of investigating vegetation dynamics has several pitfalls (Austin, 1977). In particular, the interpretation of results ought, strictly speaking, to be guided by knowledge of the reproductive biology of the species involved. In the present study, unfortunately, as in cases which deal with the vegetation of remote areas generally, the required biological information may simply be lacking. It must therefore be accepted that at best only short-term changes in dynamic status will have been detected and that the conclusions reached are provisional in nature. What the analysis has done is to provide a description of the covariance structure of the tree and seedling counterparts of the vegetation examined; the ecological interpretation of this structure points to the likelihood of the stability of the vegetation in the short term while at the same time indicating communities between which dynamic shifts in composition seem most plausible in the longer term.

10. The structure of grassland vegetation in Anglesey, North Wales

10.1 Introduction

Many ecological endeavors lead naturally to comparisons between plant communities with respect to their overall species' composition. The object of *comparative studies* of this kind is to obtain through a description of community differences insight into the processes responsible for the differences and an indication of the avenues along which more intensive study might be rewarding. In this chapter the use of canonical analysis in a comparative study is illustrated. Previous work on the composition and structure of grassland vegetation overlying a small (5 ha) area of Carboniferous Limestone in Anglesey, North Wales, showed the vegetation to be composed of three comparatively distinct communities (Gittins, 1965). The communities recognized were described as limestone, neutral and fertile grassland. With the aim of obtaining further insight into the structure of the vegetation a clearer understanding of interrelationships among the communities was sought. Examination of the sample in a geometric space chosen to emphasize both the distinctiveness of the communities and their mutual relationships would be appropriate for this purpose. Such a mapping could be achieved by a canonical analysis in which one set of variables consisted of estimates of species' representation in the stands surveyed while the variables of the second set consisted of dummy variables specifying the community affiliation of stands – that is, in other words, by a *canonical variate analysis*.

The original vegetation survey was based on a random sample of $N=45$ $10\ \text{m} \times 10\ \text{m}$ stands. Thirty-three species were encountered during the survey and stand composition was expressed in terms of estimates of the abundance of these species. The species are identified in Appendix A.3. Preliminary analysis, which included a transformation to principal components, showed that a majority of the stands could be referred unequivocally to one or other of the three communities mentioned above; six stands, however, which were transitional between limestone and neutral grassland, could not readily be assigned to either community. Before canonical analysis could be undertaken, it was necessary to allocate the transitional stands in some acceptable way among the communities. For this purpose a decision rule based on a linear discriminant function calculated between limestone and neutral grassland was defined. The details of the analysis are not essential for present purposes and are omitted; it is sufficient to mention here that application of the rule led to four of the transi-

Table 10.1. Grassland vegetation, Anglesey. Mean representation of eight selected species in three grassland communities

c	Grassland community	Sample size n_c	Species[a] x_1	x_2	x_3	x_4	x_5	x_6	x_7	x_8
1	Limestone	27	70.2	44.4	68.1	16.3	20.6	30.1	2.4	0.1
2	Neutral	16	14.7	4.7	71.5	75.0	72.7	90.4	35.5	11.0
3	Fertile	2	0.0	2.0	6.0	4.0	56.0	86.0	83.5	80.0
Average of means			28.3	17.0	14.7	48.5	49.8	68.8	40.5	30.4

[a] Species $x_1, \ldots, x_8$: *Pseudoscleropodium purum* (Hedw.) Fleisch., *Thymus drucei* Ronn., *Carex flacca* Schreb., *Agrostis tenuis* Sibth., *Trifolium repens* L., *Helictotrichon pubescens* (Huds.) Pilger, *Dactylis glomerata* L., *Phleum bertolonii* DC

tional stands being assigned to limestone grassland and two of them to neutral grassland. In this way the total sample came to be partitioned into $g=3$ mutually exclusive and exhaustive groups which corresponded to limestone, neutral and fertile grassland. Something of the composition of the communities can be obtained from Table 10.1. The table reports the mean representation of eight species chosen to emphasize the differences in composition between the communities. Comparison of the rows of the table, that is of the community mean vectors or *centroids,* reveals substantial differences in composition between the three grassland types.

Because the number of species encountered (33) was large relative to the size of the sample (N=45), the two rarest species were omitted from the analysis in order to reduce the ratio of variables to samples, if only slightly. The thirty-one remaining species were then taken to comprise one set of variables in the canonical analysis. Binary-valued dummy variables corresponding to the vegetation classification comprised the variables of the second set and were constructed in the following way. A variable, z_1, designated to correspond to limestone grassland, was established by assigning to each limestone grassland stand a value 1 on this variable while each non-limestone grassland stand received a value of 0. A second dummy variable, z_2, corresponding to neutral grassland, was created in a similar way by assigning a score of 1 to each neutral grassland stand on this variable and a score of 0 to all other stands. The variables z_1 and z_2 together completely accounted for the three-way stand classification, stands of fertile grassland being uniquely specified by a score of 0 on both z_1 and z_2. In short, the value of z_{jc} ($c=1, \ldots, g-1$) in the jth stand ($j=1, \ldots, \mathrm{N}$) was defined to be

$$z_{jc} = \begin{cases} 1 & \text{if the } j\text{th stand belonged to community } c \\ 0 & \text{otherwise.} \end{cases}$$

The canonical analysis therefore came to be based on a random sample of N=45 10 m × 10 m stands of grassland vegetation, $p=2$ binary-valued dummy variables (**z**) corresponding to $g=3$ grassland communities and estimates of the abundance of $q=31$ species (**x**). The ecological objectives were to describe community relationships and to elucidate the environmental factors or other generating processes responsible for the compositional differences between the communities. In a word, that is, to describe and account for the *structure* of the vegetation. If the communties were found to differ in composition, then considerable interest would attach to the dimensionality of the solution. A single non-zero root, in particular, would indicate not only that there were differences in composition but also that the relationship between communities was one of *collinearity*. One plausible ecological explanation for collinearity would be that the communities could be regarded as displacements along a *single* underlying environmental factor or other generating process. The results of the analysis are summarized in Tables 10.2 and 10.3 and in Fig. 10.1.

10.2 Results

10.2.1 The canonical correlation coefficients

The canonical correlation coefficients are shown in Table 10.2a. Both coefficients exceed .96. Their departure from zero is striking and entirely consistent with the sizeable differences in composition apparent in Table 10.1. For all practical purposes the correlations can be regarded as being substantially equal. The total between-community variance is obtained from the canonical roots r_k^2 of Table 10.2a by expression (4.10). We find that $\Sigma f_k/(g-1)=19.18$. Of this quantity, $f_1=r_1^2(1-r_1^2)^{-1}$ accounts for roughly twice as much (67.85%) as f_2 (32.15%).

10.2.2 Equality of community centroids

Table 10.2 reports the results of formal tests of the null hypothesis (4.21) of the equality of the community centroids. For the union-intersection test we find (Table 10.2a) that $\theta_s=.963>\theta_\alpha(2, 14, 5)=.926$ for $\alpha=.01$; for the likelihood ratio test it can be seen (Table 10.2b) that $\chi_s^2=158.95>\chi_\alpha(62)=112.2$ for $\alpha=.0001$. We are therefore led to reject hypothesis (4.21) and to conclude that the communities differ in composition at the $\alpha=.01$ level of significance at least.

10.2.3 Collinearity

The results of sequential tests of hypothesis (4.22) of collinearity for $t=1, \ldots, s$ are shown in Table 10.2. For the union-intersection test we find (Table 10.2a) that $F_s(30, 12)=4.933>F_\alpha(30, 12)=3.79$ for $\alpha=.01$. From Table 10.2b it can be seen that after r_1^2 is eliminated the residual variation is appreciable as $\chi_s^2(30)=69.94>\chi_\alpha^2(30)=67.7$ for $\alpha=.0001$. Evidently, r_1^2 and r_2^2 will both be required

Table 10.2. Grassland vegetation, Anglesey. Canonical variate analysis. Relationships between three grassland communities with respect to the representation of thirty-one constituent species. Canonical correlation coefficients and tests of equality and collinearity of community centroids

(a) The canonical correlation coefficients r_k and union-intersection tests of equality and collinearity. Approximate critical values of Roy's largest-root criterion $\theta_\alpha(s, m, n)$ are shown for $\alpha=.05$ and $\alpha=.01$ ($m=14$; $n=5$).

k	s	r_k	r_k^2	$\theta_{.05}(s, m, n)$	$\theta_{.01}(s, m, n)$	p
1	2	.981	.963	.894	.926	<.01
2	1	.962	.925	–	–	<.01 [a]

[a] From the F-distribution. See Morrison (1976, p. 178): $F_s(30, 12)=4.933>F_\alpha(30, 12)=3.79$ for $\alpha=.01$.

(b) Likelihood ratio tests of equality and collinearity. Bartlett's approximate chi-squared criterion for assessing the joint nullity of the smallest $s-k$ canonical correlation coefficients.

k	Roots	Chi-squared	df	p
0	1,2	158.95	62	<.0001
1	2	69.94	30	<.0001

in order to locate the differences between communities. Clearly, an hypothesis of the collinearity of community centroids is untenable.

10.2.4 The canonical variates

Correlation coefficients between the canonical variates u_k and v_k and the variables comprising $\mathbf{z}\,(2\times 1)$ and $\mathbf{x}\,(31\times 1)$ are reported in Table 10.3. To a large extent we shall rely on these quantities in interpreting the canonical variates.

Intraset structure correlations. From the correlations between v_1 and the species on which this canonical variate is defined it is apparent (Table 10.3, lower-left) that there are appreciable differences between species in terms of both the direction and magnitude of their correlations. Species such as *Pseudoscleropodium purum* (Hedw.) Fleisch. (.92), *Thymus drucei* Ronn. (.74) and *Helianthemum chamaecistus* Mill. (.71) vary directly with v_1, while *Agrostis tenuis* Sibth. (−.87), *Trifolium repens* L. (−.85) and *Helictotrichon pubescens* (Huds.) Pilger (−.83), among other species, vary inversely with v_1. In addition, several species are substantially uncorrelated with v_1, for example *Anthoxanthum odoratum* L. (.01) and *Galium verum* L. (.06). The second canonical variate, v_2, also differentiates the species. Observe, for example, the very different correlations of *Phleum bertolonii* DC. (−.80) and *Dactylis glomerata* L. (−.52), on the one hand, and of *Carex flacca* Schreb. (.60) and *Carex caryophyllea* Latour. (.57), on the other. Together, v_1 and v_2 account for 42% of the total variance of the species' domain. Of the two canonical variates, v_1 is appreciably stronger than v_2, accounting for 30% of the total variance ($V_1^2=.304$) while v_2 absorbs some 11% ($V_2^2=.113$).

Table 10.3. Grassland vegetation, Anglesey. Canonical variate analysis. Relationships between three grassland communities with respect to the representation of thirty-one constituent species: correlations between the original variables and canonical variates

Canonical variate	u_1	u_2	h_w^2	v_1	v_2	h_b^2
Community						
Limestone grassland	.986	.170	1.000	.967	.164	.962
Neutral grassland	−.968	.251	1.000	−.950	.241	.961
Variance extracted	.955	.046	1.001	.919	.042	.961
Redundancy	.920	.043	.963	.919	.042	.961
Canonical variate	v_1	v_2	h_w^2	u_1	u_2	h_b^2
Species						
H. chamaecistus	.710	.333	.615	.697	.320	.588
T. drucei	.744	.167	.581	.730	.161	.559
L. corniculatus	.199	.484	.274	.195	.466	.255
P. lanceolata	−.378	.494	.387	−.371	.475	.363
G. verum	.062	−.283	.084	.061	−.272	.078
C. flacca	−.071	.605	.371	−.070	.582	.344
C. caryophyllea	.402	.573	.490	.394	.551	.459
P. sanguisorba	−.456	.425	.389	−.447	.409	.367
H. pilosella	.273	.285	.156	.268	.274	.147
L. campestris	−.577	.396	.490	−.566	.381	.466
T. laevigatum	.549	.242	.360	.539	.233	.345
V. riviniana	−.175	.143	.051	−.172	.138	.049
B. media	.688	.452	.678	.675	.435	.645
K. cristata	.559	.045	.315	.548	.043	.302
A. odoratum	.013	.207	.043	.013	.199	.040
S. decumbens	.148	.150	.044	.145	.144	.042
D. glomerata	−.649	−.518	.690	−.637	−.498	.654
A. tenuis	−.871	.294	.849	−.854	.283	.809
H. pubescens	−.832	−.139	.712	−.816	−.134	.684
P. bertolonii	−.323	−.803	.749	−.317	−.772	.696
H. lanatus	−.522	.137	.291	−.512	.132	.280
C. nigra	−.623	.163	.415	−.611	.157	.398
T. repens	−.853	−.047	.730	−.837	−.045	.703
T. pratense	−.554	.117	.321	−.543	.113	.308
A. vulneraria	−.089	.066	.012	−.087	.063	.012
C. impexa	.742	.146	.572	.728	.140	.550
P. purum	.918	.268	.915	.901	.258	.878
D. scoparium	.862	.187	.778	.846	.180	.748
R. squarrosus	−.143	.258	.087	−.140	.248	.081
H. radicata	−.604	.230	.418	−.593	.221	.400
C. vulgaris	.040	.057	.005	.039	.055	.005
Variance extracted	.304	.113	.417	.291	.104	.395
Redundancy	.288	.104	.392	.291	.104	.395

The first canonical variate u_1 of the community domain is characterized by three features in terms of its correlations with limestone (.99) and neutral (−.97) grasslands – the correlations are *strong*, substantially *equal* in magnitude and of *opposite* sense (Table 10.3, upper-left). Evidently, u_1 is closely identified with both communities while nevertheless distinguishing between them with respect to the species' composite represented by its conjugate, v_1. The second canonical variate, u_2, differs from u_1 in several ways. In particular, the correlations of u_2 with limestone grassland (.17) and neutral grassland (.25) differ somewhat in size, are alike in sign and are not conspicuously strong. The lack of close affinity of u_2 with either limestone or neutral grassland is in some ways the most fundamental of these features. A further difference between u_1 and u_2 concerns their strength – u_1 is twenty times stronger than u_2, accounting for 95% of the total variance of the community domain ($U_1^2=.95$), while u_2 accounts for barely 5% ($U_2^2=.05$).

Interset structure correlations. The interset variable/canonical variate correlations appear in the right-hand side of Table 10.3. From the direction of the correlations of limestone (.97) and neutral (−.95) grasslands with v_1 it is clear that the communities differ with respect to the species' contrast represented by v_1. The correlation of limestone grassland is in the same direction as those of *P. purum* etc. while that of neutral grassland is in the same direction as those of *A. tenuis* etc. Species which vary directly with v_1 indeed prove to be essentially species of limestone grassland while species which are inversely related to v_1 similarly prove to be representative of neutral grassland. Turning to the second canonical variate, recall that in the species' domain v_2 is a well-defined composite. In the community domain, on the other hand, we find that v_2 is not closely identified with either limestone (.16) or neutral grassland (.24). Of the species which are affiliated with v_2 it is known that *P. bertolonii* and *D. glomerata* are species of relatively fertile, phosphorus rich soils (Gittins, 1965). This information, together with the lack of affinity of v_2 with either limestone or neutral grassland, enables v_2 to be interpreted ecologically as an inverse expression of *fertile* grassland.

The interset structure correlations of the plant communities with the v_k ($k=1$, 2) of the species' domain therefore yield a consistent and intuitively satisfying interpretation of relationships between the two sets of variables. In Sect. 10.3 we shall consider the ecological implications of these results further. Although we do not pursue the details here, it is worth noticing that the u_k of the community domain are open to interpretation within and across domains in comparable terms to the v_k. Interpretation of the u_k is commended as an exercise for the interested reader familiar with the species and communities in question.

Redundancy. The redundancy in the community domain attributable to the canonical variates v_1 and v_2 of the species' domain is $V^2_{\mathbf{z}|v_1}=.919$ and $V^2_{\mathbf{z}|v_2}=.042$, respectively (Table 10.3, upper-right). Together, v_1 and v_2 account for 96% of the total variance of the community domain. In other words, the total variance of the community domain is virtually fully accounted for by the two species' composites represented by v_1 and v_2. To some extent this result is attributable to the use of thirty-one variables to explain the sum of the variances of just two variables. For comparison, observe that u_1 and u_2 together account for 40% of the total variance of the species' domain, an appreciably smaller figure ($U^2_{\mathbf{x}|u_1,u_2}=.395$). The explanatory power of v_1 in the community domain is twenty-two times that of v_2 (.919/.042=21.88). This leads to a very different

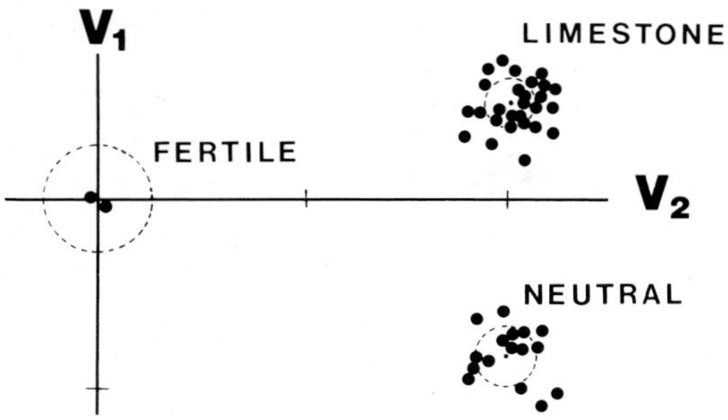

Fig. 10.1. Grassland vegetation, Anglesey. Canonical variate analysis. Relationships between three grassland communities with respect to the representation of thirty-one constituent species. N=45 stands mapped into the subspace of the first two canonical variates, v_1 and v_2, of the response variables. Circles depict the 99% confidence regions about community centroids

assessment of the relative importance of the two canonical relationships from that provided earlier by the canonical correlation coefficients (Sect. 10.2.1).

Figure 10.1 displays the N=45 stands mapped into the (v_1, v_2)-plane of species' space. The origin of the coordinate system has been placed at the contrast of fertile grassland with itself, that is at the null contrast, and the canonical variates scaled so that the average within-groups variance on each weighted by sample size is unity. This particular standardization maximizes the distances between the communities. In addition, the canonical variates have been drawn at right angles, as is customary where the original variables are not included in the representation. From the vectors of canonical weights, $\mathbf{b}_1$ and $\mathbf{b}_2$, however, it can be shown that v_1 and v_2 intersect at an angle of 119° in species' space. The circles of radius $(\chi^2\,(2)/n_c)^{\frac{1}{2}}$ are 99% confidence regions about the community centroids and provide a rough indication of the separation and uncertainty of each.

The figure shows three approximately circular clusters of sample-points corresponding to limestone, neutral and fertile grassland. The clusters differ in size, those of limestone and neutral grassland being comparable in size and considerably larger than that of fertile grassland. Apart from differences in size the most striking features of the plot are the coherence and wide separation of the clusters and the departure of their centroids from collinearity. The first canonical variate, v_1, distinguishes limestone from neutral grassland, while v_2 separates both communities from fertile grassland. Having regard for both the *direction* and the *extent* of the separation of fertile grassland from the other communities, it is clear that fertile grassland is a highly distinctive community. The group centroids are tightly enclosed by the confidence circles, a feature enhancing the mutual separation of the groups. Altogether, the canonical representation of Fig. 10.1 provides a clear description of community relationships, that is, of the *structure* of the vegetation.

10.2.5 Variable communalities

Let us first examine the species' profiles from which the intraset species' communalities are themselves obtained (Table 10.3, lower-left). The great diversity of profiles is a feature of some interest; while certain species are related, directly or inversely, to both v_1 and v_2 other species are affiliated with just one canonical variate and yet other species to neither. The communalities, h_w^2, are similarly diverse. The communalities of *P. purum* (.91), *A. tenuis* (.85) and *P. bertolonii* (.75), for example, approach unity, while those of *A. odoratum* L. (.04), *Anthyllis vulneraria* L. (.01) and *Calluna vulgaris* (L.) Hull (.01) scarcely depart from zero. It is therefore plain that the species differ greatly in the nature and the extent of their involvement with the canonical variates v_1 and v_2 defined on them. It follows that the same species' differences will exist also in relationships involving v_1 and v_2 across domains. From the table it can be seen (upper right) that the profiles of limestone and neutral grasslands with respect to v_1 and v_2 resemble one another to the extent of having a strong affinity with v_1 and a weak affinity with v_2. A fundamental difference between the profiles, however, arises from the direction of the correlation of each community with v_1. The interset communalities, h_b^2, reveal that a high proportion (>.96) of the variance of both communities is uniformly accounted for by v_1 and v_2 jointly. This is a further manifestation of the high explanatory power of the fitted model.

10.3 Ecological assessment of the results

At this point it is necessary to consider the effects of the variable/sample ratio and of possible heterogeneity of dispersion on the results obtained. The variable/sample ratio is decidedly high ($^{33}/_{45}=.73$). Almost certainly, therefore, the largest canonical correlation coefficient r_1 will be positively biased. Accordingly, the strength of at least some of the relationships described must be accepted with considerable reserve. Heterogeneity of dispersion can affect the extent of overlap between clusters and the orientation of the canonical axes as well as the validity of statistical tests. Homogeneity was not investigated prior to the canonical analysis, though it was checked afterwards. Bartlett's (1937) test of the equality of variances for the $g=3$ communities on v_1 and v_2 revealed no departure from homogeneity; for v_1, $\chi_s^2(2)=2.442<\chi_\alpha^2(2)=2.773$ for $\alpha=.250$; for v_2, $\chi_s^2(2)=0.489<\chi_\alpha^2(2)=.713$ for $\alpha=.700$. It seems, therefore, that the results of the canonical analysis may perhaps be accepted as being unlikely to have been seriously distorted by heterogeneity of dispersion.

The essential features of the analysis are to a large extent summarized by Fig. 10.1. The size, coherence and spatial relationships of the point-clusters of the figure are all readily interpreted in ecological terms. Differences in cluster size reflect differences in the size of the communities (cf. Table 10.1). The coherence and clear separation of the communities in turn suggest that the three grassland communities are discrete entities. At this point, however, it is necessary to recall and make allowance for the allocation of stands of intermediate composition prior to the canonical analysis itself. It is self-evident that this procedure will have resulted in limestone and neutral grassland appearing more discrete than they in fact are and must be borne in mind in interpreting Fig. 10.1. The proximity of limestone and neutral grasslands shows that these communities resemble one another in composition to an appreciably greater extent than either resembles fertile grassland. Similarly, the isolation of fertile grassland indicates this to be rather a distinctive community. Moreover, from departure of the community-centroids from collinearity, it is abundantly clear that the three communities can hardly be considered to represent displacements along a *single* underlying environmental or other gradient.

In attempting to explain these relationships we shall make use of field experience of the behavior of a number of species. Of those species correlated with v_1, *T. drucei* and *A. tenuis* are known to be profoundly influenced by variation in soil depth and conditions which vary with it, over the range encountered in the area surveyed (Gittins, 1965). It is also known that the species differ in their response to soil depth, *T. drucei* tending to occur in greatest abundance on shallow, well-drained soils, while the representation of *A. tenuis* tends to be greatest on deeper soils with more balanced water relationships. It follows that the change of sign associated with the differentiation of limestone and neutral grassland on v_1 is attributable largely to the differential response of species such as *T. drucei* and *A. tenuis* to variation in soil depth and its concomitants.

Phleum bertolonii and *D. glomerata* are correlated with the negative pole of v_2. Both species have been shown to be highly responsive to variation in soil fertility, their representation increasing sharply with increasing soil phosphorus in particular (Gittins, 1965). The performance of species such as *C. flacca* and *C. caryophyllea,* on the other hand, which are correlated with the positive pole of v_2, is known to be best on phosphorus depleted soils. The differentiation of fertile from limestone and neutral grasslands can therefore be attributed largely to differential species' response to variation in soil phosphorus.

The canonical variates v_1 and v_2 therefore do appear to correspond to ecological controls in terms of which the structure of the vegetation can be comprehended. Some indication of the relative importance of the corresponding factor-complexes is provided by the explanatory power of v_1 and v_2 in the community domain. In Sect. 10.2.4 we saw that the redundancy associated with v_1 was twenty-two times that of v_2, a result which might conceivably have led us to suppose that v_2 could be disregarded altogether at very little cost. It has emerged, however, that v_2 does indeed correspond to a well-defined component of the vegetation – fertile grassland – of considerable intrinsic interest notwithstanding its totally subordinate contribution to the explained variance of the community domain. Fertile grassland in fact comprises 4% of the total sample of N=45 stands, a figure which is very closely approximated by the explanatory power of v_2 ($V^2_{z|v_2}=.042$ or 4.2%) in the community domain. The direction and extent of the involvement of all thirty-one species in the differentiation of communities affected by v_1 and v_2 is specified in Table 10.3. A notable feature of the table is the remarkable fidelity with which the species' profiles accord with field experience of the affiliation or otherwise of the species with each of the three communities. This is a somewhat unexpected result in view of the exceptionally high variable/sample ratio for the analysis.

It will be instructive finally to briefly compare the results obtained with those of a principal component analysis of substantially the same 45 × 31 matrix

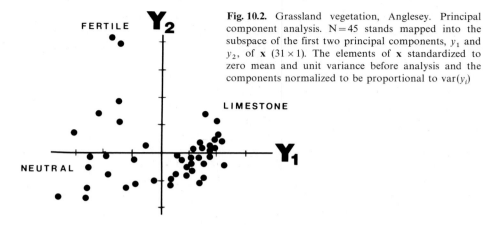

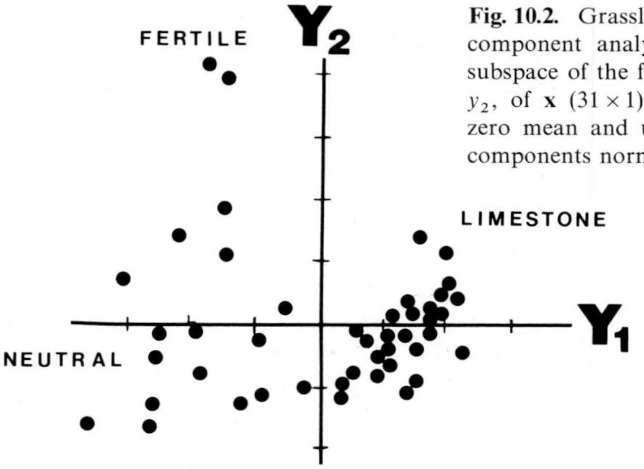

Fig. 10.2. Grassland vegetation, Anglesey. Principal component analysis. N = 45 stands mapped into the subspace of the first two principal components, y_1 and y_2, of $\mathbf{x}$ (31×1). The elements of $\mathbf{x}$ standardized to zero mean and unit variance before analysis and the components normalized to be proportional to $\mathrm{var}(y_i)$

of species' abundances as used in the canonical variate analysis. Such a comparison is illuminating because it draws attention to a distinctive property of canonical variate analysis. The comparison is conveniently made by reference to Figs. 10.1 and 10.2, which summarize salient features of the two analyses. From Fig. 10.2 it is apparent that principal component analysis discloses something of the existence, size and coherence of the three communities which together comprise the vegetation. The picture presented by canonical variate anaylsis (Fig. 10.1) is quite different. What Fig. 10.1 conveys above all is an indication of the *structure* of the vegetation as a whole – the mutual relations of its constituent communities. This aspect of the vegetation is identified and communicated with great economy and effect by canonical variate analysis. While components analysis also appears to reveal something of relations among communities the insight provided is not to be relied upon; components analysis is not formulated with features of this kind in mind. The crucial difference is that in the transformed space of a canonical variate analysis the within-group dispersions are uniform in all directions and so provide a yardstick against which distances *between* communities can properly be assessed. In components analysis, where the composite nature of the sample is disregarded and the principal components are optimized over the sample as a whole, no comparable yardstick exists.

10.4 Conclusions

The analysis provides a description of the structure of the vegetation investigated which proves amenable to explanation in ecological terms. The essential features of the results may be summarized as follows:

(a) the vegetation consists of three comparatively discrete communities, two similar in size and composition while the third is smaller and compositionally distinct;

(b) the structural relationships described can be comprehended by reference to environmental gradients in soil depth and fertility;
(c) the apparent effects of the two environmental gradients differ in magnitude, that of soil depth being some twenty times that of fertility.

These results are consistent with field knowledge of the vegetation. Furthermore, they convey insight into the structure of the vegetation with greater economy and effect than has hitherto been achieved. Differences between species in their contribution to the differentiation of the vegetation are precisely specified, the specifications according closely with expectations based on field experience despite the very high variable-to-sample ratio. The results are also useful in directing attention to lines along which further work might prove rewarding. In this regard, the identity of the remaining components of the factor-complexes of which soil depth and soil phosphorus seem likely to be part, and of the response of individual species to these components, may well repay investigation.

11. The nitrogen nutrition of eight grass species

11.1 Introduction

Identification of factors which control the occurrence and representation of plant species in the field is a prime objective of much ecological work. Preliminary indications of the identity of operative factors may be obtained either by direct observation in the field or by use of exploratory multivariate techniques in the analysis of survey data. The next step towards identification involves attempts to substantiate provisional insights of these kinds. The most direct and convincing means of substantiation is provided by examining the response of species to treatments of one or more of the indicated factors in a designed experiment. In this chapter the use of canonical analysis in the analysis of experimental data is described and illustrated. The study concerns the comparative response of eight grass species to five treatment levels of nitrate nitrogen. General field observation had indicated that the species in question differed in their response to variation in soil fertility. The experiment was designed to enable evidence for the differential response of the species to one component of the factor-complex of soil fertility, namely, nitrate nitrogen, to be assessed.

The species studied were *Lolium perenne* L., *Dactylis glomerata* L., *Phleum bertolonii* DC., *Briza media* L., *Koeleria cristata* (L.) Pers., *Festuca ovina* L., *Festuca rubra* L. and *Helictotrichon pubescens* (Huds.) Pilger. Nitrogen treatments of 1, 9, 27, 81 and 243 ppm N were obtained by varying the amount of $NaNO_3$ in culture solution, the level of other nutrients being held at or near optimum concentration. The nitrogen treatments were chosen so as to vary between what for the species was expected to be from critically low to almost toxic. Species' response to the treatments was expressed as dry weight yield in grams. The ecological objective of the experiment was to assess the evidence for differential response among the species and to characterize the differences, if present.

Individuals of each species were grown separately in pots under sand culture in an unheated greenhouse using a split-plot experimental design. Main-plots corresponded to species (8) and sub-plots to treatments (5). There were five complete replications (blocks) of the experiment. Species were randomly assigned to main-plots and treatments randomly assigned to sub-plots. The experimental unit was the individual pot and corresponded to a particular species and a particular treatment level of nitrogen. Plants were harvested and the dry weight yield of the experimental units determined after a growth period of two months. Numerical analysis was based on a logarithmic transformation (base 10) of dry weight yield.

An analysis of variance appropriate to a split-plot experimental design provided the simplest and most direct form of analysis. For present purposes, however, it will be instructive to regard the experiment from a rather different viewpoint which will enable us to pursue our principal objectives of examining the applicability of canonical analysis in ecology and of throwing light on its relationships with other statistical methods. As we shall see, by regarding the experiment as generating *vector-valued* rather than scalar responses, the results can be analyzed by both canonical variate analysis and multivariate analysis of variance. The necessary multivariate conceptualization of the experiment can be achieved quite simply by regarding the *species* rather than the individual pot as the experimental unit. The p-vector of responses associated with each species consists of the dry weight yield at each of the five nitrogen treatments separately administered to representatives (pots) of a given species. This device has the disadvantage that correlations between the responses refer to a composite representative of a species rather than to a genuine exemplar. As a result the correlations may not be well-determined. For the purpose of exposition, however, it may be permissible to overlook this point and to proceed *as if* the composite species constitutes a satisfactory experimental unit. As before, there are $b=5$ blocks each comprised of $g=8$ species. Thus, the experiment can be regarded as an orthogonal $g \times b$ multivariate randomized block design with $g=8$ conditions (species), $b=5$ blocks and $p=5$ responses. The ecological purpose of the experiment remains unchanged, namely to assess the comparative responses of the species to the treatments administered. The results obtained will also be used to verify relationships which exist between canonical analysis, multivariate analysis of variance and multiple discriminant analysis.

The multivariate analysis of variance is taken up in Sect. 11.2 followed by the canonical variate analysis in Sect. 11.3. Relationships between the two methods are examined in Sect. 11.4 and the ecological implications of the results considered in Sect. 11.5. The data underlying the analysis are reported in Table A-5 of Appendix A.2.

11.2 Multivariate analysis of variance

The species' mean vectors over the replicates in response to the five treatments are reported in Table 11.1. Differences between species are apparent. To test these differences formally, a multivariate analysis of variance may be performed. The null hypothesis addressed by the analysis is that the $g=8$ species' vector-means are equal [i.e., hypothesis (4.23)]. The results of the analysis are summarized in Tables 11.2 and 11.3.

11.2.1 Results

The multivariate analysis of variance is reported in Table 11.2. Results of likelihood ratio and union-intersection tests of hypothesis (4.23) are given in Ta-

Table 11.1. Nitrogen nutrition of eight grass species. Mean yield ($\log_{10}$ dry-weight (g) $\times$ 10) over five replicates of eight grass species grown in sand culture at five levels of nitrogen

c	Species	Sample size n_c	Mean yield[a] x_1	x_2	x_3	x_4	x_5
1	*L. perenne*	5	.9631	1.5150	1.5656	1.9683	2.0663
2	*D. glomerata*	5	.8808	1.4528	1.6527	1.8229	1.8651
3	*P. bertolonii*	5	.8082	1.4114	1.5033	1.6404	1.6850
4	*H. pubescens*	5	.7513	1.2754	1.1940	1.4057	1.3737
5	*K. cristata*	5	.6938	1.1811	0.9786	1.2695	1.1784
6	*B. media*	5	.7905	1.0702	1.1059	1.2870	1.1218
7	*F. ovina*	5	.7479	1.0589	1.1433	1.2326	1.2161
8	*F. rubra*	5	.3393	1.0426	0.8275	1.0418	1.3643
Average of means			.7469	1.2509	1.2464	1.4635	1.4838

[a] $x_1, x_2, \ldots, x_5$ mean species' response at 1, 9, 27, 81, 243 ppm nitrogen.

ble 11.3. The matrices $\mathbf{Q}_h$ and $\mathbf{Q}_e$ of Table 11.2 are mean-corrected hypothesis and error SSP matrices of order p carrying ν_h and ν_e degrees of freedom, respectively. The values of the parameters ν_h and ν_e are $g-1$ and $N-g-b+1$, respectively. $\mathbf{Q}_s$ is the constant term; it reflects arbitrary scale and location effects, and, for this reason, is not of major concern in this study. It can be seen (Tables 11.2 and 11.3a) that the likelihood ratio criterion is $\Lambda = .015691$. Bartlett's transformation of Λ for the randomized block design of the present analysis corresponding to (4.27) for the one-way layout, is given by (4.31) with parameter values $\nu_e = N-g-b+1$ and $\nu_h = g-1$. Accordingly,

$$\chi_s^2 = -\{28+7-\tfrac{1}{2}(5+7+1)\} \log_e(.015691)$$
$$= 118.4083, \text{ with } 7\cdot 5 = 35 \text{ df}.$$

As $\chi_s^2(35) = 118.41 > \chi_\alpha^2(35) = 74.92$ for $\alpha = .0001$ we reject the null hypothesis (4.23). It therefore seems that the species differ in response and that the differences warrant interpretation. We shall return to this aspect of the analysis in Sects. 11.4 and 11.5.

It is readily verified from Table 11.3b that Λ is given also by the roots f_k of the determinantal equation $|\mathbf{Q}_h - f\mathbf{Q}_e| = 0$. The table reports the roots of this equation and shows that the continued product of a simple function of the roots, $\Pi(1+f_k)^{-1}$, leads within rounding error to the same value of Λ calculated as a determinantal ratio [see Eq. 4.26)].

Although the block effects $\mathbf{Q}_b$ (Table 11.2) have little intrinsic interest, the effectiveness of the blocking is nevertheless worth examining. For this purpose we require

$$\Lambda_b = |\mathbf{Q}_e(\mathbf{Q}_e + \mathbf{Q}_b)^{-1}| = \frac{.1158460}{.3432418} = .337506.$$

Applying Bartlett's transformation we find

$$\chi_s^2 = -\{28+4-\tfrac{1}{2}(5+4+1)\} \log_e(.337506)$$
$$= 29.3267, \text{ with } 4\cdot 5 = 20 \text{ df}.$$

Table 11.2. Nitrogen nutrition of eight grass species. Multivariate analysis of variance of species' response to five treatment levels of nitrogen. x_1: 1 ppm N, x_2: 9 ppm N, ..., x_5: 243 ppm N

Source of dispersion	df	SSP (symmetric, $p \times p$)						Univariate		Multivariate		
		x_1	x_2	x_3	x_4	x_5		F	p	Λ	χ^2	p
Constant	1	22.3121					$=\mathbf{Q}_s$					
		37.3708	62.5925									
		37.2345	62.3642	62.1368								
		43.5727	72.9802	72.7140	85.0918							
		44.3289	74.2468	73.9760	86.5689	88.0709						
Blocks	4	0.1195					$=\mathbf{Q}_b$	1.91	.14	.337506	29.33	.0815
		0.0844	0.1397					1.54	.22			
		0.0517	0.0876	0.3119				2.38	.08			
		−0.0416	0.0511	−0.0026	0.1868			1.19	.34			
		0.0366	0.0722	0.1752	0.0084	0.1042		.88	.49			
Between species	7	1.1966					$=\mathbf{Q}_h$	10.92	<.0001	.015691	118.41	<.0001
		0.8732	1.2735					8.03	<.0001			
		1.5884	1.7873	3.0667				13.37	<.0001			
		1.7106	2.0525	3.1644	3.5920			13.10	<.0001			
		1.1885	2.1172	3.0431	3.5419	4.2375		20.31	<.0001			
Within species	28	0.4383					$=\mathbf{Q}_e$					
		0.0384	0.6341									
		0.0037	0.4344	0.9173								
		0.0937	0.1731	0.1046	1.0965							
		0.1853	−0.0382	−0.1439	0.3094	0.8347						
Total	40	24.0665					$=\mathbf{Q}_t$					
		38.3668	64.6398									
		38.8783	64.6735	66.4327								
		45.3349	75.2569	75.9804	89.9671							
		45.7393	76.3980	77.0504	90.4286	93.2473						

$|\mathbf{Q}_e| = 1.158460 \times 10^{-1}$; $|\mathbf{Q}_e + \mathbf{Q}_h| = 7.383006 \times 10^{0}$; $|\mathbf{Q}_e + \mathbf{Q}_b| = 3.432418 \times 10^{-1}$

Table 11.3. Nitrogen nutrition of eight grass species. Multivariate analysis of variance. Tests of the hypothesis of joint equality of species' mean vectors

(a) Likelihood ratio criterion, Λ, calculated as a determinantal ratio

$$\Lambda=\frac{|\mathbf{Q}_e|}{|\mathbf{Q}_e+\mathbf{Q}_h|}=\frac{.115846}{7.383006}=.015691$$

(b) Likelihood ratio criterion, Λ, calculated from the roots f_k of $|\mathbf{Q}_h-f\mathbf{Q}_e|=0$

k	f_k	$(1+f_k)^{-1}$
1	10.058	.090432
2	2.623	.276014
3	0.424	.702247
4	0.103	.906618
5	0.013	.987167

$$\Lambda=\prod_{k=1}^{s}(1+f_k)^{-1}=.090432\times.276014\times\ldots\times.987167=.015688$$

(c) Union-intersection criterion

$$\theta_s(s,m,n)=\frac{f_1}{1+f_1}=\frac{10.058}{1+10.058}=.910,$$

with distribution parameters

$s=\min(g-1,p)=5$, $m=\frac{1}{2}(|g-1-p|-1)=\frac{1}{2}$, $n=\frac{1}{2}(g(b-1)-b-p)=11.$

As $\chi_s^2(20)=29.33<\chi_\alpha^2(20)=31.41$ for $\alpha=.05$ we are led to conclude that variation among the blocks is insignificant. In other words, the blocking procedure used does not seem to have been effective.

The results of the union-intersection test of hypothesis (4.23) appear in Table 11.3c. As $\theta_s(5,\frac{1}{2},11)=.910>\theta_\alpha(5,\frac{1}{2},11)=.671$ for $\alpha=.01$ we reject the null hypothesis, as before. A comparable union-intersection test of the blocking effects employing the largest root f_1^* of $|\mathbf{Q}_b-f^*\mathbf{Q}_e|=0$ could also be made. It is of interest to note that univariate F tests of the contribution of the response variables x_i to the rejection of (4.23) are provided as a by-product of the multivariate likelihood ratio test of that hypothesis (Table 11.2). Judging by the univariate F-ratios, which are based on 7 and 28 degrees of freedom, all five variables contribute to the significant p-variate effect. Nevertheless, it is apparent that x_5 is the variable which best differentiates the species while x_2 is the variable in terms of which the species differ least. These univariate tests, however, are not independent and in rejecting a univariate hypothesis at the α nominal probability level, it is necessary to bear in mind that the probability of committing a Type I error may actually be greater than α.

Having shown that the species are almost surely differentiated in terms of their responses to the treatments administered, it is natural to enquire as to the direction and magnitude of the effects. In addition we may wish to further clarify which response variables or which linear combinations of them appear to be responsible for the differences. Multiple discriminant analysis is often informative for these purposes.

Table 11.4. Nitrogen nutrition of eight grass species. Discriminant analysis. Discriminant weights and coordinates of species' centroids on the discriminant functions

(a) Discriminant weights. The weights have been scaled so that the largest element in each vector $\mathbf{c}_k$ is unity.

Response, x_i	c_1	c_2	c_3	c_4	c_5
x_1: 1 ppm N	.317	1.000	.267	−.593	.729
x_2: 9 ppm N	.095	−.328	1.000	1.000	.400
x_3: 27 ppm N	.889	.206	−.801	.557	−.024
x_4: 81 ppm N	.349	.214	.184	−.267	−.999
x_5: 243 ppm N	1.000	−.621	−.146	−.529	.450

(b) Coordinates of species' centroids on the discriminant function, w_k

Species	w_1	w_2	w_3	w_4	w_5
L. perenne	1.435	−0.232	.468	−.511	.040
D. glomerata	1.175	−0.026	−.484	.197	−.133
P. bertolonii	0.666	−0.098	−.158	.442	.061
H. pubescens	−0.217	0.175	.466	.256	.095
B. media	−0.694	1.078	−.220	−.187	−.156
K. cristata	−0.799	0.261	.885	.096	−.046
F. ovina	−0.575	0.660	−.701	−.190	.172
F. rubra	−0.992	−1.819	−.257	−.103	−.034

Multiple discriminant analysis. The essential quantities for multiple discriminant analysis are the eigenvalues f_k and eigenvectors $\mathbf{c}_k$ of $\mathbf{Q}_e^{-1}\,\mathbf{Q}_h$. The discriminant functions w_k themselves may then be calculated as $w_k = \mathbf{c}_k^t\,\mathbf{x}$.

The results of the discriminant analysis are reported in Tables 11.3b and 11.4. Table 11.3b contains the eigenvalues f_k while the eigenvectors $\mathbf{c}_k$ and species' scores on the discriminant functions w_k are presented in Table 11.4a and b, respectively. We shall defer consideration of these results until Sect. 11.4.

In the multivariate analysis of variance the species constituted nominal categories which were arbitrarily arranged with respect to one another. It is frequently the case, however, that relationships among the groups in question can be postulated at the outset on the basis of substantive insight or judgement. In these circumstances a more powerful analysis is possible. By using suitably coded dummy variables or contrasts, the hypothesis sums of squares and products matrix $\mathbf{Q}_h$ of Table 11.2 can be partitioned into independent, identifiable components that exhaust all sources of variation. The procedure will be illustrated here using the results of the nutrition experiment before proceeding to the canonical variate analysis.

11.2.2 Designed comparisons

Provided relationships of interest are declared at the outset of an investigation, the analysis of Table 11.2 can be pursued further. In such cases it will generally be advantageous to select or construct contrasts which correspond to the specified relationships. In the nutrition experiment no declaration of affinities or relationships between the species was in fact made. For the sake of exposition, let us therefore assume that certain relationships had been recognized at the outset as meriting attention. Furthermore, we shall suppose that the relationships could be represented by the polynomial and Helmert contrasts introduced in Sect. 4.7.1, both of which have wide application. Orthogonal polynomials may be helpful where the species or groups are thought to be *equally spaced* with respect to some underlying metric, while Helmert contrasts are applicable where groups are believed to stand in an *ordered* relationship to one another.

Equally spaced responses. For simplicity, consider the four species *L. perenne, D. glomerata, P. bertolonii* and *H. pubescens*. Suppose grounds had existed for postulating that the overall response of the species to the treatments would be equally spaced. That is, that a linear trend existed, with *H. pubescens* being the least responsive and *L. perenne* the most responsive of the species. Even though it may not have been possible to have specified the trend with greater precision, it will be preferable to assume the spacing $-3, -1, 1, 3$ rather than ignore the supposed response structure entirely.

To test linearity formally, an orthogonal polynomial trend analysis with a specific test of the linear trend will be appropriate. A check on linearity can be made by including higher-degree polynomials; we shall assume that at most a cubic polynomial will suffice. Table 4.4a reports the orthogonal polynomials of order four. The entries of columns 2–4 contain the corresponding contrast coefficients $z_1, \ldots, z_4$ for hypothesis (4.47) in unnormalized form. The hypotheses in question are

H(1): linear trend,
H(2): quadratic trend,
H(3): cubic trend.

The matrix **U** of orthogonal estimates [Eq. (4.49)] is reported in Table 11.5a and the multivariate analysis of variance of the polynomial contrasts in Table 11.6. The successive matrices $\mathbf{Q}_h^{(i)} = \mathbf{u}_i \mathbf{u}_i^t$ corresponding to the linear, quadra-

Table 11.5. Nitrogen nutrition of eight grass species. Orthogonal estimate matrices, **U**, for multivariate analysis of variance of designed contrasts

(a) *Orthogonal polynomial contrasts.* Matrix $\mathbf{U} = n^{\frac{1}{2}} \mathbf{P}_4^t \bar{\mathbf{X}}$ of orthogonal estimates

$$\mathbf{U} = \begin{bmatrix} 3.805 & 6.322 & 6.614 & 7.644 & 7.815 \\ -.354 & -.380 & -.632 & -.935 & -1.129 \\ .028 & -.082 & -.443 & -.100 & -0.123 \\ .003 & -.058 & .038 & -.008 & -0.076 \end{bmatrix}$$

(b) *Helmert contrasts.* Matrix $\mathbf{U} = n^{\frac{1}{2}} \mathbf{H}_4^t \bar{\mathbf{X}}$ of orthogonal estimates

$$\mathbf{U} = \begin{bmatrix} 3.805 & 6.322 & 6.614 & 7.644 & 7.815 \\ -.257 & -.357 & -.736 & -.784 & -.965 \\ .216 & .151 & -.023 & .432 & .532 \\ .115 & .066 & .236 & .289 & .285 \end{bmatrix}$$

x_2: 9 ppm N, ..., x_5: 243 ppm N

Source of dispersion	df	SSP (symmetric) x_1	x_2	x_3	x_4	x_5		Univariate F	p	Multivariate Λ	χ^2	p
		14.4789										
		24.0561	39.9681									
Constant	1	25.1664	41.8129	43.7429			$=\mathbf{u}_0\mathbf{u}_0^t$					
		29.0876	48.3277	50.5584	58.4358							
		29.7376	49.4078	51.6883	59.7418	61.0769						
		.1253						8.01	.009	.290103	31.55	<.0001
		.1346	.1445					6.38	.017			
Linear	1	.2238	.2403	.3996			$=\mathbf{u}_1\mathbf{u}_1^t$	12.20	.002			
		.3310	.3555	.5911	.8745			22.33	<.0001			
		.3996	.4291	.7136	1.0557	1.2745			<.0001			
		.0008								.761666	6.94	.23
		−.0023	.0068									
Quadratic	1	−.0126	.0366	.1964			$=\mathbf{u}_2\mathbf{u}_2^t$					
		−.0028	.0082	.0442	.0100							
		−.0035	.0102	.0546	.0123	.0152						
		.0000								.977534	0.58	.99
		−.0002	.0033									
Cubic	1	.0001	−.0022	.0015			$=\mathbf{u}_3\mathbf{u}_3^t$					
		−.0000	.0004	−.0003	.0001							
		−.0002	.0044	−.0029	.0006	.0058						
		.4384										
		.0384	.6341									
Within species	28	.0037	.4344	.9173			$=\mathbf{Q}_e$					
		.0932	.1731	.1046	1.0965							
		.1853	−.0382	−.1439	.3094	.8347						
		15.0434										
		24.2266	40.7568									
Total	32	25.3814	42.5220	45.2577								
		29.5090	48.8649	51.2980	60.4169							
		30.3188	49.8133	52.3097	61.1198	63.2071						

$|\mathbf{Q}_e| = 1.158460 \times 10^{-1}$; $|\mathbf{Q}_e + \mathbf{Q}_h^{(1)}| = 3.993095 \times 10^{-1}$; $|\mathbf{Q}_e + \mathbf{Q}_h^{(2)}| = 1.520890 \times 10^{-1}$; $|\mathbf{Q}_e + \mathbf{Q}_h^{(3)}| = 1.185032 \times 10^{-1}$.

tic and cubic effects (Table 11.6) are constructed from the rows, $\mathbf{u}_i^t$, of $\mathbf{U}$. Had all eight species been used in the comparison, the sum of the partial SSP matrices, $\Sigma_i \mathbf{Q}_h^{(i)} = \Sigma\, \mathbf{u}_i \mathbf{u}_i^t (i=1, \ldots, 7)$, of Table 11.6 would be identically the matrix $\mathbf{Q}_h$ of Table 11.2, showing that the designed comparisons provide an additive decomposition of the explained sums of squares and products of the original analysis. The tests of significance in Table 11.6 are made with respect to the error matrix, $\mathbf{Q}_e$, of Table 11.2. The χ^2 values of Table 11.6 are obtained using expression (4.31) with $\nu_e = 28$ and $\nu_h = 1$. As the .05 level of $\chi^2(5)$ is 11.07, neither the cubic nor quadratic trend is significant. The linear trend, however, is significant ($\chi_s^2(5) = 31.55$, $p < .0001$). Glancing ahead to Fig. 11.1, it can be seen that the equal spacing of the four species in response to the treatments administered indicated by the test is consistent with the result of a separate analysis of the same data.

Ordered responses. Once again, for simplicity we consider just four species, *H. pubescens, L. perenne, D. glomerata* and *P. bertolonii.* Suppose there were grounds for regarding *H. pubescens* as a control, with reference to which the remaining species were known to be more responsive to the treatments administered. Further differentiation of the species may have been possible along the following lines. *Lolium perenne* might justifiably have been suspected of being the most responsive of all the species, and, in addition, is known to be a species of prime agricultural value. *Dactylis glomerata* and *P. bertolonii,* on the other hand, though responsive, have only limited agricultural value. In terms of the characteristics described, the following comparisons among species are meaningful (from the bottom upwards):

H(1): control against mean of the three responsive species;
H(2): prime agricultural species against mean of the two species of subordinate agricultural value;
H(3): subordinate agricultural species against subordinate agricultural species, i.e. *D. glomerata* versus *P. bertolonii.*

The comparisons specify a set of ordered relationships among species which is amenable to testing by Helmert contrasts. The comparisons are particular cases of hypothesis (4.47), the appropriate coefficients $z_1, \ldots, z_4$ for which appear in columns 2–4 of Table 4.4b in unnormalized form. Each species is compared with the mean of succeeding species or species-groups in a specified order, the null hypothesis at each stage being that of no difference between the species or groups compared. The matrix $\mathbf{U}$ of orthogonal estimates [Eq. (4.49)] and the multivariate analysis of variance derived from its rows are reported in Tables 11.5b and 11.7. There is no evidence of difference between the species of lesser agricultural value (Table 11.7, $\chi_s^2(5) = 5.16$, $p = .40$) or between the species of prime and lesser agricultural value ($\chi_s^2(5) = 9.42$, $p = .09$). The effect of the control versus the remaining species, on the other hand, is clear ($\chi_s^2(5) = 28.95$, $p < .0001$). The accompanying univariate F statistics show the species difference to increase fairly regularly from response at 1 ppm N to response at 243 ppm N. The results appear to support the existence of a difference between the control and the remaining species but are not consistent with the distinction proposed within the responsive species into species of greater and lesser agricultural value.

243 ppm N

Source of dispersion	df	SSP (symmetric) x_1	x_2	x_3	x_4	x_5		Univariate F	p	Multivariate Λ	χ^2	p
		14.4789										
		24.0561	39.9681									
H(0)	1	25.1664	41.8129	43.7429			$=\mathbf{u}_0\mathbf{u}_0^t$					
		29.0876	48.3277	50.5584	58.4358							
		29.7376	49.4078	51.6883	59.7418	61.0769						
		.0661						4.22	.05	.321364	28.95	<.0001
		.0918	.1274					5.55	.03			
H(1)	1	.1891	.2626	.5411			$=\mathbf{u}_1\mathbf{u}_1^t$	16.52	.0004			
		.2015	.2798	.5767	.6146			15.69	.0005			
		.2481	.3445	.7100	.7567	.9316		31.25	<.0001			
		.0469								.691230	9.42	.09
		.0328	.0229									
H(2)	1	−.0049	−.0034	.0005			$=\mathbf{u}_2\mathbf{u}_2^t$					
		.0936	.0654	−.0098	.1867							
		.1151	.0805	−.0120	.2297	.2828						
		.0132								.816652	5.16	.40
		.0075	.0043									
H(3)	1	.0271	.0155	.0558			$=\mathbf{u}_3\mathbf{u}_3^t$					
		.0331	.0189	.0682	.0833							
		.0327	.0186	.0673	.0822	.0811						
		.4383										
		.0384	.6341									
Within species	28	.0037	.4344	.9173			$=\mathbf{Q}_e$					
		.0932	.1731	.1046	1.0965							
		.1853	−.0382	−.1439	.3094	.8347						
		15.0434										
		24.2266	40.7568									
Total	32	25.3814	42.5220	45.2576								
		29.5090	48.8649	51.2981	60.4169							
		30.3188	49.8132	52.3097	61.1198	63.2071						

$|\mathbf{Q}_e| = 1.158460\times10^{-1}$; $|\mathbf{Q}_e+\mathbf{Q}_h^{(1)}| = 3.604665\times10^{-1}$; $|\mathbf{Q}_e+\mathbf{Q}_h^{(2)}| = 1.675866\times10^{-1}$; $|\mathbf{Q}_e+\mathbf{Q}_h^{(3)}| = 1.418486\times10^{-1}$.

The comparisons considered above are to some extent artificial and intended for exemplary purposes only. The principal justification for the analyses of this subsection is in enabling something of the opportunities provided by designed contrasts in multivariate analysis of variance to be sketched. The concepts involved generalize in a straightforward way to canonical variate analysis.

11.3 Canonical variate analysis

Ecological interest centers on the comparative response of the species to the nitrogen treatments administered. Canonical variate analysis could be expected to lead to a low-dimensional representation of the sample in which differences in response between species are maximized relative to variation within species. The procedure therefore seems appropriate for our purpose.

The variables on which analysis was based consisted of binary dummy variables to take account of the partition of the sample among the species, on the one hand, and of the measured responses of the species to the experimental treatments, on the other. The value of the cth dummy variable ($c=1, \ldots, g-1$) in the jth sample, z_{jc}, ($j=1, \ldots, N$) was defined to be

$$z_{jc}=\begin{cases} 1 & \text{if the } j\text{th sample belonged to species } c \\ 0 & \text{otherwise.} \end{cases}$$

Helictotrichon pubescens was arbitrarily chosen as the species not explicitly represented by a dummy variable. The response of this species could therefore be used as a datum against which the performance of other species might be assessed.

For the canonical analysis we therefore had $N=40$ samples comprising $g=8$ species which were randomly assigned to $b=5$ blocks, $p=7$ binary-valued dummy variables ($\mathbf{z}$) corresponding to the species and $q=5$ response variables ($\mathbf{x}$) corresponding to the experimental treatments. Nitrogen treatments were randomly assigned to five separate representatives (pots) of each species. The ecological objective was to clarify the comparative relationships among species with respect to their responses to the five experimental treatments considered simultaneously. Particular interest attached to the dimensionality of these relationships. Collinearity of the species' mean vectors, in particular, would imply not only that there were differences in response between species, but, furthermore, that the species could be considered to form an 'ecological series' in terms of their response to increasing concentrations of soil nitrogen. A second objective was to draw attention to connections between canonical analysis and one-way multivariate analysis of variance. To facilitate comparison, the canonical analysis was performed using sums of squares and cross products matrices rather than correlation matrices in order to put the two analyses on an equal footing. For the same reason it was also necessary to remove the block effects before embarking on the canonical analysis, the only function of the blocks

Table 11.8. Nitrogen nutrition of eight grass species. Canonical variate analysis. Canonical correlation coefficients and tests of hypotheses of equality and collinearity of species' centroids

(a) The canonical correlation coefficients r_k and union-intersection tests of equality and collinearity. Approximate critical values of Roy's largest-root criterion $\theta_\alpha(s, m, n)$ are shown for $\alpha = .05$ and $\alpha = .01$ ($m = \frac{1}{2}$; $n = 11$).

k	s	r_k	r_k^2	$\theta_{.05}(s, m, n)$	$\theta_{.01}(s, m, n)$	p
1	5	.954	.910	.605	.671	$<.01$
2	4	.851	.724	.554	.623	$<.01$
3	3	.546	.298	.482	.564	$>.05$
4	2	.305	.093	.395	.484	$>.05$
5	1	.113	.013	—	—	—

(b) Bartlett's approximate χ^2 test of the joint nullity of all s canonical correlation coefficients.

$$\Lambda = \prod_{k=1}^{s} (1 - r_k^2)$$

$$= (1 - .910)(1 - .724) \ldots (1 - .013) = 0.015610.$$

$$\chi^2 = -\{(N - b) - \tfrac{1}{2}(p + g)\}\log_e \Lambda$$

$$= -\{35 - \tfrac{1}{2}(5 + 8)\}(-4.159844) = 118.5556, \text{ with } p \cdot (g - 1) = 35 \text{ df.}$$

(c) Likelihood ratio tests of equality and collinearity. Bartlett's approximate chi-squared criterion for assessing the joint nullity of the smallest $s - k$ canonical correlation coefficients

k	Roots, r_k^2	Chi-squared	df	p
0	1, 2, 3, 4, 5	118.555	35	$<.0001$
1	2, 3, 4, 5	49.928	24	$<.005$
2	3, 4, 5	13.239	15	$>.50$
3	4, 5	3.155	8	$>.90$
4	5	0.373	3	$>.95$

being to eliminate extraneous variation. The block effects were removed by subtracting the corresponding SSP matrix, $\mathbf{Q}_b$, from $\mathbf{Q}_t$ prior to analysis (see Table 11.2). Thus the total SSP matrix used in the canonical analysis had been *adjusted* for blocks (cf. Bartlett, 1951, 1965).

The results of the analysis are summarized in Tables 11.8–11.10 and in Fig. 11.1. See also Table A-6 of Appendix A.2.

11.3.1 Results

The canonical correlation coefficients. The canonical correlation coefficients r_k ($k = 1, \ldots, 5$) are reported in Table 11.8a. It can be seen that $r_1 = .95$ and $r_2 = .85$ are substantial, the remaining correlations declining in a regular manner until $r_5 = .11$. Certainly, the magnitude of r_1 is such as to suggest the existence of differences between species in relation to the treatments administered. The total between-species variance is found from the r_k^2 and expression (4.10) to be $\Sigma f_k/(g - 1) = 13.275/7 = 1.896$, which is identically the sum of the diagonal elements of the SSP matrix $\mathbf{Q}_h$ of Table 11.2 after division by their degrees of freedom. Of the total between-groups variance, $f_1 = r_1^2(1-r_1^2)^{-1}$ absorbs 76% and $f_2 = r_2^2(1-r_2^2)^{-1}$ 20%. Thus we find that the first root r_1^2 is responsible for 3.85 times as much of this quantity as r_2^2, while together r_1^2 and r_2^2 are responsible for 96% of total variance between species.

Equality of species' centroids. The results of union-intersection and likelihood ratio tests of hypothesis (4.21) are contained in Table 11.8. For the union-intersection test (Table 11.8a) we find $\theta_s(5, \frac{1}{2}, 11) = .910 > \theta_\alpha(5, \frac{1}{2}, 11) = .671$ for $\alpha = .01$. Similarly, for the likelihood ratio test (Table 11.8b) we have $\chi_s^2(35) = 118.56 > \chi_\alpha^2(35) = 74.926$ for $\alpha = .0001$. Hypothesis (4.21) therefore is not supported by the data. It appears that there are differences between the species and that the differences warrant interpretation.

Collinearity. The results of sequential tests of hypothesis (4.22) are reported in Table 11.8. From Table 11.8a it appears that only r_1 and r_2 are statistically significant ($p < .01$), while Bartlett's test of collinearity (Table 11.8c) indicates that the three smallest roots r_k^2 $(k > 2)$ reflect only random variation. Both tests show the data to be consistent with the hypothesis $H_0(2)$: $\rho_1 \neq 0$, $\rho_2 \neq 0$, $\rho_3 = \ldots = \rho_5 = 0$. Accordingly, the dimensionality of the relationships among species may be taken as two. Further, we can anticipate that the canonical variates corresponding to r_1^2 and r_2^2 will be required in order to establish where the differences between species lie. Clearly, the hypothesis that all eight species from an 'ecological series' is untenable.

The canonical variates. Correlation coefficients between the canonical variates u_k and $v_k (k = 1, \ldots, 3)$ and the original variables are reported in Table 11.9. The intraset correlations of the first canonical variate v_1 of the response domain (Table 11.9, lower left) are noteworthy on three accounts: the correlations indicate that the response variables all contribute to v_1 in the *same direction* with respect to the linear combination of species represented by u_1, secondly that each variable makes a *sizeable contribution* ($>.67$) to the relationship, and, moreover, one which *increases systematically in strength* from that associated with x_1 (.67) to that associated with x_5 (.91). Evidently, v_1 seems to represent both species' response to N in any concentration together with a progressively increasing response to increasing dosage. Similarly, it can be seen that v_2 represents essentially a contrast between responses to the highest ($-.32$) and the lowest (.65) N treatments. Of the two canonical variates, v_1 is six times stronger than v_2, absorbing 71% of the total variance of the response domain, while v_2 accounts for less than 12%. From the interset correlations of v_1 (Table 11.9, upper-right) it can be seen that species differ appreciably with respect to the linear combination of responses represented by v_1. The correlations of three species are in the same direction as those of the response variables on which v_1 is defined, while those of the four remaining species are in the opposite direction. In particular, we find that *L. perenne* (.58) and *F. rubra* ($-.40$), for example, are affiliated in opposite senses with v_1 while *H. pubescens* is scarcely related to v_1 at all. Turning to the interset correlations of v_2, it is apparent that this canonical variate is characterized principally by the differenti-

Table 11.9. Nitrogen nutrition of eight grass species. Canonical variate analysis: correlations between the original variables and canonical variates

Canonical variate	u_1	u_2	u_3	h_w^2	v_1	v_2	v_3	h_b^2
Species								
L. perenne	.608	−.110	.347	.502	.580	−.094	.189	.381
D. glomerata	.498	−.013	−.358	.376	.475	−.011	−.195	.264
P. bertolonii	.282	−.047	−.117	.095	.269	−.040	−.064	.078
H. pubescens	−.092	.084	.344	.134	−.086	.071	.188	.048
B. media	−.294	.512	−.163	.375	−.280	.436	−.089	.276
K. cristata	−.339	.124	.656	.561	−.323	.106	.358	.244
F. ovina	−.244	.313	−.519	.427	−.233	.266	−.283	.205
F. rubra	−.420	−.863	−.190	.957	−.401	−.734	−.104	.710
Variance extracted	.143	.143	.143	.429	.130	.103	.043	.276
Redundancy	.130	.104	.043	.277	.130	.103	.043	.276
Canonical variate	v_1	v_2	v_3	h_w^2	u_1	u_2	u_3	h_b^2
Treatment response								
x_1: 1 ppm N	.671	.654	.169	.907	.604	.556	.091	.727
x_1: 9 ppm N	.818	−.033	.366	.804	.780	−.028	.200	.649
x_3: 27 ppm N	.887	.207	−.197	.868	.846	.176	−.108	.758
x_4: 81 ppm N	.904	.118	.190	.867	.862	.100	.104	.764
x_5: 243 ppm N	.912	−.318	.041	.935	.870	−.271	.022	.831
Variance extracted	.711	.117	.048	.876	.647	.085	.014	.746
Redundancy	.646	.085	.014	.746	.647	.085	.014	.746

ation of *B. media* (.44) from *F. rubra* (−.73). For the most part, the remaining species are uncorrelated with v_2.

Redundancy. The explanatory power of v_1 and v_2 across domains is expressed by the redundancy indices $V^2_{\mathbf{z}|v_1}=.130$ and $V^2_{\mathbf{z}|v_2}=.103$ (Table 11.9, upper-right). Together, v_1 and v_2 account for 23.3% of the total variance of the species' domain. Notice that while v_1 is a well-defined component of $\mathbf{x}(V_1^2=.711)$, its explanatory power in the species' domain is less than one-fifth of this value ($V^2_{\mathbf{z}|v_1}=.130$). The communalities of species with the v_k ($k=1, \ldots, 3$) tend to be small; with the exception of *F. rubra* (.71), the communalities of the remaining species are all $\leq .38$.

Figure 11.1 displays the sample of $N=40$ experimental units projected into the subspace defined by the canonical variates v_1 and v_2 of the response variables. The origin in the figure has been placed at the null contrast, that is at the contrast of *H. pubescens* with itself, and the canonical variates drawn at right angles even though they are not orthogonal in response space. Observe also that v_1 and v_2 have been standardized so that the average within-groups variance on each variate is unity, while the circles, of radius $(\chi^2(2)/n_c)^{\frac{1}{2}}$, depict the 95% confidence regions about the species' centroids. The figure is characterized principally by two features:

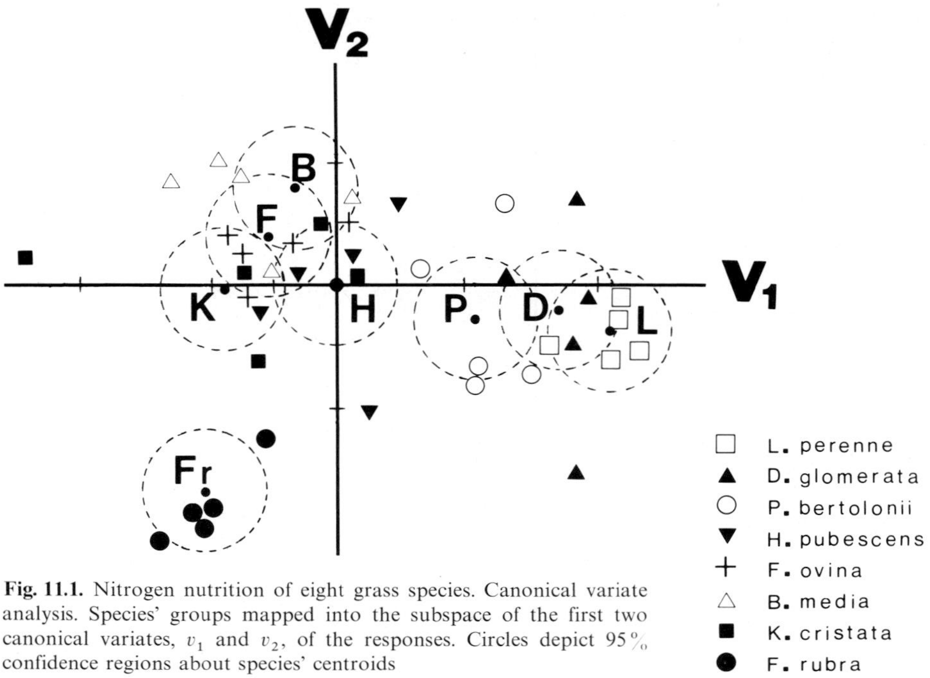

Fig. 11.1. Nitrogen nutrition of eight grass species. Canonical variate analysis. Species' groups mapped into the subspace of the first two canonical variates, v_1 and v_2, of the responses. Circles depict 95% confidence regions about species' centroids

(a) a sharp distinction between species in relation to v_1, *L. perenne, D. glomerata* and *P. bertolonii* having sizeable positive scores while the four remaining species have either zero or negative scores;
(b) a contrast among those species with zero or negative scores on v_1 in terms of their scores on v_2.

Recall that v_1 was thought likely to correspond to an expression of overall responsiveness and v_2 to a contrast in response at the extreme treatment levels. It seems, therefore, that species are clearly differentiated in terms of overall performance, *L. perenne* being the most responsive and *F. rubra* the least. With respect to v_2, it appears that *B. media* and *F. rubra* are the species most susceptible of contrasts in performance at the extreme N dosages administered, though in opposite senses. Bearing in mind the direction of v_2 (Table 11.9, lower-left), it follows that the performance of *B. media* tends to be highest at low nitrogen levels and lowest at high levels, while the converse is true of *F. rubra*. Altogether, *F. rubra* emerges as a somewhat singular species.

From Fig. 11.1 it can also be seen that there are appreciable differences between species with respect to the extent of their within-group variation, that is in their variability over replicates. *Lolium perenne* is the most homogeneous of the species and *K. cristata* and *D. glomerata* among the least.

We shall return to consider the ecological implications of these results after having compared them with those of the multivariate analysis of variance. Some further results of the canonical variate analysis which will be useful for the comparison appear in Table 11.10.

11.4 Relationships between multivariate analysis of variance, discriminant analysis and canonical variate analysis

We saw in Sect. 4.6.1 that multivariate analysis of variance, discriminant analysis and canonical variate analysis are closely related. We shall now use the results of the nutrition experiment to verify the connections described in that section.

Consider the overall hypotheses (4.21) and (4.23) of *canonical variate analysis* and *multivariate analysis of variance*, respectively. Recall that hypothesis (4.21) is itself equivalent to the hypothesis of the nullity of the $q \times p$ matrix of regression coefficients $\mathbf{B}$ in canonical correlation analysis [hypothesis (3.10)]. Table 11.3 reports the quantities required for testing hypothesis (4.23) and Table 11.8 those necessary for testing (4.21). On comparing Table 11.3a *or* b with Table 11.8b it is readily verified that, within rounding error, the two analyses lead to identical values of the likelihood ratio criterion, Λ. Reference to Table 11.3c and to the first line of the body of Table 11.8a similarly confirms the identity of the corresponding union-intersection test statistics, θ_s. Tables 11.3a and b show also that in the multivariate analysis of variance Λ can be calculated equally as a ratio of determinants or as a function of the eigenvalues f_k of $\mathbf{Q}_e^{-1}\mathbf{Q}_h$.

The interpretive indices commonly associated with *discriminant analysis* are the eigenvalues f_k and the eigenvectors $\mathbf{c}_k$ of $\mathbf{Q}_e^{-1}\mathbf{Q}_h$ together with the discrimi-

Table 11.10. Nitrogen nutrition of eight grass species. Canonical variate analysis. Canonical weights for the response variables and scores of species' centroids on the canonical variates v_k of the response variables

(a) Canonical weights. The weights have been scaled so that the largest element in each vector $\mathbf{a}_k$ is unity.

Response, x_i	a_1	a_2	a_3	a_4	a_5
x_1: 1 ppm N	.316	1.000	.267	.592	−.726
x_2: 9 ppm N	.091	−.327	1.000	−1.000	−.401
x_3: 27 ppm N	.890	.207	−.802	−.567	.025
x_4: 81 ppm N	.347	.215	.185	.267	1.000
x_5: 243 ppm N	1.000	−.621	−.147	.529	−.450

(b) Coordinates of species' centroids on the canonical variates, v_k.

Species	v_1	v_2	v_3	v_4	v_5
L. perenne	1.503	−.271	.482	.570	−.034
D. glomerata	1.227	−.048	−.497	−.174	.145
P. bertolonii	0.693	−.113	−.162	−.440	−.060
H. pubescens	−.227	.188	.482	−.271	−.096
B. media	−.729	1.134	−.223	.169	.169
K. cristata	−.836	.288	.914	−.121	.050
F. ovina	−.604	.696	−.722	.177	−.180
F. rubra	−1.029	−1.874	−.274	.090	.006

nant functions $w_k = \mathbf{c}_k^t \mathbf{x}$. For the nutrition experiment, these quantities are reported in Tables 11.3b and 11.4a and b, respectively. The corresponding indices in *canonical variate analysis* are the eigenvalues r_k^2 and eigenvectors $\mathbf{a}_k$ of $\mathbf{Q}_t^{-1} \mathbf{Q}_h$ together with the canonical variates $v_k = \mathbf{a}_k^t \mathbf{x}$. The r_k^2 are reported in Table 11.8a and the $\mathbf{a}_k$ and v_k in Table 11.10a and b, respectively. It can be seen (Tables 11.3b and 11.8a) that the f_k and r_k^2 are not identical. They are, however, connected by the simple relationships (4.37), as may readily be verified from the tables. For example, letting $k=1$ we have

$$f_1 = 10.058 = \frac{.910}{1-.910}, \qquad r_1^2 = .910 = \frac{10.058}{1+10.058}.$$

The remaining f_k and r_k^2 $(k>1)$ are similarly related.

The elements of the weights vectors $\mathbf{c}_k$ and $\mathbf{a}_k$ of discriminant analysis and canonical variate analysis are proportional. To simplify comparison the elements in each case have been scaled so that the largest absolute component in each vector is unity (Tables 11.4a and 11.10a). It can be seen that after scaling corresponding elements are identical, at least within rounding error and a possible reflection in sign. Tables 11.4b and 11.10b contain the scores of species' centroids on the discriminant functions w_k and canonical variates v_k, respectively. It is easily verified that corresponding scores differ only by a scale factor together with, in some cases, a reflection in sign. For practical purposes, therefore, the scores may be treated as equivalent.

Multivariate analysis of variance, discriminant analysis and canonical variate analysis are each generally formulated in rather different terms from one another. They differ also in the emphasis placed on particular aspects of the results. The relationships verified nevertheless show the methods to be closely related algebraically and statistically.

Having shown the results of the multivariate analysis of variance and discriminant analysis of Sect. 11.2 together to be numerically equivalent to those of the canonical variate analysis of Sect. 11.3, we shall for the remainder of the chapter confine our attention to the results of the canonical variate analysis.

11.5 Ecological assessment of the results

We shall first consider the effects of the variable/sample ratio and of possible heterogeneity of dispersion on the results obtained. The variable/sample ratio is $^{12}/_{40} = .30$. This is a comparatively large value and can be expected to have had some inflationary effect on the larger canonical correlation coefficients as well as on the interpretive indices derived from them. Accordingly, it would be well to accept the apparent strength of several of the quantities reported with some reserve. A test of the equality of the species' dispersions (Morrison, 1976, p. 252) prior to analysis showed them to be nonhomogeneous ($p < .0001$), notwithstanding the initial $\log_{10}$ transformation of the data. The equality of the species sample sizes nevertheless provided some basis for accepting the pooled within-species' dispersion matrix as the best available estimate of the corresponding population quantity (see Bock, 1975, p. 236). At the same time it was also necessary to accept that inferential features of the results would have to be treated circumspectly. In order to throw further light on the extent of the nonhomogeneity of dispersion, tests of the homogeneity of species' variance on v_1 and v_2, respectively, were performed following the canonical analysis. With respect to v_1, Bartlett's (1937) test gave $\chi_s^2(7)=12.47<\chi_\alpha(7)=14.07$ for $\alpha=.05$, while for v_2 the corresponding result was $\chi_s^2(7)=11.69<\chi_\alpha(7)=12.02$ for $\alpha=.10$. There can be little doubt, therefore, that the species' variances on each canonical variate, at least, are homogeneous. To this extent, therefore, it seems that the effects of nonhomogeneity of dispersion may not have been so severe as to completely undermine the analysis.

We are now in a position to take up the ecological implications of the results described in Sect. 11.3.1.

The differences between species with respect to the linear combinations of responses represented by v_1 and v_2 which are evident in Fig. 11.1 may be confirmed by reference back to the original data (Table 11.1). What the analysis, and, in particular, the figure, appear to have accomplished, therefore, is to have identified the salient features of the data and to have communicated these in an efficient, readily comprehended manner. The differentiation of species associated with the change of sign along v_1 is broadly consistent with existing views on the comparative ecology of the species in question. *Lolium perenne, D. glomerata* and *P. bertolonii* are well-known constituents of fertile grassland while the remaining species are known to perform best on impoverished soils (Tansley, 1939; Hubbard, 1954). In addition, the ranking of the responsive species effected by v_1 is supported by field knowledge of the behavior of these species (Gittins, 1965). Such independent substantiation of these features of the results shows canonical analysis to be capable of identifying and communicating those relationships of ecological significance about whose existence we may be reasonably sure, even though their precise specification may previously have been lacking. The point of interest concerning the subsidiary distinction among the less responsive species is the indication that, of the species studied, those of nitrogen depleted soils may be more diverse in their response to N than are species of comparatively nitrogen rich soils. This is an aspect of the comparative ecology of the species which to my knowledge has not previously been commented on. It is a minor point but one which could conceivably repay attention.

The extent to which v_1 and v_2 account for the total variance among species is modest ($V^2_{\mathbf{z}\,|\,v_1,v_2}=.233$). Thus, while the results are apparently open to meaningful ecological interpretation, and, moreover, are for the most part consistent with the results of other work, the fit of the rank 2 model is manifestly weak. It seems that species' performance is profoundly influenced by sources other than N. It may be that the plant material used in the experiment is to some extent responsible for the lack of fit. That the species' replicates are in most cases rather variable was noted earlier. The most homogeneous species in this regard is *L. perenne,* which alone was grown from the *seed* of a selected agricultural strain (s. 23). The remaining species were all grown vegetatively from *tillers* collected in the field, a conscious effort having been made to acquire material as genetically diverse as possible within the confines of the community of interest. It seems possible, therefore, that the lack of genetically uniform material could in part be responsible for the lack of fit of the model.

11.6 Conclusions

The purpose of the study was to clarify the comparative ecology of the species examined and to verify the existence of connections between canonical variate analysis and other multivariate methods.

The salient points of ecological interest to have emerged may be summarized as follows:

(a) Differences exist between species with respect to their overall responsiveness to the N treatments administered. *Lolium perenne, D. glomerata* and *P. bertolonii* are characterized by their responsiveness at *each* treatment level and by their *systematic increase* in performance with increasing dosage. *Briza media, F. ovina, F. rubra, H. pubescens* and *K. cristata,* on the other hand, are comparatively unresponsive at *all* treatment levels.

(b) A subsidiary distinction exists among the less-responsive species in relation to their performance at the extreme treatment levels. *Briza media* in particular performs comparatively poorly at the highest nitrogen dosage administered and comparatively well at the lowest, while the converse is true of *F. rubra.*

The first result is consistent with generally held views concerning the comparative ecology of the species. The results of the present analysis do, however, differ from those of previous work in specifying the relationships of interest with a precision that has not hitherto been achieved and in the ease with which they enable these relationships to be communicated. The second result suggests that, of the species investigated, those of nitrogen impoverished soils may be more diverse in their response to variation in available nitrogen than are species of nitrogen rich soils. This point may merit further study.

The results were also used to illustrate two points of theoretical interest. The results allowed us to verify that one-way multivariate analysis of variance and discriminant analysis together are formally equivalent to canonical variate analysis. Furthermore, it was shown that where relationships among the species or other entities of interest can be postulated prior to analysis, opportunities for even more incisive analysis exist. The significance of these demonstrations lies in enabling us to better perceive the domain and hence the potential role of canonical analysis in ecology.

12. Herbivore-environment relationships in the Rwenzori National Park, Uganda

12.1 Introduction

The Rwenzori National Park lies astride the equator in the Western Rift Valley of Uganda. The Park contains a variety of plant communities, including forests, swamps and several kinds of grassland. The grasslands are noteworthy in that they support the highest recorded large mammal biomass of any natural area of the world (Bourlière, 1965; Coe, Cumming & Phillipson, 1976). An estimate of 294.9 kg ha^{-1} for the average year-round standing-crop herbivore biomass is given by Field and Laws (1970). Interrelationships between the herbivore species, in particular, and their environment – vegetation, climate, soils and other animal species – have been studied by Field and Laws (1970). Herbivore distribution was found to show considerable heterogeneity in relation to the occurrence of different plant communities, the presence of standing water and the incidence of fire. Table 2 of Field and Laws' (1970) paper gives estimates of the mean density km^{-2} of nine herbivore species in ten sites or study areas. These data provided the starting point for the present investigation. Proceeding along different lines from Field and Laws, the data were analyzed in an attempt to clarify relationships within and between herbivores and study areas. Two methods were employed for the purpose – dual scaling and the contingency table analysis of E.J. Williams (1952). At a more fundamental level, the aims of the study were to illustrate the use of canonical analysis in an ecological context involving two categorial variables and to verify connections between dual scaling and contingency table analysis described in Sect. 5.5.1.

The herbivore species in question were elephant (*Loxodonta africana* Blumenbach), hippopotamus (*Hippopotamus amphibius* L.), warthog (*Phacochoerus aethiopicus* Pallus), buffalo (*Syncerus caffer* Sparrman), Uganda kob (*Adenota kob* Neumann), waterbuck (*Kobus defassa* Rüppell), reedbuck (*Redunca redunca* Pallas), bushbuck (*Tragelaphus scriptus* Pallas) and topi (*Damaliscus korrigum* Ogilby). Of these species, hippopotamus had the singular distinction of being subject to a management cropping program designed to reduce its numbers in certain areas. The sites or study areas examined were selected principally for their accessibility and representativeness of the varied grassland communities and animal associations of the Park. A summary of certain physical characteristics of the sites is given in Table 12.1. On the basis of these and other characteristics, the sites were grouped by Field and Laws into four broad categories, namely: area 2; areas 7, 9 and 10; areas 1, 3, 5 and 6; and areas 4 and 8.

Table 12.1. Rwenzori National Park, Uganda. Physical characteristics of study areas

Study area	Vegetation	Proximity to standing water	Management (cropping)
1	Short grass/thicket	Close	Moderate
2	Tall grass with thicket	Distant	None
3	Short grass/thicket	Close	None
4	Short grass/thicket	Close	Substantial
5	Short grass/thicket	Moderate	None
6	Short grass/thicket	Close	None
7	Short grass	Distant	None
8	Short grass/thicket	Moderate	Substantial
9	Short grass	Distant	None
10	Short grass	Distant	None

Before embarking on analysis, Field and Laws' data were adjusted to bring them to a more amenable form. Adjustment involved (a) rounding estimates to their nearest integral value; and (b) eliminating the entries for two species (bushbuck, reedbuck) and one site (site 2). The items eliminated were unusual in either occurring in very small numbers or in supporting a very low overall herbivore density. It was anticipated that such items might have a disproportionate effect on the outcome of the analyses contemplated (see Anderberg, 1973, p. 221; Teil, 1975). The resulting frequency table is shown here as Table 12.2. The analyses which follow are based exclusively on the $r \times c$ table for which $r=7$ and $c=9$. The omitted items were later incorporated into the analytical results obtained, so that the information conveyed by them was subsequently retrieved.

Table 12.2. Rwenzori National Park, Uganda. Density km^{-2} of nine large herbivores in ten study areas

Herbivore	Study area										Total
	1	2[a]	3	4	5	6	7	8	9	10	
Elephant	1	2	3	3	1	4	2	2	0	0	16
Warthog	1	0	3	8	1	3	1	1	4	2	24
Hippopotamus	15	0	28	1	13	21	1	4	2	6	91
Reedbuck[a]	0	0	0	0	0	0	0	0	1	0	–
Waterbuck	7	0	3	10	1	4	2	1	0	0	28
Kob	9	1	2	0	6	0	44	10	78	71	220
Topi	0	0	0	0	0	0	0	0	30	83	113
Bushbuck[a]	0	0	1	3	0	1	0	0	0	0	–
Buffalo	12	5	7	25	21	18	13	18	22	17	153
Total	45	–	46	47	43	50	63	36	136	179	645

[a] Omitted from both χ^2 analysis and canonical analysis; not entered into marginal or grand totals.
Source: Field and Laws (1970), Table 2 with modifications.

Interest centers on detecting any relationship between herbivores and study areas, and on characterizing the relationship, if such indeed exists. The contingency table analysis of Williams (1952) is identified as offering one approach to the declared ecological goal. Williams' analysis enables one to first assess the evidence for association between herbivores and study areas. If the balance of the evidence points towards the existence of association, one may then partition the overall expression of association for the table into independent components. Numerical scores can also be derived for rows and colums which may be helpful in interpreting the individual components ecologically. Canonical analysis offers an alternative approach to the same ecological goal. For this purpose, binary dummy variables corresponding to the row and colum categories of Table 12.2 are constructed as a first step. Linear composites of the dummy variables which differentiate between herbivores and between study areas to the maximum possible extent and which simultaneously maximize the association between the two, are then sought as described in Sect. 5.2. The results of analysis are often adequately summarized by the derived composites corresponding to the first two or three canonical correlation coefficients. In these cases the canonical variates enable the relationships of ecological interest to be displayed graphically. Such an analysis corresponds to one of several possible formulations of dual scaling.

Formal tests of hypotheses are scarcely possible with the data of Table 12.2 owing to the lack of statistical independence among samples and the concentration of much of the sample in a few cells with relatively large cell counts. Yet Fienberg's (1979) suggestions regarding the adequacy of asymptotic chi-squared approximations that the expected cell size be 1 or more and that for large tables the total sample size be 4 or 5 times the number of cells are both substantially met. For illustrative purposes the results of several hypothesis tests are in fact reported below. The significance levels given in connection with these tests are strictly nominal and in assessing results reliance has been placed on other less sensitive indices.

The contingency table analysis is taken up on Sect. 12.2, followed by an account of dual scaling in Sect 12.3. In Sect. 12.4 a comparison of the methods based on the results obtained is made. Ecological implications of the results are considered in Sect. 12.5.

12.2 Contingency table analysis

Table 12.3 shows the herbivore densities of Table 12.2 expressed as percentages by rows. In the absence of association the entries of each column would be equal apart from sampling effects. There is evident heterogeneity among columns. The analysis described in Sect. 5.5.1 provides a means for systematically investigating effects of this kind. Results of such and analysis for the data of Table 12.2 are summarized in Tables 12.4–12.7.

Table 12.3. Rwenzori National Park, Uganda. Densities of Table 12.2 expressed as percentages by rows. Items signified by footnote in Table 12.2 omitted

Herbivore	Study area									Total
	1	3	4	5	6	7	8	9	10	
Elephant	6.25	18.75	18.75	6.25	25.00	12.50	12.50	0.00	0.00	100
Warthog	4.17	12.50	33.33	4.17	12.50	4.17	4.17	16.67	8.33	100
Hippopotamus	16.48	30.77	1.10	14.29	23.08	1.10	4.40	2.20	6.59	100
Waterbuck	25.00	10.71	35.71	3.57	14.29	7.14	3.57	0.00	0.00	100
Kob	4.09	0.91	0.00	2.73	0.00	20.00	4.55	35.45	32.27	100
Topi	0.00	0.00	0.00	0.00	0.00	0.00	0.00	26.55	73.45	100
Buffalo	7.84	4.51	16.34	13.73	11.76	8.50	11.76	14.38	11.11	100

12.2.1 Results

Independence. The trace of the matrix **T** of departures from expectation is reported in Table 12.4. It can be seen that tr(**T**)=1.8977. To discover whether the magnitude of tr(**T**) is such as to represent a 'sizeable' departure from its expectation of unity under a null hypothesis of independence, we first calculate χ_o^2 as in Eq. (5.47):

$$\chi_o^2 = N[\mathrm{tr}(\mathbf{T}) - 1] = 645[1.8977 - 1] = 579.02.$$

Under independence χ_o^2 is distributed as a chi-squared variate on $(r-1)(c-1)$ degrees of freedom. A formal test of hypothesis (5.42) based on this quantity is inappropriate here for the reasons cited in Sect. 12.1. For the sake of completeness the test is nevertheless not without interest. Referring χ_o^2 to a table of critical values of the chi-squared distribution we find that $\chi_o^2(48) = 579.02 > \chi_\alpha^2(48) = 93.22$ for $\alpha = .0001$. The sample value so far exceeds the critical value cited as to render the hypothesis of independence scarcely tenable. In an exploratory or descriptive spirit more in keeping with the realities of the data, we shall simply tacitly accept that the departure of tr(**T**) from unity is not inconsistent with the association of herbivores and study areas. Some justification for the adoption of this position will be forthcoming when scores for rows and colums are considered shortly. Interest now turns to the

Table 12.4. Rwenzori National Park, Uganda. Analysis of association between seven herbivore species and nine study areas. $\mathbf{T} = \mathbf{GG}'$ matrix, where the elements of $\mathbf{G} = [g_{ij}]$ are given by $g_{ij} = x_{ij}/(r_i c_j)^{\frac{1}{2}}$

$$\mathbf{T} = \mathbf{GG}' = \begin{bmatrix} .0579 & .0051 & .1169 & .0686 & .0408 & .0000 & .1144 \\ .0551 & .0819 & .0887 & .0917 & .0624 & .0347 & .1335 \\ .1169 & .0887 & .3909 & .1287 & .0804 & .0318 & .2186 \\ .0686 & .0917 & .1287 & .1374 & .0426 & .0000 & .1602 \\ .0408 & .0624 & .0804 & .0426 & .4961 & .3179 & .2130 \\ .0000 & .0347 & .0318 & .0000 & .3179 & .3992 & .0968 \\ .1144 & .1335 & .2186 & .1602 & .2130 & .0968 & .3343 \end{bmatrix}$$

$$\mathrm{tr}(\mathbf{T}) = \sum t_{ii} = 1.8977$$

Table 12.5. Rwenzori National Park, Uganda. Analysis of association between seven herbivore species and nine study areas. Latent roots r_k^2 of $T = GG^t$ and the partition of χ_o^2 into additive components by the relation $\chi_o^2 = N[r_1^2 + r_2^2 + \ldots + r_s^2]$

k	r_k^2	Nr_k^2	$\% \chi_o^2$	$C_t \%$
0	.999942	–	–	–
1	.563173	363.25	62.79	62.79
2	.169158	109.11	18.86	81.65
3	.109106	70.37	12.16	93.81
4	.037283	24.05	4.16	97.97
5	.013226	8.53	1.47	99.44
6	.005011	3.23	0.56	100.00
Total	1.896899	578.54	100.00	–

dimensionality of the presumed association and to its ecological implications, if any. For these purposes the latent roots and latent vectors of **T** will be informative.

Dimensionality. The latent roots of r_k^2 of **T** are reported in Table 12.5. The first root, r_0^2, is for all practical purposes unity, as expected. Recall that this root has no bearing on the presence of association; r_0^2 expresses the contribution to tr(**T**) of the expected frequencies and could have been eliminated by working with departures from expectation rather than the observed frequencies themselves. We shall henceforth ignore this root. Of the remaining roots, r_1^2 is clearly dominant, absorbing 63% of the total association represented by χ_o^2, while r_2^2 and r_3^2 account for some 19% and 12%, respectively. The remaining roots are, by comparison, negligible ($\leq .04$). Together, r_1^2, r_2^2 and r_3^2 account for 94% of the overall association, χ_o^2, or, equivalently of tr(**T**) after removing r_0^2. These points lead us to suppose that for practical purposes, at least, the dimensionality of the relationship might well be considered to be in the vicinity of three. It is easy to verify from Table 12.5 that the terms Nr_k^2 ($k = 1, \ldots, s$) do provide an additive decomposition of χ_o^2, as Eq. (5.51) asserts.

Scores for attribute-states. The normalized latent vectors $\mathbf{a}_k$ and $\mathbf{b}_k$ ($k = 1, \ldots, 3$) of **T** are reported in Table 12.6 and the sought vectors of row and column scores $\mathbf{a}_k^*$ and $\mathbf{b}_k^*$ in Table 12.7. The score vectors here are in fact just the latent vectors after centering and scaling in accordance with the scheme specified by Eqs. (5.8) and (5.9) of Sect. 5.2.1. Accordingly, the weighted mean of the scores comprising $\mathbf{a}_k^*$ and $\mathbf{b}_k^*$ is zero while the scaling of the kth pair of vectors reflects the magnitude of the kth root, r_k^2. The sign and magnitude of the scores specify affinities or contrasts between row or column attribute-states. In this way the scores may aid interpretation of the kth component, Nr_k^2 ($k > 0$), of the global association, χ_o^2. In the absence of precise significance tests, in reaching a decision as to the number of ecologically informative components we have been guided by two considerations: the substantive meaning, if any, which could be placed on the score vectors $\mathbf{a}_k^*$ and $\mathbf{b}_k^*$ ($k = 1, \ldots, s$), and the magnitude of the goodness-of-fit criterion $C_t = \Sigma_1^t r_k^2/[\mathrm{tr}(\mathbf{GG}^t) - 1]$ of Table 12.5 for increas-

Table 12.6. Rwenzori National Park, Uganda. Analysis of association between seven herbivore species and nine study areas. Normalized latent vectors $\mathbf{a}_k$ and $\mathbf{b}_k$ corresponding to latent roots r_1^2, r_2^2 and r_3^2

a_1	a_2	a_3	b_1	b_2	b_3
−.2144	.0458	.0075	−.2675	−.0560	.0555
−.1680	.2558	.3094	−.4135	−.5271	.1464
−.5319	−.7020	−.2360	−.3224	.5966	−.5900
−.3001	.2738	.3322	−.2436	−.0223	.1151
.4334	.2120	−.5977	−.4047	−.1901	−.0378
.5259	−.4042	.6973	.1112	.3560	.5683
−.2973	.4132	.1098	−.1247	.2312	.1073
			.3416	.1556	.2676
			.5348	−.3494	−.4547

Table 12.7. Rwenzori National Park, Uganda. Analysis of association between seven herbivore species and nine study areas. Standardized scores for row $\mathbf{a}_k^*$ and column $\mathbf{b}_k^*$ attribute-states for the fitted model of rank $r=3$

Row attribute-state	a_1^*	a_2^*	a_3^*	b_1^*	b_2^*	b_3^*	Column attribute-state
Elephant	−1.0216	.1194	.0159	−0.7599	−.0871	.0693	1
Warthog	−0.6537	.4816	.5298	−1.1617	−.8117	.1812	3
Hippopotamus	−1.0628	−.7686	−.2077	−0.8964	1.0811	−.8586	4
Waterbuck	−1.0806	.5403	.5264	−0.7080	−.0356	.1472	5
Kob	0.5568	.1493	−.3380	−1.0909	−.2809	−.0449	6
Topi	0.9429	−.3970	.5502	0.2669	.4685	.6006	7
Buffalo	−0.4581	.3489	.0745	−0.3963	.4026	.1499	8
				0.5584	.1394	.1925	9
				0.7618	−.2728	−.2851	10

ing values of t $(1 \leq t \leq s)$. Taking both features into account led to acceptance of the $r=3$ rank model reported in Tables 12.6 and 12.7.

The ecological assessment of these results is deferred until after consideration of the canonical analysis, to which we now turn.

12.3 Dual scaling

Ecological interest centers on relationships between herbivores and study areas. Dual scaling provides a visual representation of an $r \times c$ array such as that of Table 12.2 in which rows and columns are simultaneously displayed as points. The hope is entertained that, in an analysis of Table 12.2, ecological relationships within and between herbivores and study areas will be reflected in the spatial relations of the corresponding points. Dual scaling therefore seems appropriate for our purpose.

We chose to regard dual scaling as that special case of canonical analysis which arises where the variables of both sets are binary dummy variables. Consequently, it was necessary to construct $p=r-1$ dummy variables to represent the rows of Table 12.2 and $q=c-1$ dummy variables to represent the columns. For the hth doubly-classified sample-observation of the table ($h=1, \ldots, \mathrm{N}$) the variables z_{hi} ($i=1, \ldots, p$) and z_{hj} were defined as follows:

$$z_{hi}=\begin{cases}1 & \text{if the } h\text{th sample belonged to species } i\\ 0 & \text{otherwise}\end{cases}$$

$$z_{hj}=\begin{cases}1 & \text{if the } h\text{th sample belonged to area } j\\ 0 & \text{otherwise.}\end{cases}$$

In this way, the x_{ij} observations of the ijth cell of Table 12.2 came to be specified uniquely in terms of the vector variables $\mathbf{z}_1$ ($p \times 1$) and $\mathbf{z}_2$ ($q \times 1$). The relationship between herbivores and study areas thus became that between $\mathbf{z}_1$ and $\mathbf{z}_2$.

The canonical analysis therefore came to be based on a biased sample of $\mathrm{N}=645$ observations, $p=6$ binary dummy variables, $\mathbf{z}_1$, corresponding to seven herbivore species and $q=8$ binary dummy variables, $\mathbf{z}_2$, corresponding to nine study areas. The ecological goal was to assess the evidence for association between herbivores and study areas and to characterize the association, if it exists. Recall from Sect. 5.3 that dual scaling supplies distance functions which define the relative positions of attribute-states in the respective row and column spaces of the analysis. The effect of these supplied functions in ecological terms here will be to equalize the contributions of rare and abundant species and of impoverished and rich sites to the analysis. Experience indicates that while this represents one acceptable weighting of information, especially where an overriding interest in a *joint* relationship has been declared, it is by no means the only possible such weighting.

Results of the analysis are summarized in Tables 12.8–12.11 and in Figs. 12.1 and 12.2.

12.3.1 Results

The canonical correlation coefficients. The canonical correlation coefficients r_k ($k=1, \ldots, s$) are reported in Table 12.8a. The correlations vary in size from $\mathrm{r}_1=.75$ to $\mathrm{r}_6=.07$. The departure of r_1 from zero in itself is such as to provide grounds for cautiously accepting that herbivores and study areas are indeed associated, at least as far as the sample data themselves are concerned. The remaining correlations r_k ($k>1$) are comparatively weak ($\leq .41$). Yet, having regard for the moderately large size of the sample ($\mathrm{N}=645$), it may be that r_2, r_3 and perhaps r_4 also are not entirely negligible. Altogether, the magnitude of the leading correlations suggests that there are grounds for guardedly concluding that herbivores and study areas are associated and for proceeding to consider what the dimensionality of the association might be.

Before doing so, let us review the results of nominal tests of an overall hypothesis of independence, that is of hypothesis (5.38). Results of union-inter-

Table 12.8. Rwenzori National Park, Uganda. Relationships between seven herbivore species and nine study areas. Dual scaling

(a) Canonical correlation coefficients r_k, goodness-of-fit C_t and nominal tests of significance. Approximate critical values of Roy's largest-root criterion, θ_α (s, m, n), are shown for $\alpha=.05$ and $\alpha=.01$; $m=\frac{1}{2}$, $n=314\frac{1}{2}$.

k	s	r_k	r_k^2	% tr(**KK**')	C_t%	$\theta_{.05}$	$\theta_{.01}$	p
1	6	.750	.563	62.83	62.83	.083	.099	$<.01$
2	5	.411	.169	18.86	81.69	.072	.085	$<.01$
3	4	.330	.109	12.17	93.86	.061	.074	$<.01$
4	3	.193	.037	4.13	97.99	.049	.061	$>.05$
5	2	.115	.013	1.45	99.44	.038	.048	$>.05$
6	1	.071	.005	0.56	100.00	.024[a]	.031[a]	$>.05$

[a] From the U-distribution: $\theta_\alpha(1, m, n)=1-U(p, 1, n)$.

(b) Bartlett's approximate chi-squared test of the hypothesis $H_0: \rho_k=0$ for all k.

$$\Lambda=\prod_{k=1}^{s}(1-r_k^2)$$
$$=(1-.563)(1-.169)\ldots(1-.005)=.306.$$
$$\chi_B^2=-\{(N-1)-\tfrac{1}{2}(r+c-1)\}\log_e\Lambda$$
$$=-\{644-\tfrac{1}{2}(9+7-1)\}(-1.184)$$
$$=753.72, \quad \text{with} \quad (r-1)(c-1)=48 \text{ df.}$$

section and likelihood ratio tests of this hypothesis are shown in Table 12.8a and b, respectively. For the union-intersection test we find that $\theta_S(6, \frac{1}{2}, 314\frac{1}{2})=.563>\theta_\alpha(6, \frac{1}{2}, 314\frac{1}{2})=.10$ for $\alpha=.01$. Turning to the likelihood ratio test we have $\chi_B^2(48)=753.72>\chi_\alpha^2(48)=93.22$ for $\alpha=.0001$. In this instance, therefore, the results of the tests agree and indicate that the hypothesis of independence would not be tenable.

Dimensionality. The index C_t of Table 12.8 is defined in Eq. (5.7). C_t expresses the goodness-of-fit of the rank t model and so is helpful in assessing dimensionality. Reference to C_t in Table 12.8 shows that for values of t in the neighborhood of 3 adequate model fit combined with a useful reduction in dimensionality could be achieved. The interpretability of the canonical variates corresponding to all six canonical roots was next evaluated. Plausible ecological interpretations were forthcoming for the canonical variates corresponding to only the three leading roots, r_k^2 $(k=1, 2, 3)$. From Table 12.8 we find that for $t=3$, $C_t=93.86$. Plainly there are grounds for believing that a model of rank 3 would provide a judicious combination of fit, parsimony and insight. These considerations led to the provisional adoption of a model of rank $r=3$. That is to say the dimensionality of the association was taken to be three. Further evidence in support of this judgement will be presented when we come to examine redundancy shortly. In order to interpret individual components of the global association, we shall require the canonical variates corresponding to the roots r_k^2 $(k=1,$

Table 12.9. Rwenzori National Park, Uganda. Relationships between seven herbivore species and nine study areas. Dual scaling. Bartlett's approximate chi-squared test of dimensionality; $m=\{(N-1)-\frac{1}{2}(r+c-1)\}=636.5$

k	Source	df[a]	$-m\ln(1-r_k^2)$	χ^2	$\%\chi_o^2$	$C_t\%$	p
1	r_1^2	13	$-m\ln(1-r_1^2)$	526.91	69.91	69.91	–
2	r_2^2	11	$-m\ln(1-r_2^2)$	117.83	15.63	85.54	–
3	r_3^2	9	$-m\ln(1-r_3^2)$	73.46	9.75	95.29	–
4	r_4^2	7	$-m\ln(1-r_4^2)$	24.00	3.18	98.47	–
≥5	r_5^2, r_6^2	8	$-m\ln\{\Lambda/(1-r_1^2)\ldots(1-r_4^2)\}$	11.52	1.53	100.00	.174
Total		48	$-m\ln\Lambda$	$\chi_o^2=753.72$	100.00	–	<.0001

[a] Degrees of freedom: $r+c-(2k+1)$.

2, 3). Sample scores on the canonical variates u_k and v_k are reported in Table 12.10 while Figs. 12.1 and 12.2 display the sample projected into subspaces associated with the u_k and v_k.

Results of nominal tests of hypothesis (3.18) of dimensionality are presented in Table 12.8a and Table 12.9. The tests would lead to acceptance of models of rank $r=3$ and $r=4$, respectively. Broadly speaking, therefore, the results are largely consistent with the estimate arrived at earlier on other grounds.

The canonical variates. The canonical variates (Table 12.10) have been scaled so as to emphasize those aspects of the relationship between herbivores and study areas which correspond to the larger canonical roots. The scaling is precisely that adopted in connection with the attribute-state scores of the contingency table analysis, and will facilitate comparison of the two analyses in Sect. 12.4. The scores shown in Table 12.10 for items omitted from the analysis (bushbuck, reedbuck, site 2) were obtained by applying the vectors of scaled, centered canonical weights to the standardized sample observations in question. In connection with Table 12.10 observe also that while the total sample size, N, is 659 (645 plus 14 'omitted' items) the lengths of the score vectors $\mathbf{u}_k$ and $\mathbf{v}_k$ are (9×1) and (10×1), respectively, rather than (659×1) as might be expected. Moreover, the elements of the vectors are identified in terms of *attribute-states* rather than samples. Table 12.10 is actually a *summary* of the sample scores on the canonical variates. The table reflects a distinctive property of the scores in dual scaling – mapped into u_k-space the $N=659$ scores fall into nine discrete groups while mapped into v_k-space they fall into ten discrete groups. Put another way, all r_i samples of the ith row (i.e., herbivore) of Table 12.2 have identical scores on u_k; similarly, all c_j samples of the jth column (i.e., study area) of Table 12.2 have identical scores on v_k. What Table 12.10 does report is the scores of sample *groups* corresponding to particular row or column attribute-states on the canonical variates. The numbers of samples comprising the groups, though not shown in the table, are simply the row and column marginal totals of Table 12.2, respectively. Little would be gained by enumerating the scores for all 659 samples individually. On the other hand, each sample-group is use-

Table 12.10. Rwenzori National Park, Uganda. Relationships between seven herbivore species and nine study areas. Dual scaling. Sample scores on the canonical variates $u_k = \mathbf{a}_k^t \mathbf{z}_1$ and $v_k = \mathbf{b}_k^t \mathbf{z}_2 (k = 1, \ldots, 3)$. Attribute-states corresponding to sample-groups are indicated

Herbivore	u_1	u_2	u_3	v_1	v_2	v_3	Site
Elephant	−1.021	.119	.016	−0.758	−.086	−.069	1
Warthog	−0.653	.481	.530	−0.629	.648	−.025	2[a]
Hippopotamus	−1.066	−.769	−.208	−1.163	−.810	−.181	3
Reedbuck[a]	0.744	.337	−.571	−0.893	.909	.722	4
Waterbuck	−1.080	.539	.526	−0.705	−.037	−.147	5
Kob	0.555	.148	−.338	−1.088	−.280	.045	6
Topi	0.946	−.399	.479	0.270	.469	−.601	7
Bushbuck[a]	−1.317	.795	1.229	−0.398	.403	−.150	8
Buffalo	−0.458	.350	.075	0.555	.140	−.193	9
				0.758	−.271	.285	10

[a] Sample scores calculated following main analysis.

fully identified by means of the attribute-state whose presence uniquely distinguishes the group. It is nevertheless worth bearing in mind that Table 12.10 does in summary form refer to all N=659 samples. Even a cursory inspection of the scores reveals evidence of the existence of affinities and contrasts within and between herbivores and study areas. More will be said about these relationships in Sect. 12.5.

Table 12.11 contains correlation coefficients between the canonical variates and the dummy variables on which they are defined, as well as other indices calculated from the correlation coefficients. The sign and magnitude of the correlations apparently indicate that considerable diversity exists among both herbivores and study areas in relation to the canonical variates. Further, the variance extracted by the canonical variates shows that all are comparatively weak components of their respective domains ($U_k^2 \leq .234$; $V_k^2 \leq .131$).

Redundancy. Interest centers principally on the variance or explanatory power of the canonical variates v_k $(k=1, \ldots, 3)$ of the study areas across domains. This property is expressed by the redundancy indices $V^2_{\mathbf{z}_1|v_1} = .080$, $V^2_{\mathbf{z}_1|v_2} = .026$ and $V^2_{\mathbf{z}_1|v_3} = .026$ (Table 12.11, upper-right). Of the three canonical variates, v_1 is appreciably stronger than the others, which are of equal strength. Notice also that while v_2 is the strongest component of the study area domain $\mathbf{z}_2$ ($V_2^2 = .131$), the explanatory power of this canonical variate in the herbivore domain is less than one-third that of v_1. The redundancies associated with u_4 and v_4 (not reported in Table 12.11) are very small, $U^2_{\mathbf{z}_1|u_4} < .004$ and $V^2_{\mathbf{z}_1|v_4} < .004$, a fact which contributed towards the adoption of the rank 3 model reported in Tables 12.10 and 12.11. Together v_1, v_2 and v_3 account for 13.2% of the total variance of the herbivore domain. The fit of the rank 3 model in these terms is therefore manifestly weak. A further expression of the lack of fit is to be seen in the herbivore communalities, all of which are small, varying between .01 for reedbuck and .29 for hippopotamus (Table 12.11, upper-right).

Table 12.11. Rwenzori National Park, Uganda. Relationships between seven herbivore species and nine study areas. Dual scaling. Correlations between dummy variables and canonical variates

Canonical variate	u_1	u_2	u_3	h_w^2	v_1	v_2	v_3	h_b^2
Herbivore								
Elephant	−.217	.076	.008	.049	−.163	.019	.003	.027
Warthog	−.171	.230	.315	.181	−.128	.095	.104	.036
Hippopotamus	−.574	−.757	−.255	.968	−.431	−.311	−.084	.290
Reedbuck[a]	.052	.078	−.206	.051	.039	.032	−.068	.007
Waterbuck	−.307	.280	.340	.288	−.230	.115	.112	.079
Kob	.534	.261	−.736	.895	.401	.107	−.243	.231
Topi	.579	−.445	.669	.981	.435	−.138	.221	.272
Bushbuck[a]	−.206	.416	.883	.995	−.154	.171	.293	.139
Buffalo	−.340	.474	.126	.356	−.255	.195	.042	.105
Variance extracted	.142	.153	.234	.529	.080	.026	.026	.132
Redundancy	.080	.026	.026	.132	.080	.026	.026	.132

Canonical variate	v_1	v_2	v_3	h_w^2	u_1	u_2	u_3	h_b^2
Study area								
1 Grass/thicket	−.277	−.057	−.057	.083	−.207	−.024	−.019	.044
2[a] Tall grass/thicket	−.125	.429	−.026	.200	−.094	.177	−.008	.040
3 Grass/thicket	−.429	−.547	−.152	.506	−.322	−.225	−.050	.157
4 Grass/thicket	−.335	.619	.613	.871	−.251	.255	.202	.169
5 Grass/thicket	−.252	−.023	−.119	.078	−.189	−.009	−.039	.037
6 Grass/thicket	−.421	−.198	.039	.218	−.316	−.081	.013	.107
7 Short grassland	.117	.375	−.598	.512	.088	.154	−.198	.071
8 Grass/thicket	−.128	.238	−.110	.085	−.096	.098	−.036	.020
9 Short grassland	.384	.175	−.301	.269	.288	.072	−.099	.098
10 Short grassland	.629	−.411	.535	.851	.472	−.169	.177	.283
Variance extracted	.120	.131	.117	.367	.068	.022	.013	.103
Redundancy	.068	.022	.013	.103	.068	.022	.013	.103

[a] Structure correlations calculated following main analysis.

Figures 12.1 and 12.2 display the sample (N = 659) projected into several different subspaces defined by the canonical variates. Figures 12.1a and 12.1b depict the sample projected into subspaces spanned by the first two canonical variates of study area and herbivore space, respectively. In Fig. 12.2 the sample has been mapped into a subspace spanned by the three leading, superimposed canonical variates of each domain, v_k and u_k ($k = 1, \ldots, 3$). In all figures the scaling adopted for the canonical variates is such as to emphasize those relationships which contribute most to the overall association between herbivores and study areas. The samples fall into discrete groups, each uniquely distinguished by the possession of a particular attribute-state. Though the mappings appear to emphasize attribute-states, more fundamentally in the context of canonical analysis they represent mappings of the sample into selected, low-dimensional subspaces. Various similarities and contrasts between study areas and herbivores are implied by the spatial relations of points in Fig. 12.1a and 12.1b. These features can also be seen in Fig. 12.2, which in addition reveals relationships

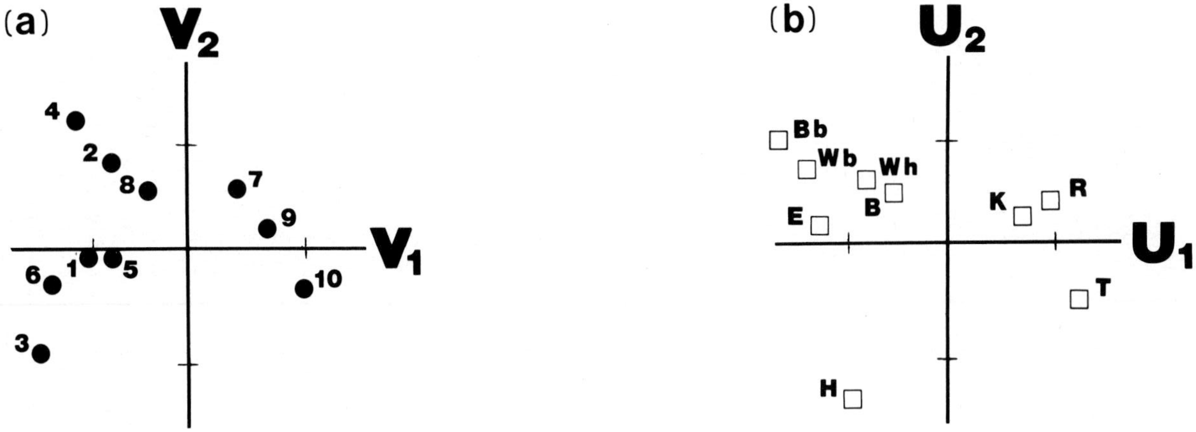

Fig. 12.1 a, b. Rwenzori National Park, Uganda. Relationships between nine herbivore species and ten study areas. Dual scaling. N = 659 samples mapped into subspaces defined by canonical variates (**a**) v_1 and v_2 of site domain; (**b**) u_1 and u_2 of herbivore domain. B: *buffalo*; Bb: *bushbuck*; E: *elephant*; H: *hippopotamus*; K: *kob*; R: *reedbuck*; T: *topi*; Wb: *waterbuck*; Wh: *warthog*

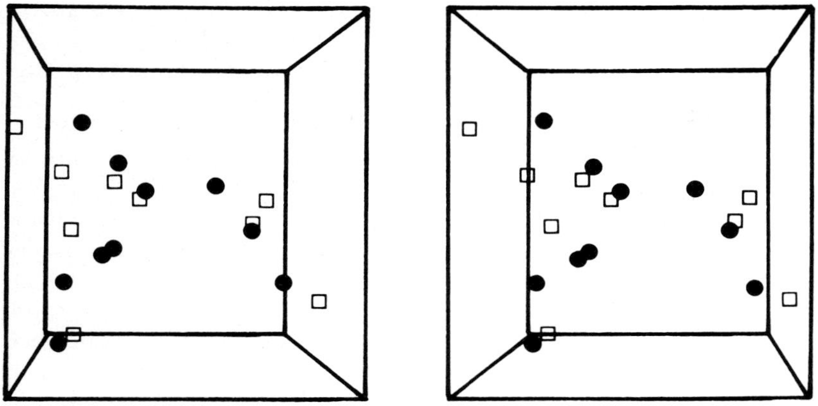

Fig. 12.2. Rwenzori National Park, Uganda. Relationships between nine herbivore species and ten study areas. Dual scaling. N = 659 samples simultaneously mapped into subspaces specified by the canonical variates u_k and v_k ($k=1,2,3$). Key to symbols as in Figs. 12.1a and b; u_1 and v_1 are coincident and comprise the horizontal axis, u_2 and v_2 are coincident and comprise the vertical axis, u_3 and v_3 are coincident and comprise the axis at right angles to the page

between attribute-states of different kinds. Recall, however, that distances between study areas and herbivores in Fig. 12.2 are to be treated circumspectly (cf. Sect. 5.3).

We shall take up the ecological implications of Figs. 12.1 and 12.2 after comparing the results of dual scaling with those of the contingency table analysis of Sect. 12.2.

12.4 Relationships between contingency table analysis and dual scaling

Connections between dual scaling and E.J. Williams' (1952) method for analyzing contingency tables were described in Sect. 5.5.1. We shall use the results presented in Sects. 12.2 and 12.3 to verify relationships to which attention was drawn there. Let us first examine the overall test of independence associated with each analysis before going on to consider the goodness-of-fit criteria and the scores for attribute-states and samples of the two analyses.

In analyzing an $r \times c$ array, hypotheses (5.38) or (5.42) are sometimes of interest. One criterion for testing hypothesis (5.38) in dual scaling is given in Eq. (5.54); similarly, the quantity specified in (5.55) provides a suitable criterion for testing hypothesis (5.42) in contingency table analysis. Numerical values of these criteria are reported in Tables 12.8b and 12.5, respectively. For the Williams' analysis we find

$$\chi_o^2 = \mathrm{N}[\sum_1^s \mathrm{r}_k^2] = 645[.8969] = 578.54,$$

and for Bartlett's approximation of Wilks' Λ,

$$\chi_B^2 = -m \log_e \prod_1^s (1-\mathrm{r}_k^2) = -636.5\,(\log_e .306) = 753.72.$$

A sizeable discrepancy exists between the two criteria. The discrepancy is attributable to two factors – use of a finite sample and to the divergence in the marginal totals of the original table (cf. Sect. 5.5.1). Despite the discrepancy, the criteria would be consistent to the extent of leading to the decisive rejection of the respective null hypotheses.

The quantities χ_o^2 and χ_B^2 may each be partitioned to yield more sensitive tests which in turn permit further comparison. The partitions are given by $\chi_o^2 = N\,\Sigma_{k=1}^6\, r_k^2$ and $\chi_B^2 = -m\,\Sigma_{k=1}^6 \log_e(1-r_k^2)$, respectively. These relationships provide the theoretical justification for Tables 12.5 and 12.9. The results are conveniently compared by means of the goodness-of-fit index, $C_t\%$. Although the distributions of $C_t\%$ differ, some similarity is apparent. The latent roots r_k^2 themselves provide a further and more direct basis for comparison. Reference to Tables 12.5 and 12.8a shows the latent roots in each case to be identical, as also is the goodness-of-fit criterion $C_t\%$ derived from them (at least within rounding error). The equivalence of the roots extends also to the scores for attribute-states (contingency table analysis) and for samples (dual scaling), derived from the corresponding latent vectors. The scores, normed to a common scale, are reported in Tables 12.7 and 12.10. For all practical purposes, the two sets of scores are the same.

We have verified the existence of identities between the results of dual scalling and contingency table analysis applied to the same data. Certain discrepancies were also noted. The discrepancies arose in connection with an identity which is known to hold only asymptotically and accordingly may perhaps be discounted. Altogether, the results presented show that dual scaling and the contingency table analysis of Williams (1952) may justifiably be regarded as being substantially at least numerically equivalent.

Having shown the results of the two analyses to be effectively the same, we have only one set to deal with. In discussing these results below, we shall employ the framework and terminology of canonical analysis.

12.5 Ecological assessment of the results

Before proceeding it will be useful to recall the salient points of ecological interest to have emerged so far. We have seen that there are grounds for believing that:

(a) herbivores and study areas are associated;
(b) the association may be resolved into three elements or components; and that
(c) of the three components, one is appreciably stronger than the remaining two, which themselves are of roughly equal strength.

Our concern now is to enquire into the ecological nature of the components. The essential features of the analysis are summarized in Figs. 12.1 and 12.2. Field and Laws' (1970) work has been drawn on extensively in interpreting the figures. The views expressed below, however, are not necessarily those of Field and Laws.

The first canonical variate of the site domain, v_1, contrasts sites 7, 9 and 10 which have positive scores, against sites 1–6 and 8 with negative scores (Table 12.10; Fig. 12.1a). Reference to the summary of site characteristics (Table 12.1) shows that sites with positive scores support open grassland at some distance from standing water, while sites with negative scores for the most part support a mosaic of grassland and woody, thicket vegetation adjacent to water. More generally, it can be deduced from Field and Laws' account that sites with negative scores on v_1 tend to be (a) close to water; (b) overgrazed; (c) fire-free; and (d) to support a mosaic of herbaceous and woody vegetation. The converse is true in each case of sites with positive scores. There is also a geographical component to this canonical variate, as can be seen on comparing the sites scores with the spatial position of the sites on Field and Law's map of the Park (Field & Laws, 1970, Fig. 1). Sites centered about the Mweya Peninsula are on the whole rather sharply differentiated by their negative scores on v_1 from sites south of the Maramagambo Forest or otherwise at some distance from standing water (sites 7, 9 and 10).

The second canonical variate, v_2, affects principally the grass/thicket sites, that is sites with negative scores on v_1. Among these, site 3 (−.81) for example is sharply differentiated from site 4 (.91). Reference to Table 12.1 and to Field and Laws' account shows site 4 to have been subjected to continuous hippopotamus cropping over a lengthy period, while site 3 has not; on the contrary, there has been an influx of hippopotamus into site 3 during the period of the survey. Moreover, the virtual disappearance of hippopotamus from site 4 has been accompanied by an increase in the density of other species in this site, especially buffalo and waterbuck. Thus, v_2 may well correspond to the effects of *management* on the herbivore complement of the sites. The scores of other sites on v_2 are broadly consistent with such an interpretation. There is evidence that dynamic changes in herbivore density have repercussions on the vegetation and other physical characteristics of sites (Field, 1968, 1972; Thornton, 1971; Lock, 1977).

The third canonical variate, v_3 (Table 12.10; Fig. 12.2), chiefly affects sites 4 (.72) and 7 (−.60); the remaining sites with the exception of site 10 (.28) have scores $<|.20|$. The ecological implications of this canonical variate, if any, are obscure. The vegetation of sites 5 (−.15), 7 (−.60) and 9 (−.19) consists of fire-climax grassland while that of site 4, which is heavily over-grazed, is largely fire free. Thus it may be that v_3 in part reflects the effects of fire.

In short, it seems that of the site characteristics examined, vegetation contributes most to the physical differentiation of the sites, while proximity to water, grazing, fire and management have more remote though perhaps still sizeable effects. Notice also that the site-configuration in the (v_1, v_2)-plane largely recovers the habitat categories into which Field and Laws initially grouped the sites; only the distinctiveness of Group 1 (comprised of site 2) is not substantiated by the results of the analysis.

The first canonical variate of herbivore-space, u_1, distinguishes two groups of species – bushbuck, waterbuck, hippopotamus, elephant, warthog and buffalo toward the negative pole, and kob, reedbuck and topi towards the positive pole (Table 12.10; Fig. 12.1b). A striking ecological difference between these species concerns their dependence on free standing water. Buffalo, elephant, hippopotamus and warthog are all dark-skinned species whose reduced hair renders them particularly vulnerable to heat. As a consequence, they depend for their survival on water-containing wallows (Field & Laws, 1970); in a similar way, waterbuck is known to require water for drinking while bushbuck has an affinity for swamps and is seldom found far from water (Walker et al., 1968, pp. 1437 and 1414; Okiria, 1980). Kob, reedbuck and topi, on the other hand, are quite unlike the species mentioned in being to a considerable extent independent of both wallows for thermoregulation and water for drinking; they obtain much of their water requirement with their food. The change of sign along u_1 therefore appears to be related to the extent of the physiological dependence of the species on standing water. For convenience we shall refer below to the two species groups as *water-dependent* and *water-independent*, respectively. Observe also from Fig. 12.1b that the disposition of species in relation to u_1 is suggestive of a partial discontinuity in the data.

Turning to u_2, we find this canonical variate to be characterized principally by a contrast within the water-dependent species. Above all, u_2 distinguishes hippopotamus (−.77) from bushbuck (.79), waterbuck (.54) and buffalo (.35). We have remarked that hippopotamus was substantially eliminated from one study area (site 4) and its numbers drastically reduced in two others (sites 1 and 8). Moreover, the cropping program also affected the densities of species not themselves directly involved, notably waterbuck and buffalo. Thus, u_2 appears to correspond to changes in the dynamic balance of the water-dependent species as a result of the cropping program directed against hippopotamus. It can be seen that u_2 also distinguishes topi (−.40) from reedbuck (.34) and kob (.15) among the water-independent species. That this distinction represents a further manifestation of management practices, however, seems somewhat remote.

The third canonical variate, u_3, is characterized principally by bushbuck (1.23), towards the positive pole, and by hippopotamus (−.21), kob (−.34) and reedbuck (−.57) towards the negative pole (Table 12.10; Fig. 12.2). Given the rarity of reedbuck (see Table 12.2) it may be prudent to regard u_3 as essentially a contrast between bushbuck, on the one hand, and simply hippopotamus and kob, on the other. One well-defined ecological distinction between these species concerns *feeding habit* – bushbuck is basically a browser, having a preference for the woody, dicotyledenous shrubs *Capparis tormentosa* Lam. and *Erythrococca bongensis* Pax (Field & Laws, 1970, Okiria, 1980), while hippopotamus and kob are exclusive grazers (Field, 1968). Waterbuck (.53) and buffalo (.10), with positive scores, have also been reported to be browsers of *Capparis* thicket (Lock, 1977, p. 399). These remarks suggest that u_3 may perhaps correspond to differences among species in feeding habit. Such, however, cannot be the whole story, as among the water-independent species u_3 differentiates kob (−.34) and topi (.48), both of which are grazers. At first sight, this finding is rather disconcerting. Bearing in mind the partial discontinuity noted earlier, it indicates that in seeking to interpret this canonical variate it may be necessary to contemplate the operation of ecological processes of different kinds on each side of the disjunction. Kob and topi are known to differ in the Rwenzori National Park in their susceptibility to predation by lion (*Panthera leo* L.), topi being twice as vulnerable as kob (Yoaciel & van Orsdol, 1981). Among the water-independent species, u_3 may conceivably reflect this difference.

At this point it will be useful to summarize our notions as to the ecological significance of the u_k. These canonical variates appear to express similarities or differences among the herbivores in terms of (a) dependence on standing water; (b) response to management; and (c) food habit or vulnerability to predation.

As a final step let us examine the canonical variates of the site and herbivore domains simultaneously. Figure 12.2 integrates the displays of Figs. 12.1a and 12.1b while extending the representation to three dimensions. The proximity of grass/thicket sites adjacent to water and water-dependent species, on the one hand, and of grassland sites at some distance from water and water-independent species, on the other, is a general expression of the association between the entities in question. Observe, for instance, the general proximity of kob, reedbuck and topi to sites 7, 9 and 10. Figure 12.2 also draws attention to effects of the contrasting management practices of sites 3 and 4; broadly speaking, positive association between hippopotamus and site 3 and negative association between this species and site 4 are implied by the spatial relationships in question (cf. also Table 12.2). Similarly, the proximity of waterbuck and buffalo to site 4 in a general way reflects the positive association between these entities, which is attributable to the elimination of hippopotamus. Notice, also, the proximity, and hence positive association, between bushbuck, which is a browser, and the fire-free grass/thicket vegetation of site 4 and that between kob and reedbuck, which are grazers, and the fire-climax grassland of sites 5, 7 and 9.

The stereogram possesses other features of interest. Notice, for example, the strong positive association between kob and site 9 and between topi and site 10 as well as the negative association between elephant and topi. The peripheral and comparatively isolated positions of hippopotamus and bushbuck indicate that these are species of narrow ecological amplitude or else species whose distribution is otherwise controlled by some rather specific circumstance. In contrast, buffalo, which is a widely and comparatively evenly distributed species, is located towards the center of the plot. Kob and topi are shown to resemble one another at least to the extent of being water-independent species of open grassland. Yet considerable spatial separation exists, suggesting that the species are differentiated by a sizeable 'ecological distance'. Reference to Table 12.2 confirms this point. The isolation of topi, in particular, is certainly suggestive of the control of this species by some overriding ecological characteristic or requirement. A further point concerns the extent of the separation of the herbivores as a whole. It can be seen that in general species tend to be well-separated. This may reflect their *ecological separation* in the field, the effect of which is to minimize competition between them and to which Field (1968) and Field and Laws (1970) have drawn attention.

The above remarks by no means exhaust the relationships of ecological interest suggested by Fig. 12.2. Sufficient has however been said to describe the salient features of the association between herbivores and study areas and to account for them, at least in a provisional way. In terms of redundancy, the fitted model accounts for only 13.2% of the total variance of the herbivore domain ($V^2_{\mathbf{z}_1|v_1, v_2, v_3} = .132$). Yet, despite the lack of fit in these terms, a coherent and reasonably detailed picture of the differentiation of herbivores and study areas emerges. For the most part, the results described are substantiated by those of Field and Laws' (1970) study, in which a quite different approach was used. The correspondence between the results of the two studies lends weight to the view expressed elsewhere in the text that canonical analysis can be used successfully to recover information of ecological value. Indeed, it might be argued that the canonical analysis provides a more penetrating and better integrated approach than that of the original study, as well as one whose results are more easily communicated. Canonical analysis unquestionably provides numerous insights of potential ecological value. It is emphasized, however, that the worth of these insights could ultimately be established only by additional field work.

12.6 Conclusions

The study had two goals – to clarify relationships between herbivores and study areas and to verify the existence of connections between canonical analysis and a particular form of contingency table analysis.

The primary mechanism of separation among herbivores appears to be the extent of their dependence on standing water. Further differentiation is attributable to the effects of management, feeding habit and predation. The essential points are as follows:

(a) *Standing water* generates or contributes in sites adjacent to water to association between water-dependent species, over-grazing and grass/thicket vegetation, and in remote sites to association between water-independent species, grassland vegetation and fire.

(b) *Management* leads to negative association between hippopotamus and sites 4 and 8, on the one hand, and to positive association between these sites and buffalo and waterbuck on the other.

(c) Differences in *feeding habit* apparently give rise to association between woody, thicket vegetation and bushbuck, and to that between fire-climax grassland and kob, while the partial dissociation of kob and topi is attributable to *predation*.

Similar conclusions were expressed by Field and Laws (1970). To some extent this consistency validates the results presented here, and lends weight to the view that canonical analysis can contribute towards the attainment of ecological goals. Much of the information provided by the canonical analysis can in fact be obtained by inspection of the original data themselves (Table 12.2). However, the information is more readily absorbed from Figs. 12.1 and 12.2, following analysis.

The results obtained were also used to illustrate a point of theoretical interest. It was shown that dual scaling and a latent root and vector method of contingency table analysis are substantially the same. The practical significance of this result to us is in showing canonical analysis to be applicable to a distinctive, new class of data, such as that exemplified by the table of frequencies analyzed. This extension represents a further widening of the applicability of canonical analysis in ecology, to which reference was made at the end of Chap. 11.

Part III
Appraisal and prospect

13. Applications: assessment and conclusions

13.1 Introduction

The analyses of Chaps. 6–12 were intended to illustrate how and to what extent canonical analysis can contribute towards the attainment of ecological goals. More specifically, the analyses were directed towards revealing something of the varied opportunities offered by canonical analysis in analyzing ecological data; the flexibility of canonical analysis which results under specialization of the variables; and the existence of connections between canonical analysis and other statistical methods widely used in ecology. The present chapter addresses the question of the worth of the analyses in ecological terms. In evaluating the analyses, appeal will be made to information outside the analyses themselves as well as to internal, data-based criteria. Scaling methods generally, of which canonical analysis is one, are difficult to evaluate except by reference to existing substantive knowledge – in other words, that is, by reference to information external to a particular analysis. This process is sometimes known as validation and has the general goal of adding conviction to the results of an analysis. The most convincing evidence for the validity of a result is that it leads to further insight. External assessment will be guided and informed by the use of internal, data-based indices. In particular, reliance will be placed on indices of variance accounted for. However, neither external knowledge nor explained variance offer entirely satisfactory means of evaluating an analysis. As the *true* relationships of interest are rarely known with certainty, in seeking external support reliance has inevitably to be placed on preconceived ideas as to what the true relationships might be. At best such notions are susceptible of bias. While the results of an analysis can sometimes be verified by experiment, more generally little can be done except to be alert to the danger of bias and to strive to minimize it. The value of explained variance in assessment is similarly equivocal. Difficulties arise because explained variance is partly a function of features of an analysis which are not directly related to its substantive worth. The issues involved have been discussed by O'Grady (1982). It seems that used circumspectly measures of explained variance are not without value.

The worth of an analysis is ultimately a matter for ecological judgement. Nevertheless, statistical as well as ecological considerations have to be weighed in the process. Essentially what is involved is the matching of theoretical and empirical knowledge.

In the following subsection an evaluation of the analyses of Chaps. 6–12 is offered, while in Sect. 13.3 conclusions concerning the role of canonical analysis in ecology are drawn based on this evaluation.

13.2 Assessment

An assessment of the worth of each of the seven analyses in relation to ecological knowledge external to the analyses themselves has already been made towards the end of the chapters in question. The conclusions expressed there may be summarized by remarking that in each case the results obtained:

(a) are open to meaningful ecological interpretation;
(b) support interpretations which are consistent with those of independent studies; and
(c) appear to represent more informative and more convenient statements of the relationships of interest that existed hitherto.

Altogether, therefore, grounds certainly exist for declaring that the analyses have merit. However, the data-based indices, to which we now turn, will cause us to qualify or refine this initial statement appreciably.

Squared canonical correlation coefficients and redundancy are the expressions of explained variance on which we shall rely. In referring to these indices our interest is soley in their value as measures of the success of a model in fitting the data – we are not concerned here with matters of statistical inference. Squared canonical correlations indicate the component of the total predictable variance which is shared by a linear composite of one measurement domain and its conjugate of the other. Redundancy expresses the shared variance of the measurement domains themselves, and, unlike a squared canonical correlation coefficient, is an asymmetric or directed quantity. Sample estimates of both quantities are biased, the bias being a function principally of sample size in relation to the total number of variables studied. Before proceeding, it will be instructive to review implications of these design characteristics on the results of an analysis.

Sample size. Table 13.1 reports the sample size, number of variables and variable to sample ratios for the seven analyses. The table shows that, with one exception, the analyses are all based on small samples ($N \leq 45$). Moreover, the variable to sample (v/s) ratios for the most part are $\geq .30$. Recommendations given by Barcikowski and Stevens (1975) and by Thorndike (1978, p. 184) suggest that v/s ratios in the range .025–.05 are desirable if firm conclusions are to be reached. This range of values provides a yardstick as to just how inadequate the samples might be in six of the seven analyses. Two consequences follow immediately:

(a) estimates of the canonical correlation coefficients, canonical roots, and redundancy will be biased, perhaps appreciably;

Table 13.1. Some sample characteristics of the analyses of Chaps. 6–12. Sample size, N, and number of variables p and q of X- and Y-domains, respectively; v/s denotes the variable/sample ratio, $v=p+q$, $s=N$

Characteristic	6 Limestone grassland spatial variation	7 Limestone grassland soil species	8 Tropical rain forest soil vegetation	9 Tropical rain forest dynamics	10 Grassland structure	11 Nitrogen nutrition	12 Rangeland herbivore environment
N	45	45	25	25	45	40	645
p	3	8	6	6	2[a]	7[a]	8[a]
q	3	6	9	6	31	5	6[a]
v/s	.13	.31	.60	.48	.73	.30	.02

[a] Binary-coded dummy variables.

(b) the fitted models will be vulnerable to sample specific variation and covariation.

It seems that the strength and perhaps the number of relationships described are to be accepted with reserve, while any attempt to extend the applicability of the conclusions reached beyond the sample data on which each analysis was based could be justified only on extra-statistical grounds.

Evidence can be adduced which suggests that the undesirable consequences of a high v/s ratio can sometimes be exaggerated. To appreciate this point, we return briefly to the analyses of Chaps. 6 and 10. In both cases, the extent of external ecological substantiation of the results (see Sects. 6.3 and 10.3) points to the applicability of the conclusions drawn beyond the samples actually analyzed, notwithstanding high v/s ratios ($\geq .13$). The investigation of the joint distribution pattern of three grassland species (Chap. 6) is particularly instructive in this regard because the canonical variates of this analysis possess a spatial distribution. Though the canonical variates cannot themselves be observed or measured directly, their spatial distribution in the field can be *mapped* and hence checked empirically. It is a simple matter to verify by direct observation in the field that the joint distribution of the species for the statistical universe (plant community in this case) as a whole is indeed precisely that of the canonical variate u_1 (cf. Fig. 6.2). Further, field examination shows that relationships among species with respect to the extent of the 'southwesterliness' of their distribution for the statistical universe itself are exactly as specified in Sect. 6.3 on the basis of sample data. The grassland structure analysis (Chap. 10) is characterized by the exceptionally large v/s ratio of .73. Yet the results of this analysis correspond with extraordinary fidelity to expectations based on some years' field experience of the statistical universe (plant community) in question. In short, despite high v/s ratios the results of both analyses are in my view demonstrably sound and represent significantly more ecologically than simply the fitting of sampling error. It therefore seems that canonical analysis may on occasion be surprisingly resistant to the presumed deleterious effects of high variable to sample ratios.

Table 13.2. Quality of the analyses of Chaps. 6–12. The canonical roots, r_k^2 ($k=1, \ldots, s$) and percentage $\mathrm{tr}(\mathbf{R}_{22}^{-1}\mathbf{R}_{21}\mathbf{R}_{11}^{-1}\mathbf{R}_{12})$ absorbed by the canonical roots individually (% tr) and cumulatively (C %). The cumulative percentage trace of the fitted models in boldface

	6 Limestone grassland spatial variation			7 Limestone grassland soil-species			8 Tropical rain forest soil-vegetation			9 Tropical rain forest dynamics		
k	r_k^2	% tr	C%	r_k^2	% tr	C%	r_k^2	% tr	C%	r_k^2	% tr	C%
1	.631	73.97	**73.97**	.851	45.75	45.75	.857	29.08	29.08	.977	24.60	24.60
2	.214	25.09	99.06	.484	26.02	**71.77**	.817	27.72	56.80	.914	23.01	47.61
3	.008	0.94	100.00	.271	14.75	86.34	.556	18.87	75.67	.894	22.51	70.12
4				.140	7.53	93.87	.427	14.49	**90.16**[a]	.760	19.13	**89.25**
5				.066	3.55	97.42	.220	7.47	97.63	.424	10.67	99.92
6				.048	2.58	100.00	.070	2.38	100.01	.003	0.08	100.00
Trace	.853	–	–	1.860	–	–	2.947	–	–	3.972	–	–

	10 Grassland structure			11 Nitrogen nutrition			12 Rangeland herbivore-environment		
k	r_k^2	% tr	C%	r_k^2	% tr	C%	r_k^2	% tr	C%
1	.963	51.01	51.01	.910	44.65	44.65	.563	62.83	62.83
2	.925	48.99	**100.00**	.724	35.53	**80.18**	.169	18.86	81.69
3				.298	14.62	94.80	.109	12.17	**93.86**
4				.093	4.56	99.36	.037	4.13	97.99
5				.013	0.64	100.00	.013	1.45	99.44
6							.005	0.56	100.00
Trace	1.888	–	–	2.038	–	–	.896	–	–

[a] For the roots r_k^2 ($k=1, 2, 3, 5$), C% = 83.14.

The above remarks are not intended as an endorsement of the use of small samples in canonical analysis. To the contrary. Indeed, it could be argued that the data on which the investigations of Chaps. 6–12 are based are scarcely sufficient to bear the weight of the analysis to which they have been subjected. Large samples are almost always likely to be advantageous and are mandatory for incisive analysis where external, confirmatory support is lacking.

Explained variance. The departure of a squared canonical correlation coefficient from zero may throw light on the worth of an analysis – broadly speaking, larger departures are consistent at least with analyses of most worth. The squared canonical correlation coefficients for the seven analyses are reported in Table 13.2. From the table it can be seen that in four of the analyses the departure of the r_k^2 $(k=1, 2)$ from zero is appreciable ($r_1^2 \geq .86$; $r_2^2 \geq .72$). In these terms, therefore, the analyses for the most part might be considered successful. The canonical roots also yield a composite index which reflects the adequacy of a fitted model as a whole. The sum of all s roots, $\Sigma_k\, r_k^2$ $(k=1, \ldots, s)$ is identically the trace of the matrix product $\mathbf{R}_{22}^{-1}\mathbf{R}_{21}\mathbf{R}_{11}^{-1}\mathbf{R}_{12}$, which represents the shared, predictable variance *common* to two domains. The percentage of this quantity which is accounted for by a fitted model of rank t, namely $[\Sigma_k^t r_k^2/\text{tr}\,(\mathbf{R}_{22}^{-1}\mathbf{R}_{21}\mathbf{R}_{11}^{-1}\mathbf{R}_{12})] \times 100$, is one index of the quality of the rank t model. Table 13.2 reports the cumulative percentage trace (C%) for different values of t for the analyses of Chaps. 6–12. The cumulative indices for the fitted models range from 72% to 100%. In terms of the percentage of predictable, shared variance which the models account for, therefore, the analyses also appear to have been largely successful. Assessment based on the canonical roots alone however is deficient in two respects – neglect of the upward bias in the r_k^2 attributable to high v/s ratios, and neglect of the within-set variance of the canonical variates themselves. These shortcomings can result in an analysis appearing to have a respectability that in reality it does not possess.

Total redundancy is a more comprehensive index of the adequacy of a fitted model than a squared canonical correlation coefficient, taking account of both the within-set and between-set variance of a canonical variate. Before evaluating the analyses in these terms, we need to first establish guidelines as to the adequacy of the index for the purpose. Table 13.3 reports the total reduncancies for

Table 13.3. Quality of the analyses of Chaps. 6–12. Total redundancies, $V^2_{\mathbf{x}|v_1 \ldots v_s}$ and $U^2_{\mathbf{y}|u_1 \ldots u_s}$, of the X- and Y-domains for the fitted models. Bias adjusted values are in italics

Total redundancy	6 Limestone grassland spatial variation	7 Limestone grassland soil species	8 Tropical rain forest soil vegetation	9 Tropical rain forest dynamics	10 Grassland structure	11 Nitrogen nutrition	12 Rangeland herbivore environment
$V^2_{\mathbf{x}\|v_1 \ldots v_s}$	.404	.382	.501	.757	.961	.233	.132
	.360	*.284*	*.202*	*.676*	*.868*	*.120*	*.124*
$U^2_{\mathbf{y}\|u_1 \ldots u_s}$	.234	.402	.391	.768	.395	.732	.103
	.178	*.269*	*.188*	*.691*	*.366*	*.673*	*.092*

all seven analyses in two forms – corrected and uncorrected for bias. The bias adjusted values were calculated using Wherry's formula (see Dawson-Saunders, 1982), and attention here is confined exclusively to the adjusted values. The quantities reported vary from .092 for the rangeland analysis to .868 for the grassland structure investigation. For the most part, however, the redundancies are $\leq .300$ and are manifestly smaller than either the canonical roots or the composite index calculated from them. Total redundancy is evidently a more conservative indicator than either of the others. Yet it seems to me that of the three indices total redundancy is the most satisfactory. To justify this claim, it is necessary to refer back to the appraisal of the analyses of Chaps. 6 and 7, both of which yielded redundancies $\leq .40$. In connection with these analyses, the following points were made (see Sects. 6.3 and 7.3):

(a) both analyses completely recovered what was considered to be all the pertinent information of ecological interest in the data;
(b) that quite small individual redundancies in the vicinity of, say, .05 can correspond to relationships of considerable ecological significance; and that
(c) variables in ecology are for the most part characterized by high unique variances.

These observations suggest that total redundancy provides an indicator of the worth of an analysis which is both sensitive to small effects and consistent with the high unique variance of ecological systems. Further, it appears that as a general rule in ecology it may be unwise to accept total redundancies much in excess of, say, .50 at face-value; while values $>.50$ certainly occur it may be that these are best treated initially as anomalies requiring close scrutiny. Let us return to the total redundancies of Table 13.3 in the light of these remarks.

Observe that four of the reported values exceed .50 and hence qualify as anomalies. As such, they invite further investigation. We shall see that in each case the unusually high value can be attributed either to identifiable and perhaps singular ecological circumstances or to design features peculiar to a particular analysis. With respect to the study of rain forest dynamics (Chap. 9), the redundancies in question are $V^2_{\mathbf{x}|v_1 \ldots v_4} = .676$ and $U^2_{\mathbf{y}|u_1 \ldots u_4} = .691$. It seems likely that the magnitude of these quantities reflects the exceptionally strong association between tree and seedling components of the forest communities investigated, to which reference was made in Sect. 9.3. This observation in itself appears in this case to adequately account for the values reported. For the grassland structure analysis (Chap. 10), the redundancy values are $V^2_{\mathbf{z}|v_1, v_2} = .868$ and $U^2_{\mathbf{x}|u_1, u_2} = .366$. The unusually high value of .868, as well as the disparity between the two quantities merit attention. Both features can be related to differences in the length of the vectors $\mathbf{z}(2 \times 1)$ and $\mathbf{x}(31 \times 1)$ on which the analysis was based – in other words, that is, to design features of the analysis. It is not altogether surprising that thirty-one variables should be able to account for 87% of the sum of the variances of just two variables. What perhaps is more remarkable is that 37% of the total variance of the thirty-one variables in $\mathbf{x}$ should be accounted for by the pair of variables comprising $\mathbf{z}$. At any rate, the length of $\mathbf{x}$ and the difference between $\mathbf{x}$ and $\mathbf{z}$ in this respect offer a convincing explanation for the results obtained. The size of $U^2_{\mathbf{x}|u_1, u_2} = .673$ of the nitrogen nutrition analysis (Chap. 11), as well as the disparity between this quantity and $V^2_{\mathbf{z}|v_1, v_2} = .120$, are amenable to substantially the same kind of explanation. The point of general interest here is that when total redundancies $> .50$ occur one is alerted to the possibility of the operation of effects or circumstances of a somewhat exceptional kind. The circumstances in question may be ecological or may arise simply from the design of the study. Efforts to distinguish between the two are likely to be worthwhile, design characteristics in particular having little or no direct substantive content. The implications of the larger redundancies of Table 13.3 are best appreciated in the light of these remarks. The ecological significance of several of the larger quantities is accordingly regarded somewhat sceptically. The remaining redundancies of Table 13.3 are small in size ($\leq .36$). Nevertheless, external substantiation of the pertinent relationships has shown these relationships to be plausible on ecological grounds.

Total redundancy therefore emerges as a sensitive though essentially conservative measure of the worth of an analysis. It is most useful in situations where ecological insight exists *and* is declared prior to analysis. It so happens that the analyses of Chaps. 6 and 7 were conducted in precisely this way, though the motivation there was rather different. Where a declaration of relationships of ecological significance was made before analysis, individual redundancy proved to be a remarkably sensitive and trustworthy indicator of the relative magnitude of a particular effect or relationship, while total redundancy provided a plausible composite index of the worth of an analysis as a whole. It may be that the redundancy values achieved could have been boosted appreciably had an attempt been made to maximize redundancy itself rather than the canonical correlation. Nevertheless, too much ought not to be expected of any single, scalar-valued quantity in adequately conveying the complexity of the dependency between two sets of variables. Moreover, where prior insight is lacking or where the design of the analysis is not fully taken into account, the danger of misinterpreting redundancy increases, especially in small samples.

In short, from the external and internal assessments jointly it seems that for descriptive purposes at least the analyses have merit, notwithstanding the low levels of explained variance reported. With some justification the results can be said to have identified the salient relationships in question, at least in a provisional way, and to have communicated these relationships simply and clearly. On the other hand, any wider applicability which the results may possess beyond the sample data themselves was not convincingly demonstrated. Appreciable shrinkage in the values of both the canonical correlation coefficients and redundancy indices could be anticipated under cross-validation. More light may well have been thrown on the stability and generality of the results by jackknife or bootstrap analysis. We shall see in Chap. 14 how this omission can be turned to useful advantage by pointing towards opportunities for improved analysis.

As well as showing how canonical analysis can contribute towards the solution of ecological problems, the analyses of Chaps. 6–12 also illustrate something of the flexibility of canonical analysis. It is to this aspect of the analyses that we now turn.

The range of substantive questions which can be addressed by canonical analysis is very wide. The varied ecological content of the analyses of Chaps. 6–12 in itself reflects something of the versatility of the procedure. Broadly, the applications are of the two kinds – field surveys and controlled experiments. In the former category, the applications exemplify studies of (a) vegetation/environment relationships; (b) vegetation/animal relationships; (c) spatially distributed vector-valued variables; (d) vegetation structure, and (e) aspects of temporal change in vegetation. In the context of a controlled experiment, canonical analysis was used to explore the comparative p-variate response of several plant species to an applied treatment regime in an RB-g design. Several other facets of the versatility of canonical analysis are revealed by the analyses. For example, use was made of variables observed on measurement scales of different kinds. Continuous and nominally-scaled variables were encountered in connection with the questions investigated while binary-coded dummy variables were introduced as a device to handle nominally-scaled variables. Binary dummy variables were shown to increase the applicability of canonical analysis very considerably. In the mineral nutrition experiment all p variables were measured in the same metric and had the same origin and unit; more generally, the response variables used in canonical analysis are qualitatively distinct, at least. In the study of spatial variation, coordinates of samples with respect to an arbitrary reference frame formed one set of variables, while in the rain forest studies some or all of the variables were principal components of observed or measured variables. The seven analyses also illustrate the use of canonical analysis for purposes of exploratory analysis and as a means of testing formal hypotheses. Graphical summary and display of the covariance structure was an essential ingredient of the exploratory analyses. In connection with hypothesis testing, we saw how hypotheses of (a) independence and of (b) rank or dimensionality can be addressed. Tests of the joint nullity of $s = \min(p, q)$ canonical correlation coefficients, of the equality of k p-variate means and of association in two-way contingency tables are all fundamentally tests of independence. Tests of rank were employed to evaluate the hypothesis that $t < s$ pairs of canonical variates account for the correlation between two sets of variables or for the separation between k p-variate means. Finally, in the nutrition experiment we saw how *a priori* biological knowledge or insight can sometimes be incorporated into and exploited in canonical analysis.

Canonical analysis also proved instructive in quite a different way – namely in providing a common framework that served to unify apparently diverse techniques. One-way multivariate analysis of variance, multiple discriminant analysis, canonical variate analysis, dual scaling and a latent root and vector method of contingency table analysis were all shown to be formally equivalent to canonical correlation analysis.

In these various ways the analyses reveal something of the domain of applicability and richness of canonical analysis. On the other hand, none of the applications in itself amounts to much in ecological terms. That is to say the scope or complexity of each of the questions addressed is comparatively simple. This is entirely natural and appropriate where the purpose is to explicate a little-known procedure and to set out to systematically investigate its performance in a particular context.

13.3 Conclusions

The principal conclusions drawn from the preceding assessment are that canonical analysis:

(a) can contribute to the description and comprehension of relationships between two sets of variables;
(b) is applicable in a wide class of ecological studies; and
(c) provides a unifying framework for comparing and relating statistical methods.

What the analyses accomplish above all is to show how the model underlying canonical analysis is fitted and how the results obtained are interpreted. In so doing the analyses also reveal the kind and extent of the insight afforded by canonical analysis. Perhaps the most perplexing aspect of investigating the dependency between two sets of variables is the difficulty of visualizing patterns in the data. We have seen that canonical analysis yields summaries of complex patterns of this kind in the form of tables and graphical displays. These devices invariably proved informative in aiding comprehension of the covariance structures of interest. In this respect canonical analysis resembles the analysis of variance. The familiar analysis of variance table provides a compact summary of patterns in data and is largely responsible for the analysis of variance being one of the most widely applied statistical methodologies. Tabular summaries of this kind can frequently be supplemented or even supplanted by graphical displays. Such displays have the capacity to transmit information on relationship and structure with great ease and efficiency. Hence, it emerges that one aspect of the contribution which canonical analysis can make is that of an *information summary*. This indeed is the primary function of the method at the present time.

The analyses also reveal that the *extent* of the opportunities offered by canonical analysis in ecology and related disciplines is very wide. Five broad areas of application seem promising: analysis of vegetation/environment relationships, analysis of plant/animal relationships, description of vegetation, description of taxonomic variation and covariation and analysis of experiments in which a p-variate response is of interest. In addition, because so many familiar methods of univariate and multivariate analysis can be embedded in canonical analysis, the method can contribute to expositions of the interrelatedness of statistical methods derived from the multivariate general linear model.

There are several aspects of canonical analysis about which the applications of Chaps. 6–12 disclose little or nothing. Among these, reference is made to problems presented by large data sets, the investigation of relationships between more than two sets of variables and the use of canonical analysis in predictive or longitudinal studies. There is obviously a great deal of scope for further work. Moreover, the question as to what extent the data themselves might be allowed to guide the course of analysis was not addressed, nor was any attempt made to assess the stability of the results obtained. Something will be said of these and other opportunities which can be expected to lead to more incisive analysis in the following chapter.

14. Research issues and future developments

14.1 Introduction

We have endeavored in the preceding chapters to draw attention to the richness and flexibility of canonical analysis. The flexibility of the underlying model, in particular, is expressed in part by the number of apparently distinct methods which prove on examination to be special cases of canonical correlation analysis. Canonical variate analysis and dual scaling were both shown abstractly to be equivalent to canonical correlation analysis. The model's flexibility is complemented by a correspondingly diverse range of application. While, in relation to the worked examples of Chaps. 6–12, it may be that the ability of canonical analysis to have parsimoniously summarized the dependence structures in question is to be attributed to the small scale of these investigations, within the class of small-scale investigations the applicability of canonical analysis was shown to be very wide. Altogether, it seems justifiable to conclude that canonical analysis offers an unusually versatile structure by means of which statistical theory, ecological theory and empirical data can be harnessed to provide insight into the organization of ecological systems.

While canonical analysis may be a useful analytical tool it is nevertheless one which must be used with the utmost care and good judgement if sensible results are to be obtained. This is partly because canonical analysis is quite sensitive to failure of the data to comply with assumptions made in fitting the model. Consequently, naive application of the method all too often leads to incorrect or meaningless results and hence to confusion rather than insight. Automated analysis is especially reprehensible in this respect, hiding the data from the analyst and making arbitrary decisions concerning the course of analysis which might better be left to the investigator. An important question to raise in applications, therefore, is that of the likely *impact of the data* on the outcome of analysis. Furthermore, the results in canonical analysis are inherently likely to be sample-dependent. That is to say the stability of results over repeated samples drawn from the same population is likely to be low. Accordingly, it is pertinent to inquire in applications as to the *quality* of the results yielded by analysis.

The present chapter is devoted to a discussion of matters such as these. Much of the relevant work derives from the areas of statistical data analysis and regression methodology, rather than canonical analysis itself. We shall draw extensively from work in both areas. The discussion is somewhat speculative in nature and no claim of completeness is made. In particular, problems which may be peculiar to canonical variate analysis, dual scaling or to the analysis of more than two sets of variables are not specifically identified or addressed. Statistical data analysis and modern regression methodology both tend to be computer-intensive – they depend heavily on the power and cheapness of modern computational means. Canonical analysis, in contrast, was developed well before the advent of modern computers and involves little more than a standard eigenanalysis or singular value decomposition. Our treatment will therefore provide us with a serendipitous glimpse of the vay in which modern computational power may be expected to impact on canonical analysis in the foreseeable future. Initial effects in this direction are already discernible. It seems reasonably clear that we may expect a shift in emphasis away from automated model fitting and a preoccupation with questions of inference towards the fitting and development of models guided by the insight acquired through a process of interactive, graphical data analysis.

In short, our purpose in this chapter is to survey issues which seem likely to influence the way in which canonical analysis will come to be used and its effectiveness enhanced during the course of the next decade. In so doing we shall find that, while the worked examples of Chaps. 6–12 may be illuminating, they fall short of good, modern statistical practice in several respects.

In Sect. 14.2 issues which arise in the design stage of an investigation prior to the collection of data are addressed. Section 14.3 focusses on procedures which are available for probing the data acquired before analysis so as to obtain some indication of their distributional properties. Special problems are posed in canonical analysis by outlying and influential observations as well as by collinear relationships among variables. These problems are taken up in Sect. 14.4. In Sect. 14.5 the statistical assessment of a fitted model is addressed, while a number of miscellaneous items are also brought together for discussion in this section. The chapter concludes with Sect. 14.6 which is devoted to a general summing up. A work of reference which will be found useful throughout the chapter is Gnanadesikan (1977).

14.2 Data collection

Suppose that canonical analysis has been properly identified as a plausible model which is well-matched to some declared ecological purpose. Then, proper identification of the model in itself is no guarantee that the analysis will prove to be informative. To appreciate this point, consider the effect of the design of a study on the outcome of analysis. Large increases in clarity and economy can reasonably be anticipated where the collection of data is guided and informed by familiarity with properties of the model to be fitted. Evidently, the design of a study is likely to repay careful attention. Two considerations which arise during the design stage concern the choice of variables for inclusion and the experimental or survey design. Fairly obviously, a computer algorithm for canonical analysis cannot in itself be expected to compensate for deficiencies at this stage.

14.2.1 Choice of variables

The need for substantive judgement in choosing variables is largely self-evident. It is plain that the omission of variables which play a significant part in the system investigated will have serious and frequently irredeemable consequences. On the other hand, the over-inclusion of variables also has harmful effects. In such cases there is a loss of precision in estimation. In other words, the overfitted model is sensitive to sampling error. Further, the power of any statistical tests which might be applied is a strictly decreasing function of the number of variables. In short, as more variables are acquired, the greater the danger that interesting relationships or effects will go undetected. Moreover, highly correlated variables add little to explanatory power and can create difficulties in the numerical estimation of the canonical weights while rendering the estimates sensitive to sampling error and vulnerable to instability. For these reasons, the need for care and judgement in selecting variables can scarcely be over-emphasized.

Once the essential variables have been identified, the question of the scale or level of their measurement arises. Although interval-scaled measurements are widely used in ecology, binary, nominal or ordinal measurements might often be more appropriate. They have the decided advantage of being easier and speedier to observe. An important consequence is that larger samples can be realized for a given expenditure of time and effort than would otherwise be the case. Large samples are advantageous in two ways. First, there is the reduction in sampling error that accompanies an increase in sample size. Second, for sufficiently large samples, the opportunity exists to subdivide the sample randomly prior to analysis and to proceed to a two-stage analysis as a means of validating the fit of the model. Efficient numerical methods for the canonical analysis of categorical data have recently become available (e.g., see Wold, 1980; van der Burg & de Leeuw, 1983). Such advantages, however, have to be weighed against the decreased stability of results which also accompanies a weakening of the measurement level employed for the variables (Perreault & Young, 1980).

14.2.2 Experimental design

A principal challenge in canonical analysis is to obtain accurate and precise estimates for the parameters of the fitted model. Hence a prime requisite of a sample is that it be representative or unbiased. In ecology the difficulty of arriving at stable and reliable parameter estimates is especially acute because of the high sampling variance which characterizes ecological systems. Two steps towards stabilizing the estimates are, however, available. In the first place, efforts to control the error variance can be made; secondly, sample size may be increased, as indicated above. In order to control unwanted variance the sources of variation must be identified and their relative magnitudes assessed. The principles of randomization, replication and stratification may then be used to reduce the inherent variability of the measurements. In connection with vegetation and other surveys, random samples, though unbiased, are not necessarily desirable nor always feasible. The advantages of systematic designs merit careful consideration in connection with vegetation surveys in particular.

Guttman (1968) has remarked that it is the task of experimental design – before data are collected – to ensure that *enough* data points are gathered with sufficient *spread* to reveal the basic lawfulness sought. With respect to sample size, it is largely axiomatic that efforts to ensure that the sample is large enough to answer decisively the research questions posed will be rewarded. The inherent sparsity of high-dimensional spaces (Gower, 1973; Friedman & Stuetzle, 1981) needs to be borne in mind here. Considerations of spread are largely the province of subject-matter knowledge, insight and experience. Two matters of a non-substantive kind, however, merit attention – namely, the desirability of (a) arriving at a sample which is substantially homogeneous, and of (b) dealing with relationships of a broadly linear kind. As homogeneity and linearity are both to some extent functions of scale, significant advantages are likely to accrue from minimizing the areal and subject-matter domains of a study as far as is consistent with realizing the overall goals in question. One has therefore to attempt to balance the requirement for an intensive and representative sample against the desiderata of homogeneity and linearity.

Further reading. Canonical analyses based on *binary, nominal or ordinal-level measurements* have been reported by Jones and Bock (1960), Maxwell (1961), Burnaby (1971), Cooley and Lohnes (1971, p. 179), Buzas (1972), Harris (1975, p. 137), Gnanadesikan (1977, p. 78), Green (1978, p. 266), Huba, Wingard and Bentler (1980) and by van der Burg and de Leeuw (1983). Gleason and Staelin (1973) have proposed a scheme for *improving the metric properties* of ordinal data. *Estimation methods* well-suited to the canonical analysis of categorical data have been developed by Wold (1966a, b, 1980), Young, de Leeuw and Takane (1976) and by van der Burg and de Leeuw (1983).

In relation to sampling matters, Kruskal and Mosteller (1979a, b, c, 1980) have discussed the concept of *representativeness* at length while Sterling (1979) has investigated *sources of bias* in a particular class of multivariate surveys. W.T. Williams (1971) has drawn attention to some consequences of bias in vegetation surveys. A study of the effects of *stratified and clustered designs* in multivariate analysis has been made by Bebbington and Smith (1977, p. 175). Beaton (1978) has examined the consequences of least-squares fitting where samples are *arbitrary* (i.e. nonrandom) and shown that useful indices of goodness-of-fit can be realized. Beaton has developed a metric for measuring the fit in such cases based on signed permutations of the residuals despite the lack of statistical independence. The question of valid *inference* from nonrandom samples has been addressed by Smith (1983). In connection with predictive applications of canonical analysis, Barcikowski and Stevens (1975) and Thorndike (1978, p. 184) have proposed guidelines as to what might constitute samples of adequate *size* relative to the number of variables of interest (i.e. relative to the subject-matter domain). Warren (1971) has discussed a dilemma which arises where attempts are made to design simultaneously for correlation (description) and regression (prediction).

14.3 Initial data exploration

Before embarking on canonical analysis it is important to gauge the appropriateness of the model itself. The quantities of fundamental interest in canonical analysis are all functions of the sample mean vector and covariance matrix. Where a sample is made up of independent, identically distributed observations from a $(p+q)$-variate normal distribution, the mean vector and covariance matrix together completely characterize the distribution. It so happens, however, that field observations in ecology are rarely multinormal. Moreover, observations are frequently not identically distributed, so that a sample may lack homogeneity. Under these conditions the adequacy of the sample mean vector and covariance matrix as summaries of the data is markedly degraded. Consequently, it is good practice to assess the homogeneity and distributional properties of a sample before proceeding to canonical analysis. The requirement for multivariate normality can in fact be relaxed in a wide class of ecological endeavors; only where tests of statistical hypotheses are contemplated is normality essential. Minimal distributional requirements for other purposes are that the data be realizations of $(p+q)$-dimensional variables from an elliptically symmetric distribution. More specifically, where the data distribution is reasonably symmetric, not too long-tailed and free from contamination, the sample mean vector and covariance matrix may be expected to summarize the data adequately for many practical purposes. Provided these conditions are broadly satisfied then canonical analysis is at least not counter-indicated. Even where homogeneity and distributional requirements are not initially satisfied, steps to remedy the deficiencies are frequently possible.

14.3.1 Homogeneity

In vegetation and other ecological surveys, the sample is frequently heterogeneous. That is to say the sample consists of observations drawn from two or more distributions which may or may not be elliptically symmetric but whose mean vectors differ. It has been well said (Webb, 1954) that variation in vegetation hovers in a tantalizing way between the continuous and the discontinuous, that is between homogeneity and heterogeneity. In general, as the geographical extent of a survey is increased so is the likelihood of heterogeneity. As a first step in gauging the appropriateness of canonical analysis, therefore, particularly in large-scale studies, it is prudent to investigate the homogeneity or otherwise of the sample.

Structure in multivariate data cannot be perceived by looking either at the data themselves or at a set of summary statistics derived from them. What can be done, however, is to examine the $(p+q)$-dimensional point cloud corresponding to the data and to base the assessment of structure on this. Perception of the sample and of its internal structure in these terms, though, for $(p+q)>3$ will still not be without difficulty. In such cases an appealing procedure is to reduce the dimensionality of the data – most simply by a well-chosen *projection*. Thus, a judicious, low-dimensional projection of the sample may be expected to be revealing with respect to structure, and, in particular, homogeneity.

Numerous methods for recognizing low-dimensional subspaces near to which high-dimensional data may be considered to lie are available. The classical procedure of *principal component analysis* is available for the linear reduction of dimensionality. Where the bulk of the data lie close to a linear subspace, components analysis offers a computationally attractive means for identifying a new coordinate system of low dimensionality. All that is required is an eigenanalysis or singular value decomposition. The new coordinates are the principal components, which themselves are linear functions of the original coordinates. The sample is projected into the subspace spanned by the components corresponding to the dimensions of greatest scatter in the $(p+q)$-dimensional cloud. In this way the internal structure of the sample is disclosed and its homogeneity can be assessed. In searching for multivariate structure in this way, two considerations need to be borne in mind. First, the effects of possible outlying observations in generating spurious structure; second, nonlinear singularities may go undetected. With respect to the second point, it is worth noticing that the class of techniques comprising *nonmetric multidimensional scaling* are better suited to uncovering curved t-dimensional subspaces $[t<(p+q)]$ close to which the data may lie. Such methods begin with a matrix of dissimilarities between pairs of data-points. A low-dimensional Euclidean representation of this matrix is then sought for which the ordered distances between points are reasonably closely monotonically related to the corresponding ordered, observed dissimilarities. A measure of the conformity of the solution to monotonicity is provided by a numerical goodness-of-fit index, the stress. The solution is in fact arrived at by explicitly setting out to minimize the stress by means of numerical optimization. By minimizing a weighted index related to stress, the resulting low-dimensional configuration can be made even more responsive to nonlinear singularities. The *parametric mapping* of Shepard and Carroll (1966) is one such procedure.

A class of procedures for the interactive, graphical exploration of multiresponse data has recently been developed. Without prior assumptions, these methods project a point-scatter into carefully chosen low-dimensional subspaces so as to shed light on whatever interesting structure may be present. These methods resemble those outlined above in seeking to uncover well-fitting, t-dimensional configurations but differ from them in being essentially computer-intensive procedures for graphical *analysis*. There will be obvious practical advantages when t is taken to be two. As before, it is the homogeneity of the sample which is of primary interest to us, or, in other words, the tendency of the points to cluster. Broadly, the new methods are of two kinds. Those in which:

(a) the data themselves determine the planes onto which they are projected;
(b) the projection planes are selected by the user.

The former are exemplified by *projection pursuit* (Kruskal, 1969; Friedman & Tukey, 1974) and *multivariate planing* (Friedman & Rafsky, 1981). Projection pursuit involves a search for structured projections – that is for projection onto new coordinates which are carefully chosen after examining the full multivariate data set. The search is accomplished by defining some suitable projection index, here related to the heterogeneity of the data, and then locating the display which maximizes the index. A numerical optimization procedure is used for this step. The great merit of projection pursuit is that it is able to quickly sift through all possible displays to find the one which best satisfies the specified index or criterion. The eye is then allowed to scan the display and identify any heterogeneity or features of related interest which may be revealed. As the internal structure of multiresponse data will rarely be reflected completely in a single projection, the structure in one projection may be removed after it has been examined and assessed, and the algorithm allowed to seek additional interesting projections. This procedure is repeated until no further structure is found. Multivariate planing offers an alternative means for uncovering multivariate structure. The objective in multivariate planing resembles that of nonmetric multidimensional scaling. Specifically, a search is made for that plane onto which sample-points in multidimensional space may be projected with least structural distortion. Multivariate planing, however, does not involve the explicit optimization of an index of fit such as stress, although it so happens that a criterion expressing the extent to which the projection onto two dimensions distorts the true distances between points in $(p+q)$-dimensional space is calculated. Further, in arriving at a two-dimensional display less weight is given to large distances than to small ones, so that the planar display concentrates on local information. As a consequence the display is especially responsive to general nonlinear features in a way that is reminiscent of parametric mapping. Multivariate planing is not only exceptionally fast to execute but planing as many as 10,000 points is perfectly feasible. The method may be used equally in connection with a completely unstructured sample as well as in situations where the existence of g subsamples or clusters is known or suspected beforehand. In either case a stagewise approach is often appropriate. Any subordinate clusters or deviant sample-points which appear in the initial display may be removed and the remaining data replaned. This sequence of operations is repeated until a final projection is arrived at which cannot be subdivided or clustered meaningfully. Planes through the hyperspace following the first are optimal for displaying the structure of the remaining data and have no memory from previous displays. As in projection pursuit, the user's eye does the work of searching for patterns in a given display.

A comparable interactive graphical system for probing high-dimensional data has been described by Stephenson, Beatty and Gentleman (1981). Like projection pursuit and multivariate planing, the system generates a sequence of two-dimensional projections. A preliminary scan of the data is provided by a frame of 100 such projections. A single projection may then be selected by the user from the scan and a variety of manipulations performed on the data as well as on the way in which they are displayed. The option also exists to select projections according to some prespecified criterion, as in projection pursuit. A variety of other graphical methods for revealing structure in data also exist. Thus, the *Fourier series* representation of Andrews (1972) can be useful for assessing the overall coherence of a sample as well as for revealing the presence of deviant observations. Gabriel's *biplot* (Gabriel, 1981; Cox & Gabriel, 1982) is another flexible graphical device which is useful in the same way.

The methods outlined are all pertinent to the task of throwing light on the homogeneity of a sample. Where a sample is shown to be substantially homogeneous, canonical analysis is at least not counter-indicated. In other cases one might first wish to remove any subordinate clusters of sample-points or aberrant individuals which may have been revealed. The nature of such peculiarities, and the mechanisms which may have given rise to them, however, deserve to be pondered carefully even though the data themselves may be excluded from subsequent analysis. On the other hand, where heterogeneity proves to be dominant, canonical analysis may still be feasible in relation to particular well-defined clusters or subsamples whose size is adequate for the purpose. The general objective of all such exploratory endeavors is to identify or create one or more essentially homogeneous samples which, in this respect at least, may be suitable for canonical analysis. Having done so attention turns to the joint distribution of the variables of interest. Before turning to consider this aspect of the data, it is worth mentioning that where a sample proves to be heterogeneous the heterogeneity itself may merit further study. In such cases, canonical variate analysis might well provide a suitable means of so doing.

14.3.2 Assessing joint distribution

The essential requirement of the data for analysis is simply that their distribution be reasonably well-behaved. More precisely, the joint distribution of variables needs to be substantially elliptically symmetric. Where, in addition, the more stringent condition of multivariate normality is met, the opportunity for a more productive analysis than would otherwise be the case exists. We propose to describe some informal graphical methods for obtaining insight as to the distribution of a multiresponse variable before proceeding to outline some formal tests for multinormality. For concreteness, we will focus on the distribution of the vector $\mathbf{x}$ $(p \times 1)$, with the understanding that the procedures to be described apply equally to $\mathbf{y}$ $(q \times 1)$ and $\mathbf{g} = [\mathbf{x}^t | \mathbf{y}^t]^t$ also.

Let $\mathbf{x}_1, \mathbf{x}_2, \ldots, \mathbf{x}_N$ be independent observations from a p-variate distribution with sample mean vector $\bar{\mathbf{x}}$ and covariance matrix $\mathbf{S}$ but one whose characteristics are otherwise unspecified. We wish to obtain an indication as to the form of the joint distribution of the variables $x_1, x_2, \ldots, x_p$. For this purpose it will be advantageous to introduce the notion of a *quantile-quantile* $(Q-Q)$ *probability plot* (Wilk & Gnanadesikan, 1968). A $Q-Q$ probability plot is a device which enables the shape of the distribution of one or more variables to be displayed and evaluated relative to that of some suitably chosen standard or reference distribution. Often, a suitable reference distribution is provided by some such theoretical probability distribution as the standard normal, uniform, chi-squared or gamma distribution. The basic idea underlying the plot is that a random sample from a population will tend to be representative in that the sample empirical cummulative distribution will approximate the theoretical distribution in the population. For observations which are multivariate, the principal steps in constructing a $Q-Q$ plot are as follows:

(a) convert each multiresponse observation to a scalar quantity;
(b) plot the ordered scalars against the corresponding quantiles of an appropriate reference distribution.

The essential property of a $Q-Q$ probability plot is that where the data and reference distributions are similar, the plotted points corresponding to sample-observations tend to fall on a *straight* line oriented towards the origin. Deviations from linearity, on the other hand, are indicative of discrepancies between the data and theoretical distributions. Two points requiring attention before a $Q-Q$ plot can be made are choices for (a) the scalars which are to represent the p-variate observations $\mathbf{x}_j$; and (b) the reference distribution against which the distribution of the scalar quantities is to be assessed.

For reasons which will become apparent, a convenient scalar quantity which can be used to represent the jth observation, $\mathbf{x}_j$ $(j=1, \ldots, \mathrm{N})$, in an acceptable way is the distance of the corresponding sample-point in p-space from the center of gravity of the point-configuration as a whole, $\bar{\mathbf{x}}$. A general expression for the square of this distance, d_j^2, say, is given by the quadratic form:

$$d_j^2 = (\mathbf{x}_j - \bar{\mathbf{x}})^t \mathbf{A} (\mathbf{x}_j - \bar{\mathbf{x}}), \qquad j=1, \ldots, \mathrm{N}, \tag{14.1}$$

where the compounding matrix $\mathbf{A}$ of order p is positive semidefinite but otherwise open to choice. By setting $\mathbf{A}=\mathbf{I}$, d_j^2 is the squared Euclidean distance between $\mathbf{x}_j$ and the sample centroid. If we let $\mathbf{A}=\mathbf{S}^{-1}$, the distance d_j^2 takes account of the possibly differing variances of $x_1, x_2, \ldots, x_p$ and of the correlations between them. The use of several different compounding matrices is generally recommended. Gnanadesikan (1977, p. 235) has offered further suggestions as to what might prove to be suitable matrices for the purpose. Turning to the selection of a reference distribution, we observe that selection is determined by what is known or can reasonably be surmised about the distribution of the scalar quantities associated with the observations $\mathbf{x}_j$ – here, that is, of the d_j^2. If for the moment we assume that the $\mathbf{x}_j$ are themselves realizations of variates from a p-variate normal distribution, then the distribution of the $d_1^2, d_2^2, \ldots, d_\mathrm{N}^2$ can be shown (Gnanadesikan, 1977, p. 233) to be approximated reasonably satisfactorily by a gamma distribution. Accordingly, under the null assumption of p-variate normality, one suitable theoretical distribution against which the data distribution can be evaluated is the gamma distribution $g(d^2; \lambda, \eta)$, where λ and η are respectively scale and shape parameters to be estimated from the d_j^2. It follows that where the $\mathbf{x}_j$ are realizations of independent p-variate normal quantities, a plot of the ordered squared distances against quantiles of the gamma distribution with appropriate choices for λ and η will give rise to an approximately linear configuration oriented towards the origin. Conversely, where the null assumption is not in accord with the data, the discrepancy will be disclosed by systemmatic or other departures from linearity in the gamma probability plot. A $Q-Q$ plot is especially valuable for discerning characteristics of the tails of the data distribution. In particular, the extremities of the plot provide information regarding the length of the tails, the symmetry or otherwise of the distribution as well as the presence of possible aberrant data points. It will perhaps be recalled that these characteristics are precisely those which bear most directly on the adequacy of the mean vector and covariance matrix as summary statistics. On the other hand, while the value of a $Q-Q$ plot for detecting discrepancies between a data and a theoretical distribution is considerable, in general such plots are not especially revealing about the precise form of the data distribution apart from the features noted.

In outlining the construction of a $Q-Q$ plot above, the p-variate normal distribution was used as a null assumption in order to develop the plot. Earlier, however, we remarked to the effect that elliptical symmetry is likely to be of greater practical importance than multivariate normality. It so happens that the multinormal distribution provides a reasonable null distribution with respect not only to multinormality but in relation to the class of elliptically symmetric distributions generally. To demonstrate this point it will be useful to introduce the density function for an elliptically symmetric variable. A random vector $\mathbf{x}$ $(p \times 1)$ is said to have an elliptically symmetric distribution, or simply an elliptical distribution, with parameters $\boldsymbol{\mu}$ $(p \times 1)$ and $\boldsymbol{\Sigma}$ $(p \times p)$ if its density function has the form:

$$f(\mathbf{x}) = c_p |\boldsymbol{\Sigma}|^{-\frac{1}{2}} h\{(\mathbf{x}-\boldsymbol{\mu})^t \boldsymbol{\Sigma}^{-1}(\mathbf{x}-\boldsymbol{\mu})\}, \qquad |\boldsymbol{\Sigma}| \neq 0,$$

where c_p is a normalizing constant and h a nonnegative function. This expression shows that an elliptical distribution retains the structure of a p-variate normal distribution, that is elliptical symmetry, while omitting the precise form of this distribution. The contours of equal density for the family of elliptical distributions therefore have the same shape as those of the p-variate normal, though the family contains long-tailed and short-tailed members relative to the normal. We have remarked that $Q-Q$ plots are well-suited to diagnosing peculiarities in the tails of a distribution. The sensitivity of the plots to elliptical distributions other than the multinormal may be sharpened by taking the cube roots of the squared distances, $d_j^{2/3}$, and plotting these against either cube roots of gamma quantiles or directly against quantiles of the standard normal distribution, N(0, 1). The gamma distribution is asymmetric and the cube root transformation has the effect of approximately symmetrizing the distribution over a wide range of values of the shape parameter, λ. The advantage of transforming is that the configuration on the resulting $Q-Q$ plot is not compressed towards the origin, as it is for the usual gamma plot. Consequently, the configuration for the transformed plot has the potential for greater flexibility in shape and a corresponding increased sensitivity to features in *both* tails of the data distribution.

Long-tailed distributions are much commoner in practice than short-tailed ones. The $Q-Q$ plot for a long-tailed elliptical distribution is distinguished by a configuration for which sample-points in the extremities are further from the center (the median) of the configuration than is the case for a multinormal distribution. The interested reader is referred to Chambers, Cleveland, Kleiner and Tukey (1983, p. 207) for an illustration of a $Q-Q$ probability plot of a long-tailed distribution of this sort. We mention also that one reason for the appeal of distance as a scalar quantity by means of which to represent the p-variate observation $\mathbf{x}_j$ is now plain – distance is a variable whose distribution is both known and tractable. Thus, the task of identifying a reference distribution is resolved. In parenthesis, it is of interest to observe that the beta distribution actually provides a somewhat closer approximation of the distribution of the d_j^2 than the gamma. Small (1978) has drawn attention to this point.

Another graphical procedure for assessing the form of a p-variate distribution is based on a modification of Andrews' (1972) method for plotting high-dimensional data. The modification involves plotting selected quantiles or percentage points of the data distribution, rather than individual samples. Such *quantile contour plots* are best suited to large samples, where they may be informative concerning the symmetry of the distribution and the presence of outlying observations.

We turn now to consider formal tests of multivariate normality. Interest in multinormality in canonical analysis is usually expressed for one of two reasons – a need to validate requirements for the proper use of significance tests, or as a means of verifying that a covariance (or correlation) matrix adequately summarizes the relationships in question. With respect to the second circumstance, it is reasonably well-known that the efficiency of the covariance matrix as a summary statistic is seriously impaired by departures from normality of two kinds, namely (a) heavy-tailed and especially skewed distributions; and (b) nonlinear dependencies. Thus, in assessing joint normality there are grounds for focussing attention on possible departures of these kinds. Numerous tests of multivariate normality have been proposed and in selecting from among these a small number for discussion below, we have been guided by the special relevance of tests which are sensitive either to the shape of the data distribution or to nonlinearity.

Two intuitively appealing test statistics for multinormality where sensitivity to the symmetry and tail characteristics of a joint distribution is required are the multivariate coefficients of *skewness* and *kurtosis* (Mardia, 1970). Suppose $\mathbf{x}_1, \mathbf{x}_2, \ldots, \mathbf{x}_N$ are a sample of N independent p-variate observations with sample mean vector $\bar{\mathbf{x}}=N^{-1}\Sigma\mathbf{x}_j$ and covariance matrix $\mathbf{S}=(N-1)^{-1}\Sigma(\mathbf{x}_j-\bar{\mathbf{x}})(\mathbf{x}_j-\bar{\mathbf{x}})^t$, respectively. Then, the sample skewness and kurtosis are defined by

$$b_{1,p}=N^{-2}\sum_{j,k=1}^{N}\{(\mathbf{x}_j-\bar{\mathbf{x}})^t\mathbf{S}^{-1}(\mathbf{x}_j-\bar{\mathbf{x}})\}^3$$

and

$$b_{2,p}=N^{-1}\sum_{j=1}^{N}\{(\mathbf{x}_j-\bar{\mathbf{x}})^t\mathbf{S}^{-1}(\mathbf{x}_j-\bar{\mathbf{x}})\}^2.$$

For a multinormal population Mardia (1970, 1974) has shown that the expected values of these statistics are $\mathscr{E}(b_{1,p})=0$ and $\mathscr{E}(b_{2,p})=p(p+1)$, respectively. Moreover, Mardia has shown that as $N\to\infty$ the distribution of these sample indices of multivariate skewness and kurtosis are given by:

$$Nb_{1,p}/6\sim\chi^2(f), \quad f=p(p+1)(p+2)/6, \quad \text{and} \quad \frac{b_{2,p}-p(p+2)}{\{8p(p+2)/N\}^{\frac{1}{2}}}\sim N(0,1).$$

Tests of multivariate skewness and kurtosis are therefore readily made by reference to tables of percentage points of the chi-squared and normal distributions. Mardia (1974, 1975) has prepared supplementary tables providing greater accuracy for tests relating to small samples for which $p=2$, 3 or 4. The null hypothesis of no skewness, $\beta_{1,p}=0$, is rejected for large values of $b_{1,p}$ and that of no kurtosis, $\beta_{2,p}=p(p+2)$, for both large and small values of $b_{2,p}$. These tests are identically tests of p-variate normality. Reservations concerning the adequacy of the normal approximation to the distribution of $b_{2,p}$ have, however, been expressed by Gnanadesikan (1977, p. 175).

A further implication of multivariate normality, in addition to the absence of skewness and kurtosis, is that the regression of one component of a p-variate vector on any other set of components is *linear*. Cox and Small (1978) have proposed tests of joint normality based on this property of the multinormal distribution. In outlining these tests below, we shall concentrate for simplicity on the case for which $p=2$. Let (x_{j1}, x_{j2}) be N independent observations ($j=1, \ldots, \mathrm{N}$) from a bivariate distribution. Then a test of the linearity of the regression of x_2 on x_1^2 is given by $Q_{2,1}$, the usual Student t statistic for the significance of the regression coefficient of x_2 on x_1^2. We define the quantity $Q_{1,2}$ in an analogous way. Under bivariate normality, the joint distribution of $(Q_{2,1}, Q_{1,2})$ is approximately bivariate normal with zero mean and unit variance. The asymptotic correlation coefficient between $Q_{2,1}$ and $Q_{1,2}$ is shown by Cox and Small to be $\mathrm{corr}(Q_{2,1}, Q_{1,2})=\rho(2-3\rho^2)$, where $\rho=\mathrm{corr}(x_1, x_2)$. We shall denote the sample estimate of ρ by r. Two composite test statistics are proposed:

(a) $\max(|Q_{2,1}|, |Q_{1,2}|)$, which for large samples may be referred to tables of the bivariate normal distribution; and

(b) $$[Q_{2,1}\, Q_{1,2}] \begin{bmatrix} 1 & r(2-3r^2) \\ r(2-3r^2) & 1 \end{bmatrix}^{-1} \begin{bmatrix} Q_{2,1} \\ Q_{1,2} \end{bmatrix},$$

which for large samples may be tested for significance by reference to tables of the chi-squared distribution with 2 degrees of freedom.

These tests extend readily to cases for which $p>2$. Details will be found in Cox and Small (1978). For $p>2$ the second test may be replaced or augmented by a quantile-quantile probability plot. For this purpose, the ordered t statistics, $Q_{r,s}^{(p)}$ ($r, s=1, \ldots, p$; $r \neq s$), are plotted against quantiles of a sample of size m from the standard normal distribution $[m=2p(p-1)]$. The linearity of the resulting configuration enables the hypothesis of p-variate normality to be assessed.

Cox and Small (1978) have suggested a third test of multivariate normality which resembles the two preceding tests in being sensitive to nonlinearity, but differs from them in being invariant under arbitrary nonsingular transformation of the p component variables. The most direct approach in constructing an invariant test is to seek that pair of normalized linear combinations of the p-variate response $\mathbf{x}$, $y=\mathbf{a}^t\mathbf{x}$ and $z=\mathbf{b}^t\mathbf{x}$, such that one has maximum curvature in its regression on the other. Now define $\eta=\gamma/[\mathscr{E}(z^4)-1-\{\mathscr{E}(z^3)\}]^{\frac{1}{2}}$, where γ is the least squares regression coefficient of y on z^2 adjusted for the linear regression on z. The square of η, η^2, expresses the curvature of the relationship of y on z – it is the proportion of the variance of y accounted for by the quadratic component in the least squares regression of y on z and z^2. Letting $\eta^2(\mathbf{b})$ be the supremum of $\eta^2(\mathbf{a}, \mathbf{b})$ over $\mathbf{a}$ for fixed $\mathbf{b}$, and writing $\hat{\eta}^2(\mathbf{b})$ for the sample value of $\eta^2(\mathbf{b})$ and $\hat{\eta}^2_{\max}(\mathbf{b})$ for its maximum, we have the following result:

$$\log \hat{\eta}^2(\mathbf{b}) \sim \mathrm{N}(\mu, \sigma),$$

where $\mu = \log(5p^2/8\mathrm{N})$ and $\sigma = \log(.53 + 3.87/p)$. Thus, the invariant test leads to a very convenient result.

A formal test of multivariate normality can also be made on the basis of a $Q-Q$ probability plot. For this test, the product-moment correlation coefficient, r_Q say, is first evaluated for the configuration. The significance of this statistic may then be assessed by reference to a table of critical values given by Filliben (1975). There are other tests of p-variate normality – for example, see Hawkins (1981), Koziol (1982, 1983) and Royston (1983). Royston draws attention to the need to examine *subsets* of variables in addition to the full set, in order to detect departure from joint normality that might otherwise go undetected.

Where the data for analysis are not multivariate normal, remedial steps to bring the data into closer conformity with Gaussianity, or, at any rate, to a form more tractable for analysis are frequently possible. We have seen that departure from multinormality can arise in several ways and insight into the nature of the departure can be useful in suggesting suitable remedies. Distributions which are longer-tailed than the normal, for example, can be made more tractable for analysis by the use of robust estimation, a topic taken up in Sect. 14.4.2 below. The exclusion of aberrant individuals in a contaminated sample, in itself, can also lead to a marked improvement in distributional characteristics. Further, skewed distributions can sometimes be symmetrized and nonlinear relationships linearized by transforming the variables appropriately. It is to transformations as a means of enhancing joint normality that we now turn.

14.3.3 Re-expressing variables

The choice of an appropriate normalizing transformation can sometimes be made on the basis of theoretical considerations. More often, however, a choice of this kind is not obvious. What nevertheless can be done is to let the data themselves suggest a transformation. The family of univariate power transformations (Box & Cox, 1964) are of this type, being indexed by a parameter λ which is estimated from the data. Data-based transformations to enhance multinormality are a direct extension of the work of Box and Cox to the p-variate case. Even if as a result of such a transformation joint normality is not achieved, transformation may go a long way towards symmetrizing and linearizing the data. For many practical purposes this in itself would represent a worthwhile improvement. Let $\mathbf{y} = [y_1, y_2, \ldots, y_q]^t$ be a q-vector of nonnegative response variables. Suppose we wish to find a vector of transformation parameters $\hat{\boldsymbol{\lambda}}(q \times 1)$ such that the transformed observations $\{g_1(\mathbf{y}_j^t; \hat{\boldsymbol{\lambda}}),$

$g_2(\mathbf{y}_j^t; \hat{\boldsymbol{\lambda}}), \ldots, g_q(\mathbf{y}_j^t; \hat{\boldsymbol{\lambda}})\}$ where $j=1, \ldots, N$, are more nearly q-variate normal, $N_q(\bar{\mathbf{x}}, \mathbf{S})$, than the original observations. A simple family of nonlinear transformations of this kind can be defined as follows:

$$g_h(\mathbf{y}^t; \hat{\boldsymbol{\lambda}}) = y_h^{(\hat{\lambda}_h)} = \begin{cases} (y_h^{(\hat{\lambda}_h)} - 1)/\hat{\lambda}_h & \text{for } \hat{\lambda}_h \neq 0 \\ \log_e y_h & \text{for } \hat{\lambda}_h = 0, \end{cases}$$

where the required $\hat{\lambda}_h$ $(h=1, \ldots, q)$ are estimated from the data by maximizing either q univariate likelihoods or a single, joint q-variate likelihood (see Andrews, Gnanadesikan & Warner, 1971). There are three variants of the procedure. We may seek to improve the *marginal* normality of the variables $y_1, y_2, \ldots, y_q$ separately, evaluating each $\hat{\lambda}_h$ in turn. Or an attempt to enhance *joint* normality can be made by finding suitable values for the elements of $\hat{\boldsymbol{\lambda}} = [\hat{\lambda}_1, \hat{\lambda}_2, \ldots, \hat{\lambda}_q]^t$ simultaneously. Estimation in this case is iterative with the starting value for $\hat{\boldsymbol{\lambda}}$ being the estimates $\hat{\lambda}_h$ $(h=1, \ldots, q)$ arrived at in seeking to improve marginal normality. Alternatively, efforts to achieve *directional* normality may be made – that is to say normality for just those directions in the space of the original variables which exhibit nonnormality. The three options differ in their computational requirements as well as in other respects. Schemes aimed at improving marginal or directional normality are most appealing computationally. On the other hand, there is evidence that the approach focussing on joint normality, though computationally demanding, yields more stable estimates of the parameters $\hat{\lambda}_h$, perhaps as a consequence of exploiting the correlations between variables. The effectiveness of transformation can be assessed by a gamma probability plot of squared distances calculated from the data before and after transformation. While there is no guarantee that joint normality will be achieved, re-expression can be a useful preliminary step before computing such statistical summaries as the vector mean and covariance or correlation matrix.

Power transformations offer a further test of multivariate normality. To see this, observe that the linear transformation specified by $\boldsymbol{\lambda} = [\lambda_1, \lambda_2, \ldots, \lambda_q]^t = \mathbf{1}$, that is the identity transformation, is the only transformation of this family which is consistent with the hypothesis that the $y_1, y_2, \ldots, y_q$ are jointly normal. A likelihood ratio test of the hypothesis $\boldsymbol{\lambda} = \mathbf{1}$ yields a test criterion which for large samples is distributed approximately as a chi-squared variate on q degrees of freedom. A test of multivariate normality is therefore readily made. Further details are given by Gnanadesikan (1977, p. 171).

The above discussion of transformations has centered on the q-variate response variable, $\mathbf{y}$. Power transformation of the predictor variable $\mathbf{x}$ $(p \times 1)$ is also possible (see Box & Tidwell, 1962; Weisberg, 1980, p. 141) and may actually be more effective in rendering the data amenable to analysis than transformation of $\mathbf{y}$. The two approaches differ in the rationale used to arrive at estimates for the transformation parameters. It has yet to be shown how such considerations may affect applications of canonical analysis in which a need for transformation arises and in which the relationship between $\mathbf{x}$ and $\mathbf{y}$ is viewed symmetrically.

The techniques described in Sects. 14.3.1 and 14.3.2 enable some appreciation of the data distribution to be acquired. Thus, the extent to which the data meet the specifications required for canonical analysis can be gauged. What is basically required is that the observations be identically and independently distributed with respect to some $(p+q)$-dimensional elliptical distribution. Where doubt is cast on the homogeneity of the sample, it may be possible to discard outlying samples or sample-groups of subordinate size until homogeneity is attained. The balance of the data may then be found, perhaps after re-expression, to be substantially elliptical or they may quite simply not conform to this specification at all. The latter case is outside the domain of canonical analysis and is not considered here. Where elliptical symmetry is demonstrated, at least approximately, one may proceed to further interrogate the data as to their suitability for canonical analysis. In Sect. 14.4 which follows, difficulties posed by outlying or influential observations, long-tailed distributions and collinear relations among variables are taken up.

Further reading. *Projection pursuit* and related methods for probing high-dimensional data by means of well-chosen, low-dimensional projections have been investigated or described by Tukey and Tukey (1981, p. 219), Friedman and Stuetzle (1982), Diaconis and Friedman (1983) and Garrett (1983). A computer program for the interactive graphical exploration of multivariate data by means of the *biplot* has been developed by Tsianco, Gabriel, Odoroff and Plumb (1981). Options for 3-dimensional displays and for the rotation of point configurations are attractive features of this program. *Quantile-Quantile probability plotting* methods for studying distributional properties of high-dimensional data have been discussed by Wilk and Gnanadesikan (1968), Gerson (1975), Gnanadesikan (1977, p. 227; 1980), and Jewell (1982). A simple procedure for obtaining a robust estimate of the shape parameter of a gamma distribution with application in constructing gamma probability plots has been developed by Kimber (1983). Wilk, Gnanadesikan and Huyett (1962), Bock (1975, p. 184) and Daniel and Wood (1980) contain standard or reference figures which are helpful in assessing $Q-Q$ plots. For an enlightening discussion of the interpretation of $Q-Q$ plots reference may profitably be made to Chambers, Cleveland, Kleiner and Tukey (1983, p. 203). The construction and use of *quantile contour plots* has been described by Gnanadesikan (1977, p. 210). The assessment of *multivariate normality* has been investigated by Andrews, Gnanadesikan and Warner (1971) and Gnanadesikan (1977, p. 168). Malkovich and Afifi (1973), Cox and Small (1978) and Mardia (1980) have reviewed work in this area. The only *exact* test of multivariate normality as opposed to tests which depend on asymptotic distribution theory, is that of Rincon-Gallardo, Quesenberry and O'Reilly (1979). In addition to providing an exact test statistic, their method can be supplemented by a probability plot which provides insight as to how and where the data may deviate from multinormality (see Dunn, 1981). The reference distribution in this case is the standard uniform distribution, $U(0, 1)$, and has the advantage of avoiding the uncertainty associated with the estimation of scale and shape parameters in distributions such as the gamma. Filliben (1975) has developed a *probability plot correlation coefficient* test for multinormality, while Mardia and Zemroch (1975) have given a Fortran subroutine for calculating multivariate *skewness* and *kurtosis*. Hawkins (1981) and Fatti, Hawkins and Raath (1982, p. 63) have described a test for multinormality and homoscedasticity of use in discriminant analysis which is also good for outlier detection.

Chmielewski (1981) has reviewed the subject of *elliptical distributions* and compiled a bibliography relating to them. *Tests for ellipsoidal symmetry* have been developed by Beran (1979). Muirhead and Waternaux (1980) have proposed tests of significance for canonical correlation coefficients where the data distribution is elliptical and their work has been extended by Tyler (1983). Andrews, Gnanadesikan and Warner (1971), Gnanadesikan (1977, p. 137), Dunn and Tubbs (1980) and Dunn (1981) have discussed *multivariate power transformations*. A quick method for choosing a symmetrizing transformation has been described by Hinkley (1977), while Henderson and Velleman (1981) have described a technique for identifying a value of the parameter λ in a power transformation which simultaneously improves symmetry and linearity in the bivariate case. It would be interesting to discover whether the latter could be extended to vector-valued variates. Cook and Wang (1983) have proposed a diagnostic procedure for assessing the extent to which individual samples or groups of samples influence the estimate of the Box-Cox transformation parameter, λ. *Splines* yield a class of smooth monotonic transformations of greater flexibility than generalized power transformations. Winsberg and Ramsay (1980) have reported an investigation of spline transformations of both the dependent and independent variables in regression analysis. Optimal transformations for multiple regression and correlation which may be useful in canonical analysis have been proposed by Breiman and Friedman (1982).

Applications of $Q-Q$ plots in biology or related areas have been made by Campbell (1978a, 1980b, 1982), Beauchamp, Begovich, Kane and Wolf (1980), Campbell and Reyment (1980), Dunn (1981), Royston (1983) and Hopper, Campbell and Caputi (1984). Malmgren (1979) has employed measures of multivariate skewness and kurtosis in testing the multinormality of data from planktonic foraminifera and has also reported the effects of certain nonlinear transformations on the distribution of these data. The power transformation test and the exact goodness-of-fit test for multinormality have been applied to taxonomic data by Dunn (1981). Applications of data-based power transformations in biological or related contexts have been made by Howarth and Earle (1979) and by Dunn (1981), while Cohen, Gnanadesikan, Kettenring and Landwehr (1977) have described an application of quantile contour plots.

14.4 Potential data problems

Throughout this section we shall be concerned with a sample composed substantially of identically distributed observations from one of the family of $(p+q)$-elliptical distributions. The restriction of our treatment to data of this kind is justified by the empirical observation that the variables of interest in ecology are not only vector-valued but are also almost invariably correlated to a greater or lesser extent. It is the correlations which are responsible for a data distribution whose probability density is ellipsoidal. While the above specification indicates that the data may be regarded as reasonably well-behaved, the naive application of canonical analysis would nevertheless be ill-advised. There are several reasons why this should be so. In the first place, the sample may contain outlying or influential observations. Second, the data distribution may be long-tailed relative to the Gaussian. Finally, one or both of the data matrices **X** and **Y** may contain near-linear dependencies. Each of these conditions can seriously diminish the worth of an analysis. Thus the data may be deficient in various ways. Accordingly, it is important to consider the likely impact of the data themselves on the outcome of analysis before proceeding. The specific goals of this aspect of the preliminary examination of the data are to uncover exceptional data points, peculiarities in the data distribution and collinear relationships between variables. Once detected, these features may be addressed directly, or, alternatively, a fitting procedure specifically designed to minimize their effects on the ensuing canonical analysis chosen. In the following section (Sect. 14.5) we shall examine the complementary issue of the impact of the model in canonical analysis and ways in which the model may prove to be inadequate.

14.4.1 Outlying or influential observations

Outlying observations are subsets of the data which appear to deviate markedly from other members of the sample of which they are part. Expressed somewhat differently, outliers are observations with high probabilities of occurring where the probability density is low, remote from the main body of the data (Hawkins, 1980). The term influential is applied to observations which individually or together with others exert a demonstrably large effect on the canonical weights and other estimates associated with a fitted model. While influential observations generally prove to be outliers the converse is not necessarily true, in other words an outlier need not be influential. Outliers merit attention in canonical analysis for several reasons – they may be interesting and informative in their own right, they frequently have drastic effects on the weights and other quantities of interest and they may lead to the adoption of models or estimation procedures that are especially designed to accommodate outliers and so result in improved estimates. Where an outlier is encountered it is always wise to first refer back to the source of the data if possible in order to verify the values. If the values are in error, they can be corrected or set aside. Where the values are correct, the question as to their significance and implications in substantive terms arises. An outlier, for example, may prove to be the most important observation in a sample. Furthermore, because of the sensitivity of least squares estimators generally to outlying observations, the adoption of a robust alternative to the usual least squares estimates of the $\{r_k;\ \mathbf{a}_k, \mathbf{b}_k;\ u_k, v_k\}$ would generally be well-advised.

Several informal graphical methods for the *detection* of multivariate outliers have been proposed. Andrews' (1972) function plot, introduced in Sect. 2.4.3, is one such method. The procedure involves mapping each data point $\mathbf{x} = [x_1, x_2, \ldots, x_p]^t$ into the function $f(t) = x_1 \sqrt{2} + x_2 \sin(t) + x_3 \cos(t) + x_4 \sin(2t) + \ldots$ and plotting over the range $-\pi < t < \pi$. Outliers are disclosed by curves which differ in shape and location from the curves corresponding to the majority of data points over the length of the plot. The flexibility of this approach is considerable; fresh insights may be obtained by permuting the variables $x_1, \ldots, x_p$ before plotting, by the use of functions other than the sine-cosine series specified above, and, for large samples, by plotting judicious combinations of selected quantiles and individual data points. The gamma probability plot described earlier is also useful in searching for outliers. For this purpose, ordered squared distances d_j^2 $(j = 1, \ldots, \mathrm{N})$ of data points from the sample centroid are plotted against quantiles of an appropriate gamma distribution, as described in Sect. 14.3.2. Aberrant observations are recognizable as points which are well separated from the remaining data points. An extension of this procedure has been proposed by Campbell (1978a). Campbell has suggested using the *influence function* associated with each sample as an alternative to the squared distance d_j^2. The sample influence function (Hampel, 1974) is a scalar random variable which is often used to assess the effect of each data point on an estimate of a parameter of interest, though this was not its original purpose. For large samples, the distribution of the sample influence function for estimates of several parameters of interest in discriminant analysis has been shown (Campbell, 1978a) to be reasonably closely approximated by a gamma variable with parameters estimated from the data. Thus, a gamma probability plot of the sample influence function for a specified parameter estimate may be useful in detecting outliers. A further development of the probability plot described at the outset is possible. The cube roots of the squared distances, $d_j^{2/3}$, are distributed approximately normally. Thus, a normal probability plot of the ordered $d_j^{2/3}$ may be made. Such a plot may be more convenient computationally than the corresponding gamma plot. Another means of exposing multivariate outliers is provided by principal component analysis. Projection of a p-variate sample onto the first few principal components will reveal outliers which may inappropriately be inflating variances and covariances or correlations and so inducing spurious relationships.

Regression analysis has been the focus of a great deal of activity directed towards the identification of data deficiencies and model inadequacies. With respect to the detection of outliers or influential data, several numerical diagnostic indices as well as graphical procedures have been developed. Many of the numerical indices are directly applicable in canonical analysis or could rather easily be adapted for the purpose. *Leverage*, which is a measure of influence related to the distance of a particular sample-point from the center of the data, is a case in point. The leverage of the jth sample is given by the jth diagonal element, h_{jj}, of the projection matrix $\mathbf{H}$, where

$$\mathbf{H} \equiv \mathbf{X}(\mathbf{X}^t \mathbf{X})^{-1} \mathbf{X}^t.$$

In this expression $\mathbf{X}$ is an $\mathrm{N} \times p$ matrix of predictor variables. Hoaglin and Welsch (1978) have discussed properties of $\mathbf{H}$. They have shown that $0 \leq h_{jj} \leq 1$ and have suggested that leverage values exceeding $2p/\mathrm{N}$ deserve attention as an indication that the jth observation may be influential. The h_{jj} are actually proportional to the so-called Mahalanobis distance of $\mathbf{x}_j$ from the center of the data at $\bar{\mathbf{x}}$. Several functions of the h_{jj} have also been proposed as indices of influence. These results are directly applicable to canonical analysis, where, in addition, the leverage $h_{jj}^{(y)}$ with respect to the projection matrix $\mathbf{H}^{(y)} \equiv \mathbf{Y}(\mathbf{Y}^t\mathbf{Y})^{-1}\mathbf{Y}^t$ similarly invites attention. We shall see in Sect. 14.5.1 that residuals from a fitted model are also useful for disclosing the presence of outliers. We remark also that the identification of *multiple* outliers poses special difficulties, detection in these cases being complicated by problems of masking and swamping.

The detection of outliers represents only a first step. Having located one or more outliers the question as to their *treatment* naturally arises. This leads to consideration of the mechanism thought likely to have generated them. In exploratory studies the options broadly speaking are twofold – outliers may be (a) *set aside* for further study; or (b) *accommodated* by robust estimation. Further study may lead to rejection, as in the case of gross errors of measurement or the inclusion of an obviously unrepresentative sample. The balance of the data may then be analyzed in the usual way. Accommodation is the domain of robust estimation and is directed towards minimizing the effects of outliers on the fitted model. Procedures for robust estimation are taken up in the following subsection.

14.4.2 Long-tailed distributions

Distributions of high-dimensional data whose tails are long relative to those of a comparable Gaussian distribution occur frequently. Short-tailed distributions on the other hand are relatively uncommon and are neglected here. Long-tailed distributions are of interest in canonical analysis because the usual least squares estimators of the coefficients of interest perform poorly where the data distribution is even slightly longer than Gaussian.

In Sect. 14.3.2 we saw that a probability plot may be useful in *detecting* long-tailed distributions. Long-tails are disclosed by curvature at both ends of the probability plot, the configuration curving sharply upwards at the right of the plot and downwards at the left. Formal test procedures for elliptical symmetry have been put forward by Beran (1979). Where long-tailed distributions are encountered they are often effectively *treated* by replacing the least squares estimators of the canonical weights and canonical correlation coefficients by robust estimators. Thus, outliers and long-tailed distributions may be dealt with in the same way. This is not altogether surprising as both conditions represent disturbances in the tails of a distribution. Robustness has to do with procedures which retain some features of methods based on assumptions of normality or multinormality but which provide more flexibility in data-fitting. Mallows (1979) has characterized robustness as consisting of three attributes: (a) *resistance*, that is insensitivity to the presence of a moderate number of bad values in the data as well as to model inadequacies; (b) *smoothness*, which requires the technique to respond only gradually to the injection of a small number of gross errors, to small perturbations in the data, and to small changes in the model; (c) *breadth*, which is the applicability of a technique in a wide variety of situations, including the initial stages of an analysis when it is appropriate to make only minimal assumptions as to what a satisfactory model may look like. A robust procedure is one which may not be optimal for any particular case, but is nearly optimal for a number of cases. There are two approaches to robustness – one may seek to modify the model on which canonical analysis itself is based, or, alternatively, the standard procedure may be applied to a matrix which is a robust estimate of a correlation matrix rather than its least squares counterpart. Here we shall be concerned largely with the second of these approaches. The basic idea of robust estimation is to consider alternatives to the usual estimators which give less weight to unusual observations and yet are also fairly efficient under multinormality.

Robust estimation of a correlation matrix. Canonical analysis is usually applied to a correlation matrix **R** even though the correlation coefficient r, whose elements comprise **R**, is known (Kowalski, 1972; Devlin, Gnanadesikan & Kettenring, 1975; Huber, 1981, p. 204) to be a fragile statistic which is sensitive to outliers and disturbances in the tails of a bivariate distribution. A single, sufficiently extreme observation (x_j, y_j) can shift r to any value in the interval $[-1, +1]$. Thus, even a single deviant observation can have a catastrophic effect on the estimate of r. An appealing strategy in these circumstances is to apply canonical analysis to a robust estimate of the correlation matrix, $\mathbf{R}^*$, say, rather than to the usual least squares estimate **R**. By a robust estimate we mean an estimate arrived at by giving full weight to an individual sample if it is a reasonable observation but downweighting it otherwise. Various robust alternatives to **R** exist. One possibility is to estimate each off-diagonal element separately by a robust correlation coefficient, r^*. Or one might seek to estimate $\mathbf{R}^*$ as a whole. We sketch some examples of these alternative approaches.

Element-wise estimation of $\mathbf{R}^*$. A robust estimate of the correlation coefficient, r^*, between two variables x_1 and x_2 can be constructed from the sum and the difference of standardized forms of the two variables. First, a robust standardization of each set of observations is made, $z_{ij}=x_{ij}/\sqrt{s_i^*}$ $(i=1,2;\ j=1,\dots,\mathrm{N})$, where s_i^* is a robust estimate of the variance of x_i. Several robust estimators of variance exist. One such is the α-trimmed variance, which is calculated in the usual way after omitting a proportion α of the largest and smallest observations and replacing the usual estimate of the sample mean, $\bar{x}_i$, by the α-trimmed mean, $\bar{x}_i^*$. In practice, choices for α of .05 or .10 are not unusual. Another robust variance estimate can be obtained from the slope of a $\chi^2(1)$ probability plot of the $\mathrm{N}(\mathrm{N}-1)/2$ squared differences between pairs of observations. The second step is to obtain robust estimates of the variance of the sum $(z_{1j}+z_{2j})$ and the difference $(z_{1j}-z_{2j})$ of the standardized variables. We shall denote these robust variances by s_+^{*2} and s_-^{*2}, respectively. The robust estimate of the correlation coefficient is then

$$r^*=(s_+^{*2}-s_-^{*2})/(s_+^{*2}+s_-^{*2}).$$

The estimate arrived at in this way lies within the admissible range $[-1, +1]$ for a correlation coefficient, although, like the usual estimate, r, it is biased unless the sample size is large. For $p \geq 2$ variables, a robust correlation matrix $\mathbf{R}^*$ can be obtained by appropriately organizing the $p(p-1)$ quantities r^*. For $p > 2$, however, $\mathbf{R}^*$ is not necessarily positive definite; consequently some or all of its eigenvalues may be negative. Procedures for adjusting $\mathbf{R}^*$ to positive definiteness have been discussed by Devlin, Gnanadesikan and Kettenring (1975). We consider next two estimation procedures for $\mathbf{R}^*$ which differ from the element-wise approach in that all p variables are manipulated simultaneously.

Multivariate trimming. Minimum covering ellipsoids (Titterington, 1978; Green, 1981) provide a means for arriving at a robust estimate of a correlation matrix. For p-dimensional problems, the minimum covering hyperellipsoid for the data is calculated, that is the smallest generalized volume or content that contains the data. The sample is then trimmed by deleting the set of observations which defines the p-dimensional convex hull peel of all the observations; these are the observations around the outside of the distribution. Titterington (1978) has shown that the number of points trimmed is at least $p+1$ and at most $\frac{1}{2}p(p+3)$. The matrix of correlation coefficients is then calculated using the balance of the data in the usual way. The appeal of the method is that it eliminates isolated data points without disturbing the general shape of the distribution. On the other hand, of the $\geq (p+1)$ points removed no more than one of them need be a contaminant.

An iterative method for multivariate trimming has been described by Devlin, Gnanadesikan and Kettenring (1981). At each step in their method the squared distances of the observation vectors $\mathbf{x}_j$ $(j=1, \ldots, \mathrm{N})$ from the current estimate of location $\bar{\mathbf{x}}^*$ in the metric $\mathbf{S}^*$, the current estimate of the covariance matrix, are calculated:

$$d_j^2 = (\mathbf{x}_j - \bar{\mathbf{x}}^*)^t \mathbf{S}^{*-1} (\mathbf{x}_j - \bar{\mathbf{x}}^*).$$

A specified percentage (e.g. 10%) of the most extreme observations is temporarily set aside and the remaining observations used to compute $\bar{\mathbf{x}}^*$ and $\mathbf{S}^*$ exactly as in computing $\bar{\mathbf{x}}$ and $\mathbf{S}$, the usual estimators of the sample mean vector and covariance matrix. The process is initiated by taking $\bar{\mathbf{x}}^*$ and $\mathbf{S}^*$ to be $\bar{\mathbf{x}}$ and $\mathbf{S}$. The covariance matrix $\mathbf{S}^*$ at each stage can be converted to a robust correlation matrix $\mathbf{R}^*$ by rescaling. Iteration ceases when the average absolute change in a simple function of the elements r_{jk}^* between two successive iterations does not exceed 10^{-3} or after the 25th step.

M-estimation. In this case instead of deleting some fraction of the data with larger d_j^2 values, weights are assigned to each vector based on these distances. The usual estimates of location and scatter $\bar{\mathbf{x}}$ and $\mathbf{S}$ are first calculated as well as the distance of each sample from the center of gravity of the sample. New weighted estimates $\bar{\mathbf{x}}^*$ and $\mathbf{S}^*$ are then computed where the weights depend on the d's. These estimates are

$$\begin{aligned} \bar{\mathbf{x}}^* &= \sum w_j \mathbf{x} / \sum w_j, \\ \mathbf{S}^* &= \sum w_j^2 (\mathbf{x}_j - \bar{\mathbf{x}}^*)(\mathbf{x}_j - \bar{\mathbf{x}}^*)^t / \sum w_j^2. \end{aligned} \tag{14.2}$$

Both $\bar{\mathbf{x}}^*$ and $\mathbf{S}^*$ are obtained iteratively. Various choices for the weights are possible. Maronna (1976) has used Huber type and maximum likelihood estimates, while Campbell (1980b, 1982) and Campbell and Reyment (1980) have used values which depend on a bounded influence function (Hampel, 1974). In the latter case, weights are inversely related to the influence of an observation with extremely deviant observations having zero weight. The required robust estimate of $\mathbf{R}^*$ is obtained by rescaling $\mathbf{S}^*$. The estimate $\mathbf{R}^*$ yielded by both procedures in which all p variables are manipulated simultaneously are positive definite except in unusual circumstances.

A refinement of multivariate trimming oriented towards interactive graphical analysis has been described by Garrett (1983). Using initial robust estimates of the sample mean vector $\bar{\mathbf{x}}^*$ and covariance matrix $\mathbf{S}^*$ a gamma probability plot of the data distribution is constructed as described earlier and displayed on a visual display unit. The cursor is then used to eliminate samples with extreme Mahalanobis distances, if such exist. An updated gamma plot based on revised estimates of $\bar{\mathbf{x}}^*$ and $\mathbf{S}^*$ is then obtained and the process repeated until a plot consistent with the sample being made up of identically distributed observations from a p-variate normal, or, at any rate, an elliptically symmetric distribution is obtained. The end result is a robust estimate of the sample mean vector and covariance matrix that can be used as the starting point for further analysis.

The estimators described do not all perform equally well. Hampel's (1974) influence curve provides a flexible tool for investigating the behavior of robust estimators. The breakdown point (Hampel, 1971) is the smallest amount of contamination that may be tolerated without an estimator taking on arbitrarily large values. The breakdown point itself is confined to the closed interval $[0, 1]$. The usual estimate of the correlation matrix, $\mathbf{R}$, has a breakdown point of zero, indicating its unacceptability as a robust estimator. Multivariate M-estimators of scatter have a disappointingly low breakdown point which is no better than $1/(p+1)$. Accordingly, they are increasingly unattractive as p increases. Moreover, M-estimators do not perform well in the presence of asymmetric contamination. It is therefore not altogether surprising that M-estimators have been found to yield only moderate improvement (e.g., see Beauchamp, Begovich, Kane & Wolf, 1980; Hill & Dixon, 1982; Portnoy, 1982). The breakdown point of the multivariate trimming approach of Devlin, Gnanadesikan and Kettenring (1981) is less dependent on p and so yields an estimate which has greater merit in these terms. The appeal of the element-wise estimator of $\mathbf{R}^*$ increases substantially where there are missing values. Interesting proposals for robust estimators of location and scatter have recently been put forward by Donoho and Huber (1983) which incorporate concepts developed in projection pursuit. The first step is to find for each data point $\mathbf{x}_j$ ($j = 1, \ldots, N$) that *one-dimensional* projection for which $\mathbf{x}_j$ is most outlying:

$$r_j = \sup_{|\mathbf{u}|=1} \frac{\mathbf{u}^t \mathbf{x}_j - \mathrm{MED}(\mathbf{u}^t \mathbf{X})}{\mathrm{MAD}(\mathbf{u}^t \mathbf{X})}.$$

In this expression MED and MAD are robust estimates of location and scatter, respectively; MED stands for the median and MAD is the median absolute deviation from the median. The vector $\mathbf{u}$ at which the supremum occurs is the direction in which $\mathbf{x}_j$ is most outlying while r_j is the extent of its outlyingness in that direction. Then, $\mathbf{x}_j$ is weighted by its outlyingness

$$w_j = w(r_j)$$

in the expressions for robust location and scatter given in (14.2). The breakdown point of the resulting estimators is $(N-2p+1)/(2N-2p+1)$.

Robust estimators of the correlation matrix offer a useful complementary approach in canonical analysis to the standard analysis. The advantage is that the influence of atypical observations or heavy-tailed distributions are reduced. Where the distribution is well-behaved in the sense that there is no asymmetry or heaviness and no aberrant observations, the robust and usual estimators will be the same. Almost any robust estimate may be worthwhile, though ideally the choice should match the type of disturbance present. The theory of robust estimation is still in a developing state, however, and firm guidelines for a choice are not always possible.

Data problems of a different kind arise in canonical analysis where the variables of either or both sets are highly correlated. It is to such problems that we now turn.

14.4.3 Collinearity

We use the term collinearity to refer to the occurrence of a dependency of an exactly or an almost exactly linear kind between two or more variables. Exact linear relationships occur only infrequently in practice and we shall not refer to them further. Collinearity has been intensively studied in regression analysis, where it is usually referred to as multicollinearity and where its effects on least squares estimation, inference, variable selection and prediction are well known. In canonical analysis, on the other hand, collinearity has been little studied. Yet, in canonical analysis where there are two equally susceptible sets of variables, the problem is likely to be at least as acute as in regression. Indeed, not only are two sets of variables involved in canonical analysis but we may well be faced with attempting to interpret as many as $s=\min(p, q)$ conjugate pairs of linear composites of them. In multiple regression only one such linear composite arises. Nevertheless, in our treatment of collinearity we shall rely heavily on the regression work. The significance of collinearity in canonical analysis is analogous to that in regression – in its presence the quality of the usual least squares estimates of the canonical weights $a_{1k}, a_{2k}, \ldots, a_{pk}$ and $b_{1k}, b_{2k}, \ldots, b_{qk}$ $(k=1, \ldots, s)$ is likely to be degraded. The precision of these estimates is indicated by their variance. In regression it is well known (e.g., Weisberg, 1980, p. 174) that the variances of those regression weights which are affected by a dependency in the data are inflated. The loss of precision in these estimates is manifest in several ways – the coefficients are frequently large and their signs may be inconsistent with known theoretical relationships among the variables. Further, the importance of variables corresponding to the affected coefficients appear to be enhanced or suppressed, forecasts in predictive studies may be unacceptably poor while variable selection procedures are often rendered ineffectual. In short, collinearity leads to problems in estimation, interpretation and prediction. Yet, collinearity is not always responsible for these conditions nor need they necessarily arise where collinearity is present. Nevertheless, there is a marked tendency for most of them to occur where collinearity is found. In attempting to apply these regression results to canonical analysis a difficulty is encountered in that simple expressions for the estimated variances of the weights vectors, var$(\mathbf{a}_k)$ and var$(\mathbf{b}_k)$, comparable to the familiar regression result that $\text{var}(\mathbf{b}) = \sigma^2(\mathbf{X}^t\mathbf{X})^{-1}$ are not available. Consequently, the effect of collinearity on the estimated variances of the individual canonical weights $a_{1k}, a_{2k}, \ldots, a_{pk}$ and $b_{1k}, b_{2k}, \ldots, b_{qk}$ cannot be assessed directly. Experience suggests, however, that the difficulties associated with collinearity in canonical analysis are similar in kind and severity to those encountered in regression. Thus, in the presence of collinearity in either or both sets of variables, it seems safe to anticipate that there will be an inability on the part of the least squares estimators to yield satisfactory estimates for the canonical weights.

Collinearity can occur for a variety of reasons. We distinguish three principal causes: collinearity which is (a) inherent in the variables studied; (b) attributable to the specification of the model fitted; and (c) a consequence of the way in which the data for analysis were collected. Collinearity which is inherent in the variables calls for little elaboration – the condition reflects the existence of near-linear dependencies among variables which are an intrinsic property of the population or system under study. An example of a model whose specification can give rise to collinearity is provided by polynomial regression. That the predictor variables in models of this kind tend to be highly correlated is well known. As a second example, consider a system for which a judicious selection of variables for inclusion cannot be made simply because the necessary substantive insight is lacking. In such cases it is common practice to incorporate a large number of variables in the hope that no pertinent variable will be overlooked. Of the variables included, it often happens that two or more are parallel forms of some single underlying variable or construct and so are highly intercorrelated. Collinearity attributable to both of the above causes will be manifest in every investigation of the system in question or in each use of the model. This is in sharp contrast to collinearity of the third kind, which arises where the data are confined to a subspace of either or both of the p-dimensional and q-dimensional spaces generated respectively by $\mathbf{X}$ and $\mathbf{Y}$. The dependency in such cases is a property of the data set in question and would not be expected in further, comparable data sets. Two special cases of data-induced collinearity may be mentioned – that arising from the occurrence of an outlying data point and that associated with an over-defined model. Outlier-induced collinearity refers, as the term implies, to a near-linear dependency manifest in the presence of one or more samples which are inconsistent with the bulk of the data, and which disappears when these samples are removed. An over-defined model is one in which the sample is too small to permit all the potential model coefficients to be estimated. In cases of both kinds, the collinearity need not be indicative of any true redundancy among the variables.

Numerous procedures for *detecting* collinearity have been put forward. Recent reviews of the available techniques have been given by Gunst (1983), Hocking (1983) and Hocking and Pendleton (1983). Here, we shall confine our attention to just one of these techniques, namely, that based on singular value decomposition of the matrices **X** and **Y**, a procedure developed by Belsley, Kuh and Welsh (1980). Consider the $N \times p$ data matrix **X** which we shall assume is nonsingular. As a first step let the columns of **X** be normalized to length one. Further, let the singular value decomposition of this matrix be $\mathbf{X} = \mathbf{PDQ}^t$, where $\mathbf{P}^t\mathbf{P} = \mathbf{Q}^t\mathbf{Q} = \mathbf{I}_p$ and $\mathbf{D} = \text{diag}(d_1, \ldots, d_p)$ is a diagonal matrix with nonnegative singular values, d_k. Then, the occurrence of a singular value which is small relative to the magnitude of the largest singular value is indicative of linear dependence. The quantity

$$\varkappa(\mathbf{X}) \equiv \frac{d_{\max}}{d_{\min}} \geq 1$$

is the *condition number* of **X**. A condition number which is strictly equal to unity shows **X** to be columnwise orthonormal (the variables are linearly independent), while increasingly large values indicate increasing degrees of linear dependence. It is also useful to define

$$\eta_k(\mathbf{X}) \equiv \frac{d_{\max}}{d_k} \geq 1, \qquad k = 1, \ldots, p,$$

as the kth *condition index* of **X**. The quantity η_k expresses the severity of the degeneracy, if any, associated with the kth singular value, d_k. Broadly speaking, values of η_k in the vicinity of 5–10 indicate mild dependency and values in the range 30–100 moderate to strong dependency. The condition of the conjugate data matrix **Y** may be investigated in the same way. Where collinearity in **X** or **Y** is detected it is natural to inquire as to the identity of the variables responsible. Again, singular value decomposition provides a means of pursuing this aspect of the problem.

In regression analysis it is well known (e.g., Weisberg, 1980, p. 252) that the covariance matrix of the least squares estimator of the regression coefficients $\mathbf{b}(p \times 1)$ is given by $\sigma^2(\mathbf{X}^t\mathbf{X})^{-1}$, where σ^2 estimates the common variance of the elements composing the vector of disturbances, $\boldsymbol{\varepsilon}(N \times 1)$. Belsley, Kuh and Welsch (1980, p. 105) have shown that the variance of the ith regression coefficient, b_i $(i = 1, \ldots, p)$ can be decomposed into a sum of p terms each of which is associated with a particular singular value, d_k. To see this, first notice that from the singular value decomposition $\mathbf{X} = \mathbf{PDQ}^t$ we may obtain $\mathbf{X}^t\mathbf{X} = \mathbf{QD}^2\mathbf{Q}^t$. Now write

$$\begin{aligned} \text{var}(\mathbf{b}) &= \sigma^2(\mathbf{X}^t\mathbf{X})^{-1} \\ &= \sigma^2\mathbf{QD}^{-2}\mathbf{Q}^t. \end{aligned}$$

The second equality enables the variance of the ith component of **b**, $\text{var}(b_i)$, to be written

$$\text{var}(b_i) = \sigma^2 \sum_{k=1}^{p} \frac{q_{ik}^2}{d_k^2}, \qquad i = 1, \ldots, p,$$

where d_k is the kth singular value of $\mathbf{X}$ and $\mathbf{Q}=[q_{ik}]$. The individual terms in this sum decompose the variance of b_i and show the effects of collinearity indicated by small d_k on the variance of each regression coefficient. Where an unusually high proportion of the variance of two or more coefficients, b_i, is concentrated in terms associated with the *same*, diminutive singular value, the variables corresponding to these regression coefficients are implicated in the collinearity. To simplify this aspect of the investigation, Belsley, Kuh and Welsch (1980, p. 106) recommend constructing the variance-decomposition matrix, $\mathbf{\Pi}=[\pi_{ki}]$, where

$$\pi_{ki}=\frac{q_{ik}^2/d_k^2}{\sum_{t=1}^{p} q_{tk}^2/d_t^2}.$$

The rows of $\mathbf{\Pi}$ correspond to singular values and the columns to regression coefficients. Variables which are responsible for the linear relationship associated with a specified η_k of small magnitude are disclosed by the occurrence of two or more large elements (say $\pi_{ki}>.5$) in the corresponding row of $\mathbf{\Pi}$. The identity of the collinear variables corresponds to the regression coefficients in whose columns the inflated values occur.

The results outlined above were developed in the context of regression but may also be productive for investigating collinearity in canonical analysis. Collinearity can also be revealed by Andrews' (1972) sine-cosine or quantile contour plots. Dependencies or near dependencies in the data are disclosed in such plots by curves which simultaneously pass through or close to a particular point for one or more values of t. Alternative procedures for detecting collinearity and assessing its impact in canonical variate analysis have been developed by Campbell and Reyment (1978), Campbell (1980a) and in canonical correlation analysis by Campbell and Palmer (1984).

The *treatment* of collinearity depends on the mechanism which is diagnosed as having given rise to the condition. Where collinearity is inherent either in the system generating the data or in the model fitted, collinearity will always exist irrespective of the sampling procedure or experimental design used. Under these circumstances there may be no compelling reason to retain all the variables. One or more variables which make no contribution either in a practical sense or statistically may then be eliminated. It is well accepted that in fitting a model which has an exact collinearity, one of the variables must be eliminated. This is precisely what happens in canonical analysis where the variables of either or both sets are indicator variables corresponding to different levels of a categorical variable. In such cases, where the categorical variable has g levels one common approach involves fitting $g-1$ rather than g indicator variables in order to circumvent the exact collinearity which would otherwise result. Variable selection will in general lead to reasonable estimates for the weights of the retained variables in instances of this kind. The situation in relation to data-induced collinearity is more complex. Here, variable selection is counter-indicated as the collinearity may be unique to a particular sample and hence would not be encountered in any future sample. Moreover, eliminating variables in data-induced collinearity has been shown (Gunst, 1983) to lead to substantial bias in coefficient estimates and predictions. Two alternative procedures exist – data augmentation and biased estimation. The preferred solution is data augmentation; additional, well-conditioned data can in themselves break up collinearity. Unfortunately, renewed data acquisition all too often proves impracticable. In such cases, biased estimates of the canonical weights may be substituted for the least squares estimators. If it is accepted that it is the large variance of the least squares weights in the presence of collinearity that is distasteful, then the appeal of estimators which, though biased, have smaller variances will be clear. Biased estimates can be obtained by adding shrinkage or ridge-type constants to the diagonal elements of the matrices $\mathbf{S}_{11}$ and $\mathbf{S}_{22}$ before evaluating the weights, as indicated in Sect. 3.3.2. Suitable values for the constants may be provided by simple multiples of the singular values of these matrices. The resulting canonical ridge weights are suitably dampened and have been found to yield more satisfactory interpretations, while leaving the canonical correlation coefficients themselves substantially unchanged. At this point it is convenient to mention that Lee (1981) has proposed a novel and apparently effective means for combatting model-induced collinearity in one of the two sets of variables in most predictive surface mapping, a special case of canonical correlation analysis.

Collinearity is only one among several potential sources of instability in the canonical weights. To some extent instability is an inherent part of the estimation procedure itself. To see this, notice that even in the absence of collinearity several different sets of weights can yield essentially the same fit of a model for a given set of data. Thus, one may legitimately inquire as to the sensitivity of a fitted model to perturbations in the parameter estimates. A procedure for this purpose based on response surface methods has been developed by DeSarbo, Hausman, Lin and Thompson (1982). A solution to the problem posed by unstable canonical weights has been suggested by the same workers. DeSarbo, Hausman, Lin and Thompson (1982) have considered restricting the weights to simple integral values such as $\{-1, 0, 1\}$ and maximizing the canonical correlation coefficient under this constraint. This procedure has been found in certain cases to enhance the stability of the weights as well as the interpretability of the results as a whole.

Further reading. *Outliers* have been the subject of general discussions by Barnett (1978a, 1983b), Barnett and Lewis (1978), Hawkins (1980) and by Beckman and Cook (1983). Issues which bear on the detection and treatment of *multivariate outliers* have been examined by Gnanadesikan and Kettenring (1972), Gnanadesikan (1977, p. 258), Barnett (1978b, 1979, 1983a, b), Barnett and Lewis (1978, p. 208), Hawkins (1980, p. 104) and Chernick (1983). Koenker and Bassett (1978) have suggested the use of regression quantiles for the detection of multivariate outliers. Techniques for outlier detection in specifically *discriminant analysis* and *canonical correlation analysis* have been proposed by Gnanadesikan and Kettenring (1972) and Gnanadesikan (1977, p. 284). A formal *test* for multivariate normal outliers which leads to Mardia's multivariate sample kurtosis as a test criterion has been derived by Schwager and Margolin (1982). In the context of *regression*, problems posed by outliers and influential observations have been discussed by Belsley, Kuh and Welsch (1980, p. 6), Weisberg (1980, p. 113), Draper and John (1981), Cook and Weisberg (1982), Hocking (1983) and Hocking and Pendleton (1983). Atkinson (1983) has shown that outliers can sometimes be reconciled with the body of the data by a suitable *transformation*, while the literature on *multiple outliers* has been reviewed by Beckman and Cook (1983). Campbell (1978a) has described the use of the *influence function* for outlier detection and given an example from biology. Campbell (1980b) and Campbell and Reyment (1980) have employed the cube-root transformation of a gamma variable to Gaussian form in using $Q-Q$ probability plots to search for outliers. The procedure is illustrated by examples from biology and palaeontology.

A basic reference work devoted to *robustness* is Huber (1981). Other contributions to the subject have been made by Hampel (1977), Huber (1977), Hogg (1977, 1979), Mallows (1979) and Joiner and Hall (1980). The notion of a *breakdown point* in robustness studies has been the subject of a recent review by Donoho and Huber (1983). Stigler (1977) and Hill and Dixon (1982) have examined the performance of several robust estimators when the data distribution is *not well-behaved*, while Kafadar, Rice and Spiegelman (1983) have proposed a robust estimator of location applicable to small samples with asymmetric contamination based on one-sided trimming. Robust estimation of the *correlation coefficient* has been considered by Gnanadesikan and Kettenring (1972), Bebbington (1978), Titterington (1978) and Green (1981). The question of robustness in specifically *multivariate contexts* has been addressed by Gnanadesikan and Kettenring (1972), Devlin, Gnanadesikan and Kettenring (1975, 1981), Gnanadesikan (1977, p. 127), Maronna (1977), Campbell (1980b, 1982) and Huber (1981, p. 199). Robust estimation has been employed by Campbell (1980b) for the purpose of *outlier detection* while Campbell (1982) has described a robust version of *canonical variate analysis* in some detail. A robust version of canonical variate analysis has also been proposed by Digby and Gower (1981). Illustrations of the *application* of robust multivariate procedures in the biological or earth sciences have been given by Campbell (1980b, 1982), Campbell and Reyment (1980), Beauchamp, Begovich, Kane and Wolf (1980), Harner and Whitmore (1981) and by Patterson, Pirkle, Johnson, Bement, Stablein and Jackson (1981).

Collinearity has been the subject of a detailed study by Belsley, Kuh and Welsch (1980). Other contributions to the topic in addition to citations which appear in Sect. 3.3.2 have been made by Rozeboom (1979) and Fearn (1983). The reviews of linear regression methodology by Gunst (1983), Hocking (1983) and Hocking and Pendleton (1983) incorporate summaries of numerous procedures for detecting and treating collinearity in the context of *regression*. Velleman and Welsch (1981) have described efficient procedures for computing regression diagnostics generally, including diagnostics for collinearity. Mandel (1982) has discussed the use of singular value decomposition in elucidating the nature of collinearity and as a means of rectifying the condition. Investigations which deal with collinearity specifically in *canonical analysis* have been made by Vinod (1976), Campbell and Reyment (1978), Campbell (1980a), Lee (1981), Vinod and Ullah (1981), DeSarbo, Hausman, Lin and Thompson (1982) and

Campbell and Palmer (1984). Examples of the *application* of methods for combatting collinearity in biological or similar uses of canonical analysis are to be found among these papers.

14.5 Statistical assessment

The usual goals of canonical analysis are either to extract and explicate the informational content of a body of data consisting of joint observations on two sets of variables **x** and **y**, so as to clarify the relationship between them, or to predict in a least squares sense one set of variables from the other. The purpose of the model fitted during analysis is accordingly either to identify and summarize the systematic component of the relationship or to facilitate prediction. In fitting a model several assumptions about the nature of the relationship between **x** and **y** are made and sometimes also about variation in **x** and **y** that is not accounted for by the variables of the other set. A further, even more basic assumption is that sample-units are identically and independently distributed. Assumptions are made in order to get the analysis under way. It so happens that the assumptions are often too optimistic, with the result that fitting cannot properly be regarded as complete until the validity of the assumptions has been examined. Information concerning inadequacies in a model due to failure of the assumptions is contained in the discrepancies or residuals between observed and fitted values. Thus, while the model or *fit* summarizes the relationship of interest, the residuals are instrumental in exposing inadequacies in the fitting procedure and specifically in the failure of assumptions. After reviewing the calculation of residuals in canonical analysis we shall go on in Sect. 14.5.2 to consider their use in evaluating a fitted model. While in many applications a good fit to the data may be all that is required, in other cases once a satisfactory fit has been achieved it is natural to inquire as to the applicability of the results beyond the data analyzed. On occasion, therefore, the accuracy and stability of the results merit attention. We shall see that the resampling schemes to be described in Sect. 14.5.3 are informative in this way.

14.5.1 Residuals in canonical analysis

Canonical correlations and variates are determined by the solutions of Eqs. (2.14) and (2.16), or, equivalently, by the singular values and vectors of the matrix **K** defined in Eq. (2.20). Together, Eqs. (2.14) and (2.16) are symmetric in **x** and **y**, as is Eq. (2.20). Thus, in canonical analysis it is not necessary to think of either vector as dependent on the other, as in regression, but rather to think of **x** and **y** as associated. This is why reference to error or residual variables in canonical analysis is infrequent. The regression formulation of canonical analysis, however, though rarely encountered, leads to the explicit recognition of residuals. We shall see that multiresponse residuals can in fact be defined quite naturally with respect to both formulations.

In the usual development of canonical analysis, normalized linear combinations of $\mathbf{x}$ and $\mathbf{y}$, $u = \mathbf{a}^t\mathbf{x}$ and $v = \mathbf{b}^t\mathbf{y}$, are sought which maximize the canonical correlation coefficient, $r(u, v)$. There are in all q pairs of such linear composites, u_k and v_k, and corresponding canonical correlation coefficients, r_k $(k = 1, \ldots, q)$. Maximizing the r_k can be shown (Brillinger, 1981, p. 370) to be equivalent to minimizing the sum of squared residuals $\Sigma_j \Sigma_k \hat{e}_{jk}^2 = (u_{jk} - v_{jk})^2$ for $j = 1, \ldots, N$ and $k = 1, \ldots, q$. The corresponding residual sum of squares and cross-products matrix is

$$\hat{\mathbf{E}}^t\hat{\mathbf{E}} = [(\mathbf{Z}^{(x)}\mathbf{A} - \mathbf{Z}^{(y)}\mathbf{B})^t(\mathbf{Z}^{(x)}\mathbf{A} - \mathbf{Z}^{(y)}\mathbf{B})],$$

where $\mathbf{Z}^{(x)}(N \times p)$ and $\mathbf{Z}^{(y)}(N \times q)$ are data matrices in standard score form and $\mathbf{A}(p \times q)$ and $\mathbf{B}(q \times q)$ are matrices of canonical weights. Thus, the $N \times q$ matrix $\hat{\mathbf{E}} = (\mathbf{Z}^{(x)}\mathbf{A} - \mathbf{Z}^{(y)}\mathbf{B})$ may appropriately be regarded as a matrix of residuals. The jth row of $\hat{\mathbf{E}}$, $\hat{\mathbf{e}}_j = [\hat{e}_{j1}, \hat{e}_{j2}, \ldots, \hat{e}_{jq}]^t$, is a multidimensional residual of length q whose elements are the discrepancies $(u_{jk} - v_{jk})$ for the jth observation.

A regression specification of canonical analysis was given in Sect. 2.3.3. Suppose now that we wish to predict in a least squares sense the original variables in $\mathbf{y}$ (in standard score form) from the canonical variates of $\mathbf{x}$ – that is we wish to regress $\mathbf{y}$ on the canonical variates of $\mathbf{x}$. Then, from (2.28) we find that $\hat{\mathbf{Z}}^{(y)} = \mathbf{Z}^{(x)}\mathbf{A}\mathbf{D}\mathbf{B}^{-1}$. In practice it may in fact be more realistic to predict the original variables in either set from some subset $t < q$ of those canonical variates of the other which correspond to the larger canonical roots. In this case we may write

$$\underset{(N \times q)}{\hat{\mathbf{Z}}^{(y)}} = \underset{(N \times p)}{\mathbf{Z}^{(x)}} \; \underset{(p \times q)}{\mathbf{A}} \; \underset{(q \times q)}{\mathbf{D}^*} \; \underset{(q \times q)}{\mathbf{B}^{-1}},$$

where $\mathbf{D}^* = \mathrm{diag}(d_1, \ldots, d_t, 0_{t+1}, \ldots, 0_q)$ has $t < q$ nonzero elements, the remainder having been set to zero. Similarly, it may be shown that

$$\underset{(N \times p)}{\hat{\mathbf{Z}}^{(x)}} = \underset{(N \times q)}{\mathbf{Z}^{(y)}} \; \underset{(q \times q)}{\mathbf{B}} \; \underset{(q \times q)}{\mathbf{D}^*} \; \underset{(q \times p)}{[\mathbf{A}^t]^{-1}}.$$

Matrices of multiresponse residuals, $\hat{\mathbf{E}}^{(x)}(N \times p)$ and $\hat{\mathbf{E}}^{(y)}(N \times q)$, are then given by $\hat{\mathbf{E}}^{(x)} = \mathbf{Z}^{(x)} - \hat{\mathbf{Z}}^{(x)}$ and $\hat{\mathbf{E}}^{(y)} = \mathbf{Z}^{(y)} - \hat{\mathbf{Z}}^{(y)}$, respectively. The matrix $\hat{\mathbf{E}}^{(x)}$ is analogous to the residual matrix in the multivariate regression of $\mathbf{x}$ on $\mathbf{y}$ while the same may be said of $\hat{\mathbf{E}}^{(y)}$ with respect to the regression of $\mathbf{y}$ on $\mathbf{x}$.

Having obtained several matrices of multivariate residuals we shall now consider how these may be analyzed so as to throw light on the validity of assumptions made during fitting.

14.5.2 Does the model fit?

Predicted or fitted values in canonical analysis are represented by the canonical variates, u_k and v_k associated with the kth root, r_k^2 $(k = 1, \ldots, t)$. The canonical variates are random variables and as such their realizations are generally conceived of as a set of numerical values. For the purpose of assessing the fit of a model there is much to be said for displaying the canonical variates graphically and studying the configuration. Thus, scattergrams of u_k against v_k for $k=1,\ldots,t$ reveal considerably more about the quality of a model than a statement of the corresponding numerical values. A scattergram not only characterizes the fit represented by the joint distribution of (u_k, v_k) more completely and with greater effect than the corresponding canonical correlation coefficient, r_k, but in addition is able to disclose possible defects in the form of the model.

Information on the *lack of fit* is conveyed by discrepancies or residuals between observed and fitted values or between a conjugate pair of fitted values. The study and analysis of residuals, therefore, is an essential part of model assessment. A fitted model may be misspecified in a number of ways. Among these, mention is made of misspecification due to a nonlinear functional relationship between **x** and **y**, to nonnormality in the error component of the model, to lack of independence among the errors, to the presence of outlying observations and to the omission of essential variables. Regression methodology is particularly rich in techniques for extracting information from residuals with the result that model failure in regression can often be identified with a specific cause. Techniques for residual analysis in canonical analysis are not nearly as well developed. The account which follows therefore is necessarily somewhat sketchy. Yet much can be done and the methods developed for use in regression provide a useful starting point. For simplicity, attention is concentrated on model failures of just three kinds – nonlinearity of the relationship between **x** and **y**, nonnormality of the error component and lack of independence among the errors.

Linearity. Canonical analysis is specified by a linear model. The model itself is most explicit in regression formulations (Robinson, 1973; Izenman, 1980; Brillinger, 1981; Muller, 1982), where an error term and sometimes an error distribution are defined. The linear form of the model in regression is expressed differently in classical and conditional versions of the model. In the classical version, **y** is treated as a random variable whose mean depends on fixed values of the predictors, **x**, and is assumed to be a linear function of the regression coefficients. More importantly for present purposes, in the conditional regression model **x** and **y** are both random and have a joint distribution. If the joint distribution is assumed to be multinormal, the best linear predictor of **y** turns out to be the conditional expectation of **y**, given **x**, which is itself a linear regression function with exactly the same coefficients as would result in fitting the classical model. In the standard formulation of canonical analysis (Sect. 2.2) reference to an error term is suppressed and the model is less obvious. Nevertheless, there is an underlying model, the linear form of which is implicit in a requirement for the linearity of both conditional expectations, that is of the conditional expectation of **y**, given **x**, and of the conditional expectation of **x**, given **y**. The linearity of the model is of course quite apparent in the search for linear functions of **x** and **y**, $u=\mathbf{a}^t\mathbf{x}$ and $v=\mathbf{b}^t\mathbf{y}$, where the elements of $\mathbf{a}=[a_1, a_2, \ldots, a_p]^t$ and $\mathbf{b}=[b_1, b_2, \ldots, b_q]^t$ are limited to quantities of the first degree. Linearity of the model restricts the usefulness of canonical analysis to situations in which the point-clouds corresponding to the data in x-space and y-space each contain, at least approximately, a t-dimensional linear manifold with $t<\min(p, q)$. Where the point-clouds have a simple structure but lie close to nonlinear manifolds a discrepancy between the model and the data arises and canonical analysis will be less successful and perhaps misleading. Accordingly, it is important to examine the adequacy of a fitted model.

In regression, the appropriateness of a linear model is sometimes investigated by plotting the response variable y against the predicted values, $\hat{y}$, and studying the display. In canonical analysis, where there are two sets of predicted values represented by the canonical variates u_k and v_k $(k=1, \ldots, t)$, a comparable check on linearity may be made by plotting u_k against v_k. Where a linear model is appropriate this will be reflected by a configuration which is substantially linear. A curvilinear display, on the other hand, would betray the inadequacy of the model. A bivariate plot of (u_k, v_k) is easily made and is also informative in other ways. Thus, the shape of the joint distribution is revealed and attention drawn to the presence of any deviant observations or to other, smoother departures from regularity in the disposition of individual points. A more direct extension of the regression plot of y against $\hat{y}$ may be feasible. Analogs of y and $\hat{y}$ in canonical analysis are represented by the observed and fitted quantities $\mathbf{y}$ $(q \times 1)$ and $\mathbf{v}$ $(q \times 1)$, both vector-valued. The shape of the distribution of some suitable scalar function of the multivariate response $\mathbf{y}$ and of a corresponding scalar function of the fitted quantity $\mathbf{v}$ could be assessed relative to one another by means of an *empirical* $Q-Q$ probability plot (Wilk & Gnanadesikan, 1968). Where the distributions are similar, an approximately straight-line configuration with slope equal to 1 would be expected. Such a configuration would represent the ideal or *null-plot* – the pattern observed when the specified model is correct. Evidence of systematic departure from the null configuration indicative of nonlinearity or other model failure might then be sought. The usual need to re-express the data prior to plotting if simple relationships are not to remain hidden, however, must not be overlooked. A comparable $Q-Q$ plot of $\mathbf{x}$ $(p \times 1)$ against $\mathbf{u}$ $(q \times 1)$ might also be worthwhile.

In regression analysis the systematic search for possible failures in a fitted model is normally conducted in terms of residuals rather than observed values. Residuals of course represent observations with the fit removed. Displays of residuals $\hat{e}$ against predicted values y are an important diagnostic tool, not only for checking linearity but also in connection with possible model misspecifications of other kinds. In canonical analysis opportunities of this sort are even more varied in the sense that there are two sets of vector-valued indices which may be regarded as counterparts of $\hat{e}$ and $\hat{y}$, namely the rows of the matrices $\hat{\mathbf{E}}^{(y)}$ and $\mathbf{V}$, and $\hat{\mathbf{E}}^{(x)}$ and $\mathbf{U}$, respectively. In principle, empirical $Q-Q$ plots appear to offer a straightforward means of extending the regression-based procedures to canonical analysis. Unfortunately, such an approach seems unlikely to be productive. This is because of the difficulty of specifying for the $Q-Q$ plots a counterpart of the null configuration represented by the horizontal, structureless point-configuration against which departures are assessed in regression analysis. Evidently, what would be useful is some equivalent or alternative $Q-Q$ plot for which the null configuration is *linear*. We shall encounter the residual matrices $\hat{\mathbf{E}}^{(x)}$ and $\hat{\mathbf{E}}^{(y)}$ again below and will find that they are amenable to investigation, though in ways less specifically directed towards diagnosing model misspecification due to nonlinearity.

A special case of a residual plot which is much used for assessing linearity and other aspects of a fitted model in regression is the *adjusted variable* plot (Chambers, Cleveland, Kleiner & Tukey, 1983, p. 268). A plot of this kind yields information about a single regression coefficient in graphical form. An analog of the plot in canonical analysis is much needed. In effect, an adjusted variable plot is a plot of a response variable y against the explanatory variables x_i $(i=1, \ldots, p)$ with the influence of all explanatory variables excepting the ith, $x_{1 \cdot 2 \ldots (i-1)(i+1) \ldots p}$, removed. The construction of such a plot may be summarized as follows:

(a) adjust y for the effects of all explanatory variables except the ith to obtain a new, adjusted response variable $y_{\cdot [i]}$;
(b) adjust the ith explanatory variable for the effects of all other explanatory variables to obtain an adjusted explanatory variable $x_{i \cdot [i]}$.

The adjusted variable plot corresponding to the coefficient b_i is given by plotting $y_{\cdot [i]}$ against $x_{i \cdot [i]}$. The slope of the configuration is equal to b_i. Further, the residuals represented by the scatter of points about the regression line are identically the residuals in the *full* multiple regression of y on all the x_i. Thus the essential features of an adjusted variable plot are (a) the fitted, systematic effect of y on x_i isolated from the effects of the $p-1$ remaining predictors; and (b) the residuals in the full multiple regression of y on $x_1, x_2, \ldots, x_p$. In addition the plot can convey interesting information of several other kinds including nonlinearity and the influence of individual data points and collinear relationships on the estimation of b_i. A counterpart of the adjusted variable plot for each of the canonical weights $a_{1k}, a_{2k}, \ldots, a_{pk}$ and $b_{1k}, b_{2k}, \ldots, b_{qk}$ would be valuable in canonical analysis even though the number of potential plots would be considerably greater than in regression. Gnanadesikan (1977, p. 264) has remarked that multivariate analogs of the partial residual plots of Larsen and McCleary (1972), which are related to adjusted variable plots, may be feasible. Little work towards extending the principles involved specifically to canonical analysis, however, appears to have been undertaken. One may anticipate that in attempting to extend the adjusted variable plot in this way it would be necessary to adjust each response variable y_h $(h=1, \ldots, q)$ for the effects of the $q-1$ remaining responses before proceeding to step (a) above. Landwehr (1983) has used partial residual plots to detect nonlinearity in multiple regression.

Where endeavors of the kind described cast doubt on the linearity of the relationship between $\mathbf{x}$ and $\mathbf{y}$, the adequacy of the fitted model is suspect. Nonlinearity, however, if revealed, may nevertheless still be exploited. Procedures for exploiting nonlinearity in canonical analysis are discussed in Sect. 14.5.4.

Normality of error. Assumptions regarding the error distribution in canonical analysis arise in two contexts. The linearity of the conditional expectations which is implicitly assumed in fitting the model requires that the observation vector $\mathbf{g}=[\mathbf{x}^t|\mathbf{y}^t]^t$ be multinormal. Hence, the distribution of the corresponding error term, $\boldsymbol{\varepsilon}^{(g)}=[(\boldsymbol{\varepsilon}^{(x)})^t|(\boldsymbol{\varepsilon}^{(y)})^t]^t$ will similarly be multinormal if the model fits, for $\boldsymbol{\varepsilon}^{(g)}$ is a linear combination of $\mathbf{g}$. An assumption concerning the error distribution is also made where tests of statistical hypotheses are made. In this case it is the multinormality of either $\boldsymbol{\varepsilon}^{(x)}$ or $\boldsymbol{\varepsilon}^{(y)}$ which is crucial. Following analysis, therefore, there is a need to examine the error distribution in order that possible model failures of the kinds described may be disclosed. As it is the normality of the distribution which is at issue in both contexts, the context may be disregarded and attention focussed simply on the normality or otherwise of the errors. In applications of an exploratory or descriptive kind, it may be that the requirement for normality is unnecessarily restrictive and can justifiably be replaced by the weaker assumption of elliptical symmetry. In any event, our goal is to acquire as much insight as possible into the shape of the error distribution. We shall rely on the residual matrices $\hat{\mathbf{E}}$, $\hat{\mathbf{E}}^{(x)}$, $\hat{\mathbf{E}}^{(y)}$ and $[\hat{\mathbf{E}}^{(x)}|\hat{\mathbf{E}}^{(y)}]$ for this purpose. For simplicity of exposition, however, only the matrix $\hat{\mathbf{E}}$ is referred to below with the understanding that all four matrices are amenable to analysis in the same way. Recall that the jth row of $\hat{\mathbf{E}}$, $\hat{\mathbf{e}}_j^t$ ($j=1, \ldots, \mathrm{N}$) is a multidimensional residual of length q.

As residuals are linear composites of observed variables they tend to be more normal in distribution than the observations from which they derive. One may turn this property to useful advantage in searching for a suitable null distribution against which the empirical distribution of residuals may be evaluated. Evidently, the gamma or normal distributions would provide such a standard. The comparison itself is readily made by means of a $Q-Q$ plot. The basic idea is to plot the ordered values of a suitable scalar function of the jth residual $\hat{\mathbf{e}}_j^t(q\times 1)$ against quantiles of either a gamma distribution with parameters estimated from the data or the standard normal distribution. A scalar quantity related to the size of the jth residual ($j=1, \ldots, \mathrm{N}$) which is suitable for the purpose is given by the quadratic form

$$d_j^2=(\hat{\mathbf{e}}_j-\hat{\mathbf{e}}_{\bullet})^t\,\mathbf{A}(\hat{\mathbf{e}}_j-\hat{\mathbf{e}}_{\bullet}),$$

where $\hat{\mathbf{e}}_j=\mathbf{a}_k^t\mathbf{z}_j^{(x)}-\mathbf{b}_k^t\mathbf{z}_j^{(y)}$ ($k=1, \ldots, q$) and $\hat{\mathbf{e}}_{\bullet}=\mathrm{N}^{-1}\Sigma\hat{\mathbf{e}}_j$. The compounding matrix $\mathbf{A}$ ($q\times q$) is positive semi-definite but otherwise arbitrary. Suitable choices for $\mathbf{A}$ are the identity matrix of order q, $\mathbf{I}_q$, the inverse of a robust covariance matrix $\mathbf{S}^*$ calculated from the residuals, $\mathbf{A}=(\mathbf{S}^*)^{-1}$, or a diagonal matrix $\mathbf{D}$ whose non-zero elements are reciprocals of the diagonal elements s_{ii}^* of $\mathbf{S}^*$. No unique measure of the size of the residuals exists and their multidimensional nature indicates that several choices of $\mathbf{A}$ are likely to be productive. The ordered, squared quantities $d_1^2\leq d_2^2\leq\ldots\leq d_{\mathrm{N}}^2$ may then be plotted against corresponding quantiles of a gamma distribution, $g(d^2;\lambda,\eta)$, with scale and shape parameters λ and η estimated from a subset of the smaller d_j^2. Where one is interested in assessing the overall shape of the residual distribution rather than specifically its normality, a better idea might be to first symmetrize the distribution of the d_j^2 by taking cube roots, $d_j^{2/3}$, and then plotting the cube roots against normal quantiles. Alternatively, the $d_j^{2/3}$ might be plotted against cube root gamma quantiles. The resulting plot is likely to be more revealing with respect to the shape of *both* tails of the residual distribution than the corresponding, initial gamma plot. Under null conditions, that is to say where the fitted canonical model was properly specified, the configurations for gamma and normal plots would in each case be reasonably linear. In relation to elliptical symmetry (normal and cube root gamma plots only), however, the configuration, while linear in the center, would curve sharply upwards at the right and sharply downwards at the left. Departure in a configuration from the corresponding null specification, such as might be represented by piecewise linearity or points widely separated from the mass of the data, would be indicative of non-normality or non-elliptical symmetry in the residual distribution, and hence of model misspecification. The precise nature of the departure may be such as to suggest appropriate remedial measures directed towards improving the shape of the residual distribution. The indicated measures might then be implemented, the model refitted and the residual analysis repeated.

Residuals of another kind arise in canonical analysis. Such a matrix, $\hat{\mathbf{E}}^{(t)}$, was defined in Sect. 3.4.5 associated with the rank t least squares approximation of the matrix $\mathbf{K}=\mathbf{S}_{11}^{-\frac{1}{2}}\mathbf{S}_{12}\mathbf{S}_{22}^{-\frac{1}{2}}$. The residuals in this case differ from those just considered in that they are unidimensional rather than multidimensional, and, furthermore, cannot be identified with individual samples. It is the elements $\hat{e}_{ij}^{(t)}$ of $\hat{\mathbf{E}}^{(t)}$ $(p \times q)$ rather than its rows which constitute residuals; $\hat{e}_{ij}^{(t)}$ is the discrepancy between the ijth element of the standardized scalar-products matrix $\mathbf{K}$ and the corresponding element of its rank t least squares estimate $\mathbf{K}_t$. The investigation of such residuals can nevertheless be motivated by the observation made earlier that residuals generally tend to be more normally distributed than the data from which they derive. Accordingly, a $Q-Q$ probability plot again proves useful for studying the residual distribution. Since in this case the residuals are scalars, the corresponding order statistics are plotted directly against quantiles of the standard normal distribution. The resulting configuration is interpreted in broadly the same way as the earlier residual plots.

Multivariate residuals are amenable to analysis by procedures other than probability plots. Sine-cosine or quantile function plots, for example, may be used to reveal outliers, clustering and singularities as well as to obtain insight as to the symmetry and other shape characteristics of the residual distribution as a whole. Projection pursuit, principal component analysis and a graph-theoretic approach devised by Friedman and Rafsky (1983) offer further avenues of approach in searching for structure in multivariate residuals.

Independence of errors. Where observations are ordered in space or time adjacent observations are likely to be affected by similar external conditions. Ecological *surveys* are inherently likely to give rise to data which exhibit systematic spatial effects of this kind. More generally, one may be faced with autocorrelation in both time and space. Spatial or temporal relationships among samples frequently cause the standard assumption that the errors are statistically independent, which is sometimes implicit in fitting a canonical model, to be violated. Consequently, both the observations and the errors are autocorrelated. Study of the effects of autocorrelation in canonical analysis has not progressed far. Work in related disciplines suggests that two important consequences may be anticipated:

(a) the variance of parameter estimates will be under-estimated, while, in applications involving statistical inference, the true probability of a type I error will in general tend to *exceed* the nominal level;
(b) the results of analysis will be *scale-dependent*.

As a consequence of the first point, where autocorrelation is present the results will appear to be more precise and more significant than they really are. Scale-dependence has far-reaching consequences, rendering the concept of correlation itself elusive. In the presence of autocorrelation, correlation between variables is partly a function of the size of the sampling unit employed. Hence, in relation to the arbitrary sampling units (quadrats) of ecological surveys, the familiar notion of correlation, as well as of a single correlation coefficient, is erroneous. The practical implications of the scale problem are twofold. First, consideration has to be given to the question as to what is the most appropriate 'scale' for an investigation. Second, attempts to disentangle scale-effects from results of genuine scientific worth may be necessary.

Methodology for detecting and treating autocorrelation in canonical analysis is not well-developed. In regression, scatter plots of residuals against spatial or temporal order are an important tool for diagnosing autocorrelation. Further, several numerical indices of univariate spatial autocorrelation have been proposed (see Cliff & Ord, 1981a, b). The Durbin-Watson statistic d is one widely used index of this kind. Analogs of such graphical procedures or numerical indices would be beneficial in canonical analysis. One possibility which might be worth investigating would be to plot some scalar-function of the multidimensional residuals against spatial or temporal order. Two-dimensional spectral analysis (Bartlett, 1975; McBratney & Webster, 1981; Renshaw & Ford, 1983) offers a further possible means of uncovering structure in the multidimensional residuals of ecological surveys.

Several remedial steps for dealing with autocorrelation have been proposed. Broadly, these are of two kinds – the model to be fitted may itself be modified to allow for the lack of independence or the autocorrelation may be removed from the data befor applying canonical analysis in the usual way. Of the two alternatives the second is much the simplest. The use of *difference operators* for eliminating spatial autocorrelation in the univariate case has been described by Cliff and Ord (1981a, b). Where tests of statistical hypotheses are of interest, the opportunity to use randomization tests exists. In such cases sampling distributions taking account of the autocorrelation would first be generated from the data and the significance of test criteria then assessed in terms of these empirical distributions. With respect to scale effects we have remarked that the notion of a single correlation coefficient for variables observed in sample units whose size is arbitrary is not well-founded. With this point in mind two-dimensional multivariate spectral or cross-spectral analysis is appealing. Analysis of this kind would allow the variance or correlation structure of interest to be partitioned into components attributable to different spatial scales. In this way spatial information which is masked in the standard application of canonical analysis might be disclosed.

Where residual analysis reveals that a fitted model is misspecified, remedial steps are called for. Often, the procedures described above for diagnosing model failures are themselves useful for suggesting the nature of appropriate measures to improve the fit. Once appropriate steps have been taken, the model may be refitted, residuals calculated and analyzed as before. Thus, model fitting is properly an iterative process which is terminated only when the residuals show the current model to be correctly specified.

14.5.3 Stability of results

Once a satisfactory fit has been achieved the quality of the results obtained merits attention. Where the purpose of analysis is prediction, the quality of the prediction is plainly fundamental to success of the venture. In exploratory or descriptive applications, on the other hand, a good fit may be all that is required. Here, the need is to assess the quality of the fit. Even in descriptive studies, however, it is often hoped that the results obtained will have at least some degree of generality beyond the sample which yielded them. Thus, the question as to extent of the applicability of results is one of general interest.

The results yielded by canonical analysis are optimal in the sense that the canonical correlation coefficient is maximal over all possible normalized linear combinations of $\mathbf{x}$ and $\mathbf{y}$ for the sample analyzed. The logic of maximization is that it facilitates prediction. However, results which are optimal with respect to a given sample may lack good general validity simply because of idiosyncrasies in the data. Hence, while a fitted model may yield what appear to be sensible results, the results may be so severely biased as to be misleading. As protection against over-optimistic assessment, therefore, some indication of the average fit of a model is needed. In other words, not only are estimates of the quantities involved in fitting a model required, but also a statement concerning the accuracy of the estimates themselves. Accordingly, we are led to inquire as to the bias and standard error of each of the quantities $\{\mathrm{r}_k; \mathbf{a}_k, \mathbf{b}_k; u_k, v_k\}$ as well as to the prediction error associated with each data point for the fitted model. A class of nonparametric methods for assessing statistical errors of these kinds has been developed. As well as being distribution-free, these methods are appealing in that they require only the sample-data initially available, being based on computer-intensive resampling schemes from this sample. We briefly consider three such procedures - cross-validation, the jackknife and the bootstrap.

Cross-validation. Assessment of a fitted model by cross-validation was outlined in Sect. 3.3.4. What is involved is an attempt to validate the model by confirmation against independent observations. Where the sample is large, it may be convenient to proceed by subdividing the sample into two equal parts. One subsample - the development sample - is used for estimation and the other - the test or validation sample - for assessment of the estimates. The performance of the model is evaluated by comparing parameter estimates and predicted values for the test sample against those of the development sample. The initial fit is generally found to be over-optimistic and often appreciably so precisely because the model has been selected to fit the data of the development sample. Subdivision of the sample into equal parts is not an essential part of cross-validation. The opposite extreme is represented by the situation in which the validation sample consists of a single data point. The model is fitted to the remaining $N-1$ data points and its effectiveness in predicting the excluded point assessed. The discrepancy between the observed and the predicted value is the prediction error. This procedure is repeated N times, once for each data point. The average of the prediction errors is the cross-validated measure of prediction error. The process of omitting samples simulates the prediction of future observations from the universe in question. The essential idea is to ensure that each data point is not used in both the fitting and the assessment stages, while nevertheless using as much of the data as possible for prediction. Refinements of this scheme (Stone, 1974; Geisser, 1975; Eastment & Krzanowski, 1982) enable cross-validation to be employed for model *selection* rather than simply evaluation.

Cross-validation has the considerable merit of permitting an evaluation of a fitted model as a whole as opposed to assessment of a particular parameter estimate or subset of estimates. On the other hand, the procedure suffers from the weakness that the test sample often resembles the development sample much more closely than it resembles a typical sample from the universe sampled. Moreover, in the classical form represented by division of the sample into two subgroups of equal size, subdivision can induce the instability it is designed to detect. Subdivision in this way can also lead to the isolation of influential samples and to the creation of collinearity in either or both subsamples.

The jackknife. The jackknife (Quenouille, 1949; Tukey, 1958) is a nonparametric technique for reducing bias in the estimation of parameters and for estimating the standard errors of these estimated parameters. Let the sample value of any one of the quantities of interest in canonical analysis $\{r_k; \mathbf{a}_k, \mathbf{b}_k; u_k, v_k\}$ be denoted by $\hat{\theta}$, where $\hat{\theta}$ is permitted to be vector-valued. Jackknife estimates of $\hat{\theta}$ are arrived at by deleting the jth sample observation $\mathbf{g}_j = [\mathbf{x}_j^t | \mathbf{y}_j^t]^t$ sequentially for $j = 1, \ldots, N$ and computing $\hat{\theta}$ on the basis of the remaining $N-1$ samples. We denote the value of $\hat{\theta}$ obtained when $\mathbf{g}_j$ is omitted by $\hat{\theta}_{(j)}$ and the average of all N $\hat{\theta}_{(j)}$, $N^{-1}\Sigma_j\hat{\theta}_{(j)}$, by $\hat{\theta}_{(\cdot)}$. Then the jackknife estimate of the bias of $\hat{\theta}$, $\hat{\beta}_J$, say, is

$$\hat{\beta}_J = (N-1)(\hat{\theta}_{(\cdot)} - \hat{\theta}),$$

where $\hat{\theta}$ is the estimate yielded by all N observations. The bias corrected estimate of the parameter $\hat{\theta}$, $\hat{\theta}_J$ say, is obtained on subtracting $\hat{\beta}_J$ from $\hat{\theta}$:

$$\hat{\theta}_J = N\hat{\theta} - (N-1)\hat{\theta}_{(\cdot)}.$$

An attractive property of jackknife estimates is that they tend to eliminate bias. The jackknife is therefore particularly useful in connection with analyses based on small samples.

The jackknife estimate of the standard error of $\hat{\theta}$, $\hat{\sigma}_J$, is

$$\hat{\sigma}_J = \left[((N-1)/N) \sum_{j=1} (\hat{\theta}_{(j)} - \hat{\theta}_{(\cdot)})^2\right]^{\frac{1}{2}}.$$

The bootstrap. The bootstrap (Efron, 1979a, b) offers an alternative to the jackknife for estimating and assessing the accuracy of the interpretive indices in canonical analysis. The bootstrap method for estimating the *distribution* of $\hat{\theta}$ consists of taking random samples with replacement from the population defined by the empirical distribution of the original sample. From B such Monte Carlo replications the empirical sampling distribution of the bootstrap estimate of $\hat{\theta}$, $\hat{\theta}^*$, its mean, and its standard error are determined. The bootstrap estimate of the bias of $\hat{\theta}^*$, $\hat{\beta}_B$, is

$$\hat{\beta}_B = B^{-1} \sum_{b=1}^{B} \hat{\theta}^{*b} - \hat{\theta} = \hat{\theta}^{*\cdot} - \hat{\theta}$$

while the standard error of $\hat{\theta}^*$ is

$$\hat{\theta}_B = \left[\sum_{b=1}^{B} (\hat{\theta}^{*b} - \hat{\theta}^{*\cdot})^2 / (B-1)\right]^{\frac{1}{2}},$$

where $\hat{\theta}^{*\cdot} = B^{-1} \Sigma \hat{\theta}^{*b}$. The standard error $\hat{\sigma}_B$ is the average amount by which the observed statistic $\hat{\theta}$ for a random sample of size N differs from the true population quantity, θ.

The usual purpose in estimating standard errors in applications is to set confidence intervals for an unknown parameter. The jackknife and bootstrap estimates $\hat{\sigma}_J$ and $\hat{\sigma}_B$ can both be used in this way. In general, jackknife and bootstrap parameter estimates are related and lead to equivalent results. However, the two are not necessarily computationally equivalent for a given set of data, the bootstrap being much the more demanding of the two computationally. Notice that the bootstrap involves the tacit assumption that the empirical sampling distribution of $\hat{\theta}^*$ reliably represents the population sampled. Where the sample size is small and the true distribution has long, thick tails this is a considerable assumption. Under these conditions, it is pertinent to ask how stable the empirical sampling distribution might be from one trial to another. Limitations of a similar kind apply to the jackknife.

14.5.4 Miscellanea

In this section five unrelated topics which do not fit conveniently elsewhere are dealt with. Only the first of these, devoted to exploiting nonlinearity, is developed at any length.

Exploiting nonlinearity. In practice relationships between **x** and **y** are frequently nonlinear. Yet the model fitted in canonical analysis is linear in the sense that linearity of both conditional expectations is assumed. Hence, a discrepancy between the specification of the model and the data to which it is fitted may arise. Provided the nonlinearity is not marked, canonical analysis may be applied in the usual way without serious harm. The consequences of strong nonlinearity, on the other hand, cannot be dismissed lightly. Severe nonlinearity poses a challenge, for it indicates that the data contain more information than can be extracted by canonical analysis in the ordinary way. Provided suitable measures are taken, however, the performance of canonical analysis in the presence of nonlinearity can be improved considerably. The available measures are of two kinds – data transformation and linearization of the nonlinear problem. Transformations enable nonlinearity to be accommodated by the model to be fitted and increase the flexibility of canonical analysis quite considerably. Moreover, a nonlinear problem may be linearized by the application of one of the standard techniques for the nonlinear reduction of dimensionality. Thus, the remedial measures are comparatively straightforward. Recall that steps can also be taken during the design stage to circumvent a need to deal with nonlinearity later. To develop this point slightly, notice that while essentially all real problems are nonlinear, severe nonlinearity is frequently partly a function of scale – the larger the scale of an investigation, the greater the likelihood of nonlinearity. This observation may be turned to useful advantage; it suggests that nonlinearity may be controlled by the local replacement of a nonlinear problem by one that is more nearly linear. Thus, an extensive ecological survey, for example, may be designed and analyzed as two smaller, largely independent surveys. Similarly, benefits may be expected where the subject-matter and areal domains of a study are minimized as far as is consistent with the realization of overall objectives. The point of general interest here is that in setting research goals measures to control nonlinearity are often feasible.

Data transformation. Canonical analysis involves a search for functions of $\mathbf{x}$ and $\mathbf{y}$, $u_k = \mathbf{a}_k^t \mathbf{x}$ and $v_k = \mathbf{b}_k^t \mathbf{y}$, which are linear in the coefficients. No such restriction, however, is placed on the elements $x_1, x_2, \ldots, x_p$ and $y_1, y_2, \ldots, y_q$ of $\mathbf{x}$ and $\mathbf{y}$ themselves. Any arbitrary functions of the x's and y's are admissible. This freedom imparts appreciable flexibility to canonical analysis in applications involving nonlinear and interactive relationships. Hence data transformation represents an important strategy for exploiting nonlinearity in canonical analysis. In certain applications the nature of the curvature may suggest appropriate remedial action. In straightforward cases nonlinearity will arise either from a few outlying data points or from a consistent curvature. The treatment of outliers has already been addressed (Sect. 14.4.1). Where curvature is encountered and has a consistent form this in itself may suggest a suitable linearizing transformation. Curvature may also be associated with just one or a few variables. In such cases it may be possible to treat the balance of the data as substantially linear, while the dependency of the anomalous variables on the remainder is described separately. Examination of the data in these ways, however, may not always be productive or even possible. The data themselves may then be used to prescribe a linearizing transformation. The data-based power transformation towards normality described in Sect. 14.3.3 may sometimes have a linearizing effect. Another use of data-based transformations is directed rather more specifically towards linearity. The method of *canonical optimal regression* (Young, de Leeuw & Takane, 1976) is based on such a transformation. Canonical optimal regression is one of a class of alternating least squares methods with optimal scaling features developed to allow the efficient analysis of nonmetric (binary, nominal, ordinal) data by linear models. Specifically, in canonical optimal regression transformations of the x's and y's are sought which minimize the least squares criterion $\mathrm{tr}[\mathbf{XA}-\mathbf{YB})^t(\mathbf{XA}-\mathbf{YB})]$. Minimizing the trace of this matrix product is equivalent to maximizing the canonical correlation coefficient. Optimization is accomplished by iterating between two alternating phases or operations – an estimation phase and an optimal scaling phase. In the estimation phase, estimates for canonical weights are sought which minimize the stated criterion while the data transformations are held constant; in the optimal scaling phase, estimates for the transformations are sought which minimize the least squares criterion while the weights are held constant. By alternating between the two phases, the analysis converges to the canonical correlation coefficient. Notice that unlike power transformations, the transformations in canonical optimal regression are optimized *under* the model. The type of admissible transformation for a particular variable is determined by the measurement level of that variable. Arbitrary, monotonic and linear (including polynomial) transformations are applicable to variables of different levels. Canonical optimal regression has been developed and refined to give *nonlinear canonical correlation analysis* (van der Burg & de Leeuw, 1983). The name of this method derives from its property that results are invariant under certain nonlinear transformations of the original variables. This distinctive property contrasts sharply with the classical form of canonical analysis, for which results are invariant under only linear transformations of $\mathbf{x}$ and $\mathbf{y}$. In nonlinear canonical analysis, weights' matrices $\mathbf{A}$ and $\mathbf{B}$ are sought exactly as in the classical version but in addition transformations of the data are simultaneously sought so as to optimize the canonical correlation. The convergence properties of the algorithm used in implementing nonlinear canonical correlation are superior to those of canonical optimal regression. Moreover, applied to the same categorical data, nonlinear canonical analysis has been found to yield more informative results than the standard linear analysis.

As an alternative to the data-based transformations and methods outlined above, prespecified transformations of the data may also be helpful in exploiting nonlinearity. The most widely employed transformations of this kind involve the use of polynomial or sine-cosine functions of one or other set of variables. The procedure is standard practice in certain areas of the application of canonical analysis, notably in *canonical trend surface analysis* (Lee, 1969; Burwash & Culbert, 1976; Webster, 1977) and *most predictable surface mapping* (Lee, 1981). Spline transformations (Wold, 1974; Smith, 1979; Winsberg & Ramsay, 1980, 1983) represent an important extension of the opportunities afforded by polynomials in data re-expression. Spline functions are piecewise polynomials whose flexibility and local properties make them unusually promising in rendering nonlinear relations more amenable to analysis. An application of spline transformations in dual scaling has been reported by van Rijckevorsel (1982).

Linearizing nonlinear problems. The preceding discussion has focussed on efforts aimed at improving the ability of canonical analysis to exploit nonlinearity by transformation of variables. A class of methods directed towards linearizing a nonlinear problem offer an alternative approach. The basic idea is to linearize **x** and **y** by applying a standard method for nonlinear reduction of dimensionality before going on to perform canonical analysis with the linearized data in the usual way. Opportunities for nonlinear reduction of dimensionality are offered by *nonmetric multidimensional scaling* (e.g., see Kruskal & Wish, 1978), *parametric mapping* (Shepard & Carroll, 1966) and *nonlinear mapping* (e.g., see Kakusho & Mizoguchi, 1983). Starting from a matrix of similarities or dissimilarities (proximities) between samples, nonmetric multidimensional scaling seeks a Euclidean representation of the proximity matrix for which interpoint distances between sample-points are monotonically related to the empirical proximities. That is to say, the larger the dissimilarity between two samples the larger the distance separating them in the Euclidean representation should be. There is no requirement or assumption of linearity. Improved recognition and description of severe nonlinearity may be achieved by replacing the requirement for monotonicity by a requirement for 'smoothness'. Smoothness is defined analytically by an index $\varkappa$ which prescribes a desired relationship between the coordinates of sample-points in the sample space determined by **X** or **Y** and the corresponding low-dimensional representation. The immediate goal of parametric mapping is to optimize $\varkappa$.

Once low-dimensional summaries of **X** and **Y** in Euclidean space are obtained by some such procedure as nonmetric multidimensional scaling or parametric mapping, canonical analysis may be applied to the summary matrices in the usual way. An alternative to canonical analysis as a means of comparing the two linearized representations may be worth mentioning – this is *Procrustes analysis* (Gower, 1975; Ten Berge, 1977; Sibson, 1978; Stenson, 1982). There is a further strategy for exploiting nonlinearity in canonical analysis. This approach requires only the ability to partition the sample into subgroups which are inherently meaningful or interesting in substantive terms. By defining binary dummy variables to correspond to the resulting subgroups or classes, *canonical variate analysis* may be used to explore relationships between the subgroups and the multivariate response of either **x** or **y**. The procedure is appealing in that it is both effective and easily implemented. Some cost in terms of adjustments to the initial, overall objectives of a study, however, may be incurred. An extension of the procedure would involve canonical variate analyses of both **x** and **y**. Relationships between the two sets of resulting canonical variates might then be investigated by means of canonical correlation analysis in the usual way.

From the foregoing it is evident that the linearity of the model in canonical analysis is not nearly as restrictive as might at first appear. Nevertheless, it would be a serious error to allow the flexibility of canonical analysis to deflect attention from the desirability of developing more realistic models for the analysis of intrinsically nonlinear systems. The challenge in so doing is to invent methods that increase the versatility of models as much as possible with the fewest additional parameters. With the power of modern computational tools, the formulation of such models is perfectly feasible. Steps in this direction have already been taken. Friedman and Rafsky (1983), for example, have proposed a completely distribution-free, computer-intensive method based on a model of this sort for investigating multivariate association and prediction. Not only is their method completely distribution-free but it is less specifically directed towards linearity than the comparable classical methods of canonical analysis and multivariate regression. Heiser and Meulman (1983b) have put forward a comparable proposal based on constrained multidimensional scaling. In this case, the underlying notion is to seek a single, suitably constrained space which is *common* to the spaces generated by $m > 2$ data or dissimilarity matrices.

Variable selection. The initial selection of variables for inclusion in a canonical analysis is a matter for subject-matter judgement. Where knowledge of the system investigated is limited there is a tendency to include as many variables as possible in the hope that no important variable will be overlooked. Yet a model which uses fewer variables is often more appealing than a model with many variables, partly because a smaller model is usually easier to understand and because a model based on a subset of variables generally yields more precise results. Thus, an interest in models based on subsets of the original variables is often expressed. Variable selection arises most naturally in the context of prediction. Here, the aim is unambiguous, namely to minimize the mean square error of prediction using the fewest predictor variables. Where the analysis is directed towards exploratory or descriptive modelling the need for a parsimonious model is less clear. However, precise parameter estimates are unquestionably desirable even in cases of this kind. The single most useful tool in selecting a subset of variables for use as a model is the investigator's knowledge of the system investigated and of the role of each variable in the system. In favorable circumstances, such insight will be sufficient to allow the expected sign and magnitude of each coefficient to be declared prior to analysis, at least for those canonical variates which correspond to salient components of the overall relationship. A second, reasonably straightforward procedure is to define variables which are combinations of the original variables.

Variable selection has often to be based on the data themselves. For concreteness, consider an analysis for which the immediate goal is to obtain precise parameter estimates, a situation arising in predictive and descriptive studies. In such cases there is usually no single model which is best; rather, several subset models can be found each of which performs well relative to the full model. The first question which arises in deriving adequate subsets of variables is that of a suitable criterion for deciding when one subset is better than another. One such index proposed by Mallows (1973) for use in regression is C_p. Where the response is q-variate ($q \geq 2$), C_p is a matrix of order q related to the mean square error of prediction. Let $\mathbf{X}$ be $\mathrm{N} \times k$ and $\mathbf{Y}$ be $\mathrm{N} \times q$. Then the multivariate C_p, $\mathbf{C}_p$ (Mallows, 1973; Sparks, Coutsourides & Troskie, 1983), is given by

$$\mathbf{C}_p = (\mathrm{N} - k)\,\mathbf{S}_e^{-1}\,\mathbf{S}_h + 2(p - k)\,\mathbf{I},$$

where

$$\mathbf{S}_e = \mathbf{Y}^t(\mathbf{I} - \mathbf{X}(\mathbf{X}^t\mathbf{X})^{-1}\mathbf{X}^t)\mathbf{Y}, \qquad \mathbf{S}_h = \mathbf{Y}^t\mathbf{A}\mathbf{Y} - (\mathrm{N}-k)\hat{\mathbf{V}},$$
$$\mathbf{A} = \mathbf{I} - \mathbf{X}_p(\mathbf{X}_p^t\mathbf{X}_p)^{-1}\mathbf{X}_p^t, \qquad \hat{\mathbf{V}} = \mathbf{S}_e/(\mathrm{N}-k)$$

and $p<k$ is the number of predictor variables which remain after r variables are deleted from the model ($p+r=k$). Desirable subsets are those which minimize some scalar function of $\mathbf{C}_p$ such as the trace, $\mathrm{tr}(\mathbf{C}_p)$, or largest root, $\lambda_1(\mathbf{C}_p)$. A nice feature of this method is that it lends itself to graphical display. Convenient displays result on plotting $\mathrm{tr}(\mathbf{C}_p)^*$ or $\lambda_1(\mathbf{C}_p)^*$ against p ($p=1, \ldots, k-1$), where

$$\mathrm{tr}(\mathbf{C}_p)^* = 1 + \{[\mathrm{tr}(\mathbf{C}_p) - p\,q]/(k-p)\,q\}$$

and

$$\lambda_1(\mathbf{C}_p)^* = 1 + \{[\lambda_1(\mathbf{C}_p) - p]/(k-p)\}.$$

Good subsets are those for which $\mathrm{tr}(\mathbf{C}_p)^*$ or $\lambda_1(\mathbf{C}_p)^*$ are closest to or smaller than 1. For small problems ($k \leq 8$) $\mathbf{C}_p$ may be calculated for all possible subsets of the original variables. As k increases, however, the procedure rapidly becomes computationally prohibitive and a good computational alternative to computing all possible subsets is required. This task represents a second question which has to be addressed in many cases of variable selection. Furnival and Wilson (1974) have considered this question and have developed an algorithm that, while computing only a small fraction of all possible regressions, does in fact find those few models which are best in the present sense.

The question of subset selection specifically in canonical variate analysis and discriminant analysis has been examined in detail by McKay and Campbell (1982a, b). For canonical variate analisis these authors advocate an all subsets procedure which isolates a number of subsets which may be considered adequate in the sense of providing essentially the same group separation as the full set. Sparks, Coutsourides and Troskie (1983) have derived the multivariate $\mathbf{C}_p$ and given worked examples of its use from multivariate regression. Their results appear to extend more or less directly to canonical analysis. A solution to the vexing problem of how best to proceed when there is little in the way of substantive insight to guide the initial selection of variables may be found in the partial least squares formulation of canonical analysis. Wold (1980, p. 70) has commented favorably on the performance of partial least squares models under the inclusion of large numbers of variables in precisely these circumstances.

Incomplete data. Observations may be incomplete in that some vectors are only partially observed during sampling. Missing observations are a source of difficulty in implementing canonical analysis since the method is applicable only to full data sets. Thus, special procedures are required before analysis can be undertaken. We distinguish two cases of incomplete data – those in which the missing data are missing at random and those in which the pattern of missing values is closely tied to the values of one or more variables. An important task where data are incomplete is to determine which of these mechanisms is responsible for the missing observations. The random case is the more tractable in that the process or processes responsible for the missing values can be ignored. On the other hand, where the missing values are influenced by the values of variables, then the cause of the missing data must be incorporated into the model. Inspection of the pattern of missing values can assist in differentiating the two cases. Our remarks are confined to values which are missing at random. Missing observations can be dealt with either by deleting partially observed samples or variables, or by obtaining estimates for the unobserved values. Eliminating samples is usually the most straightforward procedure. Once the incomplete data have been either eliminated or estimated canonical analysis can be applied in the usual way.

An intuitively appealing notion for filling in missing values is to use all the available data to obtain plausible estimates for them. A general maximum likelihood approach of this kind is provided by the expectation-maximization (EM) algorithm (Dempster, Laird & Rubin, 1977). The EM algorithm is an iterative, two-stage procedure with prediction and estimation steps.

(a) *Prediction step:* given some estimate $\hat{\theta}$ of the unknown parameters, predict the contribution of any missing observation to the sample sufficient statistics;
(b) *Estimation step:* use the predicted sufficient statistics to compute a revised estimate of the parameters.

The calculation alternates between these steps until stable values for the estimates result. The use of the EM algorithm to estimate a mean vector and covariance matrix from incomplete multivariate data has been described by Little and Rubin (1983), who also provide an introduction to recent literature on the topic.

Maximum likelihood estimation of missing values has also been considered by Hocking and Smith (1968) and Hocking and Marx (1979) and the performance of their method in canonical correlation analysis examined in some detail by Smith and Riggs (1982). Computation of the mean vector and covariance matrix with incomplete vectors from several populations with common dispersion has been studied by Huseby, Schwertman and Allen (1980). The adequacy of the often-made assumption that missing values occur at random has been questioned by Rao (1983). Little and Rubin (1983) have similarly remarked that the assumptions generally required when estimating multivariate incomplete data are unrealistic.

Observational data and prediction. Canonical analysis is usually applied to observational as opposed to experimental data. The analysis of observational data is challenging under even the most favorable circumstances. Difficulties are associated with errors of measurement in the x's and y's, correlations among variables, curvilinear relationships, inadequate sampling of the domain of interest and omission of important variables. Thus, canonical analysis at best is, or ought to be, a painstaking and time-consuming process. One key to productive analysis is provided by considering model fitting in the case of data from designed experiments. In response surface designs, for example, provision is almost always made for curvilinear and interactive effects. The comparison suggests that it may be prudent to consider fitting curvilinear (x_i^p) and interactive ($x_i x_k$) terms as a general rule in canonical analysis.

Most applications of canonical analysis are exploratory or descriptive in nature. Predictive applications, however, do occur in meteorology, geology and palaeoclimatology. Such applications pose a number of additional challenges over and above those identified above. In the first place, the true form of the functional relationship between variables is not known; instead, the canonical prediction equation is used, the coefficients of which are estimated quantities. A second point is that the parameter estimates are artifacts of the equation fitted, depending not on the true functional relationship between variables but on the data available for analysis. Thus, if the sample is to satisfactorily represent the population of interest it is most important that the sample be unbiased. Even so, the fitted model is selected for the data analyzed and will not perform as well with new data. Another issue is that of the range of validity for predictions. Since the prediction equation represents an empirical rather than a functional relationship, the prediction equation may perform adequately only over a limited range of values for the predictors. The minimum covering ellipsoid (Titterington, 1975) provides a useful guide as to the range of valid interpolation. For a discussion of some of the difficulties posed by observational studies and an introduction to the recent literature, reference may be made to Rosenbaum (1984).

Extensions of canonical analysis. At several places in the text reference has been made to extensions of canonical analysis. For the most part these are directed towards either improving interpretation or widening the class of problem that can be investigated. We summarize the principal extensions here while taking the opportunity to mention recent developments not referred to elsewhere in the text.

With respect to proposals for improving interpretability, we refer to the *shrunken* or *ridge-type estimators* for the canonical weights, *constrained canonical analysis, most predictable surface mapping* and the *rotation* of canonical variates. In constrained canonical analysis, the model is fitted under the restriction that the canonical weights be confined to the integers $\{-1, 0, 1\}$. Efficient procedures for the canonical analysis of *nonmetric* data have been developed. The *partial least squares* and *alternating least squares* approaches of Wold (1980, 1982) and of Young, de Leeuw and Takane (1976) have this objective. These are distribution-free procedures characterized by iterative, nonlinear methods of parameter estimation. Not only is the analysis of data expressed on very weak measurement scales made possible by these methods but in addition little or no prior knowledge as to the identity of pertinent variables is called for. The alternating least squares approach has been further extended to give *nonlinear canonical correlation analysis* (van der Burg & de Leeuw, 1983). *Redundancy analysis* (van den Wollenberg, 1977), unlike the preceding developments, incorporates a modification to the model itself. The goal of redundancy analysis is to maximize predictable variance rather than the shared variance common to a pair of linear composites, one from each domain. The concept of redundany analysis has been unified and extended by Tyler (1982). *Cross-validation* permits the goodness-of-fit of all aspects of a fitted canonical model to be assessed, while the *jackknife* and the *bootstrap* enable estimates of the precision of individual coefficients comprising the $\{r_k;\ \mathbf{a}_k, \mathbf{b}_k;\ u_k, v_k\}$ to be obtained.

Several further extensions of canonical analysis have been proposed. Meredith and Tisak (1982) have described procedures for the analysis of *longitudinal and repeated measures* data. The extension of canonical analysis from the usual situation in which the data comprise observations on a finite number of variables to data for which the observations are *continuous functions* of time or space has been sketched by Ramsay (1982). Jewell and Bloomfield (1983) and Jewell, Bloomfield and Bartmann (1983) have generalized the notion of canonical correlations and variates to the analysis of certain *past and future time series*.

Further reading. Opportunities for studying *residuals* in canonical analysis by means of gamma probability plots were described some years ago by Brillinger (1967, p. 236). Gnanadesikan (1977, p. 258) has discussed the analysis of multivariate residuals from a more general viewpoint. Chambers, Cleveland, Kleiner and Tukey (1983, p. 48) have given a nice introduction to the construction and use of *empirical quantile-quantile* plots. Much the greater part of the work on analyzing residuals has been carried out in the context of regression. For an introduction to this work reference may be made to Belsley, Kuh and Welsch (1980), Gunst and Mason (1980), Dempster and Gasko-Green (1981), Cook and Weisberg (1982) and Neter, Wasserman and Kutner (1983). *Adjusted variable* plots have been described and used under the name of partial residual or partial regression plots by Gunst and Mason (1980), Henderson and Velleman (1981) and Velleman and Welsch (1981). Questions of temporal and spatial *autocorrelation* in residuals have been addressed respectively by Johnston (1972, p. 246) and by Cliff and Ord (1981a, b) and are beginning to receive attention in canonical analysis (Lee, 1981, p. 240). Ecological applications of two-dimensional *spectral analysis* have been described by McBratney and Webster (1981) and Renshaw and Ford (1983).

Formal accounts of *cross-validation* have been given by Stone (1974, 1977a, b), Geisser (1975), Efron (1979b, 1982) and Efron and Gong (1983). The use of cross-validation in canonical analysis has been illustrated by Thorndike and Weiss (1973, 1983), Herman, Dunham and Hulin (1975), Weinberg and Darlington (1976), Huba, Wingard and Bentler (1980) and Lee, McCabe and Graham (1983). The *jackknife* has been discussed by Miller (1974, 1978), Mosteller and Tukey (1977, p. 133), Efron (1979a, b, 1981, 1982) and Efron and Gong (1983). Clarkson (1979) and Gifi (1981) have discussed the jackknife in multivariate contexts related to canonical analysis. A modified jackknife applicable in canonical analysis has been proposed by Dempster (1966), who has also provided a nice illustration of the use of a jackknife in canonical variate analysis (see Dempster, 1969, p. 257). Parr and Schucany (1980) have prepared a bibliography of the jackknife, while Matloff (1980) has published a Fortran subroutine for computing the jackknife. The *bootstrap* has been described by Efron (1979a, b, 1981, 1982) and by Efron and Gong (1981, 1983). Diaconis and Efron (1983) have employed the bootstrap to assess the accuracy of the weights and of the linear composites yielded by them in principal component analysis, a procedure which might also be informative in canonical analysis. Efron and Gong (1981) have used the bootstrap in discriminant analysis to assess the performance of a prediction rule and as a guide in variable selection. The bootstrap has also been used by Gong (1982) to assess the variability of a model prescribed by a particular variable selection technique in regression analysis. Connections between cross-validation, the jackknife and the bootstrap have been discussed by Efron (1979b, 1982). An example of the use of the bootstrap in canonical analysis has been given by van der Burg and de Leeuw (1983) and in a related context by Heiser and Meulman (1983a, b).

Canonical analysis is not widely used for *prediction*. Examples of its use for this purpose are, however, to be found in meteorology (Robert Miller, 1962; Glahn, 1968), palaeoclimatology (Webb & Bryson, 1972; Bryson & Kutzbach, 1974; Sachs, Webb & Clark, 1977; Webb & Clark, 1977) and geology (Lee, 1981).

14.6 Concluding remarks

Canonical analysis is known and appreciated more for its elegant mathematical properties than for its utility. Convincing practical examples specifically of canonical correlation analysis remain comparatively few. We set out to examine the worth of canonical analysis in ecology, describing several applications of the procedure with this purpose in mind. None of the substantive questions addressed amounts to much in purely ecological terms. Nevertheless, the worked examples serve to show how the model is fitted and results are interpreted in canonical analysis. In addition, the examples reveal something of the flexibility of the underlying model and of the correspondingly wide range of applicability of the method.

The view was expressed on the basis of the results obtained that as an *information summary* canonical analysis can contribute towards the attainment of objectives in a broad class of ecological endeavors. General considerations nevertheless indicate that canonical analysis is unlikely to offer solutions to all problems involving two or more sets of variables. Far from it. That boundaries exist beyond which canonical analysis is likely to be ill-advised and unproductive is largely axiomatic. The applications described represent an initial step in efforts to establish just what can and what cannot be achieved by fitting a model in canonical analysis. Beginning with two sets of data for which the dependency structure was believed in advance to be topologically simple, the conclusion was reached that canonical analysis can reliably identify structure in ecological data provided the structure in question is relatively simple. The requirement for simplicity was next relaxed somewhat and the merits of canonical analysis examined in relation to a wider class of more realistic investigations. The results lent weight to our initial conclusion that canonical analysis can serve to reliably identify simple structure while disclosing something of the versatility which is a salient characteristic of the method. Later, in Chap. 14, attention was drawn to recent theoretical and practical developments which promise to extend the applicability of canonical analysis well beyond small-scale investigations of the kind dealt with in this study, for which the dependency structure is inherently likely to be topologically simple. At the time of writing few applications of canonical correlation analysis to large data sets have been reported. The effectiveness of canonical variate analysis and dual scaling in large-scale investigations, on the other hand, has been well-established for some considerable time (e.g., see Mahalanobis, Majumdar & Rao, 1949; Barnett & Lewis, 1963; Lebart, Morineau & Tabard, 1977; Lebart & Morineau, 1982; Greenacre, 1984; Lebart, Morineau & Warwick, 1984). One large-scale and apparently worthwhile application of canonical correlation analysis may be mentioned. This is the study by Lee, McCabe and Graham (1983), for which $N = 1972$, $p = 26$ and $q = 5$.

It is instructive to consider the domain of applicability or utility of canonical analysis from a different viewpoint. Canonical analysis is one of a class of methods characterized by a search for linear composites of observed variables that have certain desirable properties. Features common to this class of methods are parsimony, wide applicability and the ready incorporation of probabalistic elements. In canonical analysis, linear composites are sought which reduce the correlation structure between two sets of variables to its simplest possible form. Two consequences of this strategy merit attention. One consequence is that it leads to the focus of attention being shifted from the original variables with which the investigator is familiar to new, unobserved variables – the canonical variates. Canonical variates are specified by mathematical rather than substantive considerations; they are linear composites of observed variables which are maximally correlated subject to the constraint of biorthogonality. Prospects for the substantive interpretation of canonical variates are accordingly not assured. Underlying these remarks is the implication that canonical analysis is essentially descriptive and interpretive in nature rather than an expression of substantive theory. The usual result of analysis is a description of the structure of what is often a dynamic system rather than of the processes involved in it. To go beyond this point stronger models based on some conception of the actual processes underlying the data are required. For these reasons canonical analysis is best suited to the exploratory phase of research where the need is simply to identify the presence of certain variables and the direction of their effect. The results provided may nevertheless be indispensable as a first step in the development of a dynamic model of the system or process in question. Canonical analysis may indeed yet be shown to have a more substantial contribution to make to the study of dynamic systems. Ramsay's (1982) remarks concerning the opportunities for canonical analysis in a Hilbert space where the data are continuous functions as opposed to a finite set of points are especially pertinent in this connection. In such cases it is natural to conceive of the data dynamically, rather than in a static way, and to analyze them accordingly. A further consequence of the search for linear composites in canonical analysis is that it has had the effect of confining attention almost exclusively to functions of a set of linear and quadratic statistics, to the neglect of all other aspects of the data. Some justification for this practice is forthcoming from the observation that for data which are p-variate normal all the sample information in the data matrix is contained in the mean vector and covariance matrix, regardless of the size of the sample. The mean vector and covariance matrix are *sufficient statistics* for a multivariate normal density. Nevertheless, such a cutting operation is to be viewed with suspicion if only because p-variate normality is far from universal in real data. Where the data distribution happens to be other than multivariate normal, the canonical correlation coefficients and other indices which depend on the data only through the sufficient statistics will not accurately reflect the quality of a fitted model in all regions of observation space. Fortunately, recent work, through the development of techniques for studying quantities associated with *each* data point, such as fitted values, residuals and measures of influence, enables shortcomings of this kind to be overcome.

The analyses reported deal with substantive questions of a comparatively simple kind. For the purpose of investigating the performance of a neglected method such questions provide a natural starting point. As a corollary, however, it follows that the analyses reveal little or nothing of the ability of canonical analysis to deal with more complex issues, such as might be presented for example by large data sets, $m > 2$ sets of variables, prediction or longitudinal studies. The analyses are deficient in another way – they fail to make the fullest possible use of the data either in model fitting or in assessing the precision of the coefficients estimated. Despite these limitations, the results obtained enable several conclusions to be drawn and stated unequivocally. The principal conclusions drawn from the study are as follows.

(a) Canonical analysis has a remarkable ability to parsimoniously capture the dependency structure between two sets of variables in cases where this structure is topologically simple.
(b) Much can be done to widen the class of problem whose analysis can profitably be undertaken beyond that for which the dependency structure is initially topologically simple.

Recent advances in the statistical data analysis of multivariate observations and in regression methodology, as well as refinements to canonical correlation analysis itself (e.g., ridge-type estimation, nonlinear canonical analysis), promise to make the analysis of dependency in systems of appreciably greater complexity than those dealt with in this study perfectly feasible. Among the challenges likely to be met in analyzing complex systems are heterogeneity, data distributions which are not well-behaved, collinearity and nonlinearity. Measures for diagnosing and treating these conditions were discussed in earlier sections of the present chapter. Accordingly, it seems justifiable to conclude that, provided proper attention is given to matters of the kind identified, opportunities for the profitable use of canonical analysis in ecology are considerable.

Canonical analysis is a delicate tool which must be used with care and good sense if worthwhile results are to be obtained. This is why naive reliance on an automated computer program in implementing analysis all too often leads to misleading or incorrect results and to confusion rather than insight. The preferred approach is otherwise. The first requirements are to assess the homogeneity of the sample and the shape of the data distribution. Where the sample is shown to be homogeneous and the data distribution well-behaved one may proceed to fit the model. Alternatively, a need to first induce homogeneity or to re-express the data may be revealed. At this stage the question as to whether to employ a robust estimate of the covariance or correlation matrix also arises. Fitting a model is itself properly treated as an interactive process in which data, residuals and the judgement of the investigator combine to guide the course of analysis from one step to the next. Graphical displays are indispensable aids throughout the fitting procedure. Having arrived at a satisfactory fit, the final step is to assess the quality of the model. For this purpose jackknife or bootstrap estimates of the precision of quantities estimated during fitting will be informative. Associated with these developments in the conduct of analysis are changes in attitude towards the purpose of analysis. Increasingly, it is accepted that insight can accrue without the need for tightly-specified and frequently over-optimistic distributional requirements of the data. Awareness that re-expression has a most valuable contribution to make in rendering the shape of a data distribution more amenable for analysis is also increasing. At the same time less emphasis is being placed on inferential aspects of an analysis than hitherto. The goal in many cases is rather to arrive at a well-fitting model to be valued either for its capacity to yield an information-rich graphical summary of the relationships of interest, or, alternatively, for its predictive power. Thus, there is a freeing of analysis from the distributional requirements associated with parametric tests as well as from the extreme form of summarization which test statistics themselves represent.

In ecology the successful use of canonical analysis calls for an ability to match statistical theory with empirical knowledge of the structure and function of ecosystems. Empirical knowledge of this sort is to be obtained above all by observation and study in the field. The importance of fieldwork in ecological applications of canonical analysis can scarcely be overemphasized. This follows because it is only through careful fieldwork that the insight necessary for a wise choice of variables for inclusion as well as for the incisive interpretation of results is to be acquired.

Appendices

A.1 Multivariate regression

1. The multivariate linear regression model

Consider two sets of jointly distributed variables $\mathbf{x}$ $(p \times 1)$ and $\mathbf{y}$ $(q \times 1)$. Suppose it desired to predict the y's simultaneously from the x's. This is the problem addressed by multivariate multiple regression analysis. The linear model for multivariate regression may be specified as follows:

$$\underset{(\mathrm{N} \times q)}{\mathbf{Y}} = \underset{(\mathrm{N} \times m)}{\mathbf{X}} \; \underset{(m \times q)}{\mathbf{B}} + \underset{(\mathrm{N} \times q)}{\mathbf{E}}, \tag{A1.1}$$

where $m = p + 1$, $m + q \leq \mathrm{N}$ and

Y: consists of N independent observations on q response variables,
X: consists of N independent observations on p regressor variables,
B: is a matrix of unknown parameters to be estimated,
E: is a matrix of unobserved random disturbances.

The first column of **X** consists of unities, so that $\mathbf{X} = [\mathbf{1} | \mathbf{X}_1]$, where $\mathbf{X}_1$ is $\mathrm{N} \times p$, allowing for an overall mean effect. The minimum error specification for (A1.1) is that the jth row of **E**, $\mathbf{e}_j^{\mathrm{t}}$ $(j = 1, \ldots, \mathrm{N})$, has a q-variate distribution with zero mean vector and finite covariance matrix, denoted by $\mathscr{E}(\mathbf{e}) = \mathbf{o}$ and $\mathscr{V}(\mathbf{e}) = \mathbf{\Sigma}$, respectively. The rows of **Y**, **X** and **E** correspond to particular samples and the columns of **Y** and **E** to particular responses. The matrix $\mathbf{B} = \{\beta_{ik}\}$ of *partial regression coefficients* consists of q columns of weights, $\boldsymbol{\beta}_k$ $(k = 1, \ldots, q)$ for predicting the kth response $\mathbf{y}_k(\mathrm{N} \times 1)$ from the $p+1$ antecedents in the rows $\mathbf{x}_j^{\mathrm{t}}$ $(j = 1, \ldots, \mathrm{N})$ of **X**. The model (A1.1) is described as linear because of the linearity of the parameters β_{ik}.

There are several versions of (A1.1), each distinguished by a particular specification for **Y**, **X**, **B** and **E** (see Press, 1972, p. 209). Of the different forms, we refer to just two: in the *fixed* version, $\mathbf{X}_1$ is a matrix of known constants while in the *random* version $\mathbf{X}_1$ consists of realizations of random variables. The theory pertaining to the fixed model is best developed, an important special case being the experimental design model where **X** consists only of 0's and 1's. The random model becomes more tractable if all likelihoods and expectations are interpreted as *conditional* on **X**. This stipulation is equivalent to asserting that **X** and **E** are independent.

The principal objectives of analysis are to obtain point and interval estimates for the entire matrix **B** and to determine whether the x's alone or collectively predict the q-variate response **y**.

2. Estimation

An important task in multivariate regression is to estimate the parameter matrix **B**.

B may be estimated in a variety of ways, the estimator proving to be the same for the fixed model and for the conditional form of the random model. For optimal prediction the matrix of coefficients **B** should be chosen so as to minimize the elements of $\mathbf{E}(\mathrm{N} \times q)$, or some function of them. The method of least squares yields an estimator which minimizes the sum of squared disturbances for each of the responses. The sum of squared disturbances for the kth response is the kth diagonal element of $\mathbf{E}^t\mathbf{E}$. The sum of these diagonal elements over all responses is the trace of $\mathbf{E}^t\mathbf{E}$. In this case, therefore, we wish to minimize

$$\begin{aligned}\sum_j \sum_k e_{jk}^2 &= \mathrm{tr}(\mathbf{E}^t\mathbf{E}) \\ &= \mathrm{tr}\{(\mathbf{Y}-\mathbf{XB})^t(\mathbf{Y}-\mathbf{XB})\}. \qquad \text{(A1.2)}\end{aligned}$$

Differentiating (A1.2) with respect to variation in **B** and equating the derivative to the null matrix **0** of order $m \times q$ leads to the *normal equations*

$$(\mathbf{X}^t\mathbf{X})\mathbf{B} = \mathbf{X}^t\mathbf{Y}, \qquad \text{(A1.3)}$$

for which the ordinary least squares (OLS) solution $\hat{\mathbf{B}}$ is

$$\hat{\mathbf{B}} = (\mathbf{X}^t\mathbf{X})^{-1}\mathbf{X}^t\mathbf{Y}. \qquad \text{(A1.4)}$$

Provided that $\mathbf{X}^t\mathbf{X}$ is nonsingular, a unique solution can be found. $\hat{\mathbf{B}}$ is an unbiased estimator of **B**, $\mathscr{E}(\hat{\mathbf{B}}) = \mathbf{B}$. Other properties of $\hat{\mathbf{B}}$ are summarized by Press (1972, p. 213), Finn (1974, p. 110) and Mardia, Kent and Bibby (1979, p. 172).

From Eqs. (A1.2)–(A1.4) it can be seen that estimation in the multivariate case is a direct extension of the estimation procedures of multiple regression. In particular, the elements of the kth column, $\hat{\boldsymbol{\beta}}_k$, of $\hat{\mathbf{B}}$ are precisely the weights which would result from a multiple regression of the kth response, y_k, on $x_1, \ldots, x_m$.

The standard errors of the $\hat{\beta}_{ik}$ are helpful in assessing the individual coefficients. Moreover, provided certain distributional requirements are met, confidence bounds for the β_{ik} individually and jointly can be obtained. We shall not pursue the details here. For a thorough discussion of interval estimation in multivariate regression reference may be made to Bock (1975, p. 227).

3. Hypothesis testing

Up to this point it has been assumed only that the disturbances $\mathbf{e}_j^t$ are q-variate quantities with zero mean vector and finite covariance. For the purposes of hypothesis testing and establishing the precision of the point estimate $\hat{\beta}_{ik}$, it is necessary to specify an error distribution. For the fixed model it is usual to require that the distribution of $\mathbf{e}_j^t$ be multivariate normal; more specifically it is required that

$$\mathbf{e}_j^t \sim \mathrm{N}_q(\mathbf{o}^t, \mathbf{\Sigma}). \tag{A1.5}$$

The kth column of $\mathbf{E}$, $\mathbf{e}_k(\mathrm{N} \times 1)$, consists of the disturbances for all N observations on the kth response ($k=1, \ldots, q$). It is usual to assume that $\mathbf{e}_k \sim \mathrm{N}_N(\mathbf{0}, \sigma_k^2 \mathbf{I})$, where σ_k^2 is the variance of y_k. The requirement (A1.5) is equivalent to assuming that the distribution of the jth response is $\mathbf{y}_j^t \sim \mathrm{N}_q(\mathbf{x}_j^t \mathbf{B}, \mathbf{\Sigma})$. For the random model the distribution of $\mathbf{y}^t$ given $\mathbf{X}$ must be q-variate normal with conditional mean vector $\boldsymbol{\mu}^*(q \times 1)$ and conditional covariance matrix $\mathbf{\Sigma}^*(q \times q)$:

$$\mathbf{y}_j^t | \mathbf{X} \sim \mathrm{N}_q(\boldsymbol{\mu}^*, \mathbf{\Sigma}^*).$$

With these assumptions hypothesis tests can be made and confidence bounds for the β_{ik} established. We shall see that, although estimation in multivariate regression and multiple regression are essentially the same, in connection with hypothesis testing differences arise. We shall outline first a test for the independence of $\mathbf{x}$ and $\mathbf{y}$ and then proceed to a test of the partial contribution of an additional predictor which may be of interest where $\mathbf{x}$ and $\mathbf{y}$ can be considered to be dependent.

3.1 Independence

A test of the linear relationship between $\mathbf{x}$ and $\mathbf{y}$ is generally first performed. The procedure is identical to that described in Sect. 3.4.1. The hypotheses of interest are:

$$\begin{aligned} &\mathrm{H}_0: \mathbf{B}=\mathbf{0}, \\ &\mathrm{H}_1: \mathbf{B} \neq \mathbf{0}. \end{aligned} \tag{A1.6}$$

Several criteria are available for testing (A1.6), all of them functions of the roots λ_k ($k=1, \ldots, q$) of the determinantal equation $|\mathbf{Q}_h - \lambda \mathbf{Q}_e| = 0$, where $\mathbf{Q}_h = \mathbf{S}_{21}\mathbf{S}_{11}^{-1}\mathbf{S}_{12}$, $\mathbf{Q}_e = \mathbf{S}_{22} - \mathbf{S}_{21}\mathbf{S}_{11}^{-1}\mathbf{S}_{12}$ and the $\mathbf{S}_{ij}$ are uncorrected SSP matrices calculated from $\mathbf{x}$ and $\mathbf{y}$. Wilks' likelihood ratio statistic Λ is one such criterion, where

$$\Lambda = \frac{|\mathbf{Q}_e|}{|\mathbf{Q}_e + \mathbf{Q}_h|} = \prod_{k=1}^{q} (1+\lambda_k)^{-1}, \qquad 0 \le \Lambda \le 1.$$

The quantity Λ has a Wilks' $\Lambda(u, \nu_h, \nu_e)$ distribution (Timm, 1975, p. 147) with parameters $u=q$, $\nu_h = p$, $\nu_e = \mathrm{N} - p - 1$. A convenient approximation to the distribution of $\Lambda(u, \nu_h, \nu_e)$, at least for moderately large N, is given by Bartlett's (1938) approximation

$$\chi^2 = -\{(\mathrm{N}-1) - \tfrac{1}{2}(p+q+1)\} \log_e \Lambda.$$

Under the null hypothesis χ^2 is distributed approximately as a chi-squared variate on pq degrees of freedom. Hypothesis (A1.6) is rejected with confidence $(1-\alpha)$ if

$$\chi^2 > \chi_\alpha(p\,q),$$

where $\chi^2_\alpha(p\,q)$ is the upper 100α percentage point of the chi-squared distribution with $p\,q$ degrees of freedom. Otherwise, H_0 is maintained. The test of (A1.6) is summarized in the *multivariate analysis of variance* of Table A1.1.

Table A1.1. Multivariate multiple regression. A test for H_0: $\mathbf{B}=\mathbf{0}$ in the regression of $\mathbf{y}(q\times 1)$ on $\mathbf{x}(p\times 1)$

Source	df	SSP$(q\times q)$	$\Lambda(u, \nu_h, \nu_e)$
Regression of $\mathbf{y}$ on $\mathbf{x}$	$p+1$	$\mathbf{Q}_h=\mathbf{S}_{21}\mathbf{S}_{11}^{-1}\mathbf{S}_{12}=\hat{\mathbf{B}}\mathbf{S}_{11}\hat{\mathbf{B}}^t$	$\lvert\mathbf{Q}_e\rvert/\lvert\mathbf{Q}_e+\mathbf{Q}_h\rvert$
Error	$N-p-1$	$\mathbf{Q}_e=\mathbf{S}_{22}-\mathbf{S}_{21}\mathbf{S}_{11}^{-1}\mathbf{S}_{12}$	
Total	N	$\mathbf{Q}_h+\mathbf{Q}_e=\mathbf{S}_{22}$	

Rejection of (A1.6) indicates that at least one of the $x_1, \ldots, x_p$ is useful for predicting at least one of the $y_1, \ldots, y_q$. In such cases more specific tests of the contribution of particular antecedent variables are likely to be of interest. A test procedure of this kind was described in Sect. 3.4.3. That procedure does not provide independent tests of the information gain attributable to individual predictors. We shall therefore consider here an alternative which does yield independent tests. The procedure is based on a decomposition of the regression SSP matrix, $\mathbf{Q}_h$, into orthogonal component matrices of order q, each of which is uniquely associated with a particular antecedent variable.

3.2 Stepwise test of the partial contribution of an additional predictor

Independent tests for predictors are not directly obtainable from the rows of $\hat{\mathbf{B}}$. Nevertheless, if on substantive grounds the antecedents can be *ordered* in a meaningful way, it is possible to separate the contribution of each antecedent to the explained variation of the responses represented by $\mathbf{Q}_h$. Stepwise, single-degree-of-freedom tests of the unique contribution of each predictor can be made after first transforming the predictors to new, uncorrelated variables in a specified order. More precisely, the linear function or portion of the ith predictor x_i which is uncorrelated with the preceding predictors $x_1, x_2, \ldots, x_{i-1}$ is substituted for x_i in $\mathbf{X}$. In other words, the new variables are those obtained by partialing out the effects of the earlier predictors in the set.

It is desirable to order the predictor variables in such a way that the least plausible appear last. A series of hypotheses is then of interest. The hypotheses focus sequentially on the last row, the penultimate row and so on of $\mathbf{B}$ and are equivalent to asserting that the pth, p-1th, p-2th, ... predictor makes no contribution to the variation of the responses. The procedure begins with the Gram-Schmidt factorization of $\mathbf{X}$:

$$\underset{(N\times m)}{\mathbf{X}} = \underset{(N\times m)}{\mathbf{X}^*}\ \underset{(m\times m)}{\mathbf{T}}, \qquad \mathbf{T}: \triangleleft \tag{A1.7}$$

where $\mathbf{X}^*$ is columnwise orthonormal. Substituting for $\mathbf{X}$ in (A1.1) we have

$$\mathbf{Y}=\mathbf{X}^*\mathbf{T}\mathbf{B}+\mathbf{E}. \tag{A1.8}$$

In this expression $\mathbf{TB}$ is the matrix of population regression coefficients for the ordered, uncorrelated predictors in $\mathbf{X}^*$. The least squares estimate of $\mathbf{TB}$ is given by

$$\begin{aligned}\mathbf{T}\hat{\mathbf{B}} &= ([\mathbf{X}^*]^t\mathbf{X}^*)^{-1}[\mathbf{X}^*]^t\mathbf{Y}=[\mathbf{X}^*]^t\mathbf{Y}\\ &= \underset{(m\times q)}{\mathbf{U}}, \quad \text{say.}\end{aligned}$$

$\mathbf{U}$ is known as the matrix of *orthogonal estimates* or of *part* or *semipartial* regression coefficients and is an unbiased estimator for $\mathbf{TB}$. The ith row of $\mathbf{U}$, $\mathbf{u}_i^t$, contains the regression coefficients for x_i eliminating all preceding antecedents. Moreover, the rows of $\mathbf{U}$, unlike those of $\hat{\mathbf{B}}$, are uncorrelated. Computationally, the orthogonal estimates are readily obtained by means of the Cholesky decomposition of $\mathbf{S}_{11}$:

$$\mathbf{S}_{11}=\mathbf{S}_{11}^{\frac{1}{2}}\mathbf{S}_{11}^{\frac{1}{2}}. \qquad \mathbf{S}_{11}^{\frac{1}{2}}: \triangleleft$$

$\mathbf{S}_{11}^{\frac{1}{2}}(m\times m)$ is identically the triangular factor of (A1.7). Then

$$\begin{aligned}\mathbf{U}&=\mathbf{T}\hat{\mathbf{B}}\\ &=\mathbf{S}_{11}^{\frac{1}{2}}\hat{\mathbf{B}}=\mathbf{S}_{11}^{\frac{1}{2}}\mathbf{S}_{11}^{-1}\mathbf{S}_{12}=\mathbf{S}_{11}^{-\frac{1}{2}}\mathbf{S}_{12}.\end{aligned}$$

The rows $\mathbf{u}_i^t$ of $\mathbf{U}$ yield an additive decomposition of the regression SSP matrix $\mathbf{Q}_h$ of Table A1.1:

$$\begin{aligned}\mathbf{Q}_h&=\hat{\mathbf{B}}^t\mathbf{S}_{11}\hat{\mathbf{B}}=\hat{\mathbf{B}}^t[\mathbf{S}_{11}^{\frac{1}{2}}]^t[\mathbf{S}_{11}^{\frac{1}{2}}]\hat{\mathbf{B}}\\ &=\mathbf{U}^t\mathbf{U}=\mathbf{u}_1\mathbf{u}_1^t+\mathbf{u}_2\mathbf{u}_2^t+\ldots+\mathbf{u}_m\mathbf{u}_m^t.\end{aligned} \tag{A1.9}$$

Each outer product $\mathbf{u}_i\mathbf{u}_i^t$ in (A1.9) is a symmetric matrix of order q carrying a single degree of freedom which is uniquely associated with the ith predictor. To test the contribution of x_i to $\mathbf{Q}_h$ we proceed as follows. An hypothesis SSP matrix for x_i, $\mathbf{Q}_h^{(i)}$, on one degree of freedom is defined as follows:

$$\mathbf{Q}_h^{(i)}=\mathbf{u}_i\mathbf{u}_i^t,$$

the corresponding error SSP matrix being

$$\mathbf{Q}_e^{(i)}=\mathbf{Q}_t-\sum_{j=1}^{i}\mathbf{u}_j\mathbf{u}_j^t.$$

Then

$$|\mathbf{Q}_h^{(i)}-\lambda\,\mathbf{Q}_e^{(i)}|=0$$

has just one non-zero root, λ_i, readily computed as

$$\lambda_i=\mathbf{u}_i^t[\mathbf{Q}_e^{(i)}]^{-1}\mathbf{u}_i,$$

and

$$\Lambda = (1 + \mathbf{u}_i^t [\mathbf{Q}_e^{(i)}]^{-1} \mathbf{u}_i)^{-1} \sim \Lambda(u, \nu_h, \nu_e),$$

where $u = q$, $\nu_h = 1$ and $\nu_e = \mathrm{N} - p - 1$. Probability points for Λ are approximated by those of the chi-squared variate

$$\chi^2 = -\{\nu_e + \nu_h - \tfrac{1}{2}(\nu_h + q + 1)\} \log_e \Lambda,$$

which has q degrees of freedom. In this way a test of the partial contribution of x_i over and above the contributions of $x_1, x_2, \ldots, x_{i-1}$ can be made. The stepwise analysis is summarized in the *multivariate analysis of variance* of Table A1.2. We stress that the *order* of the predictor makes a great deal of difference to the increment associated with a particular predictor.

Table A1.2. Multivariate regression. Stepwise analysis of the partial contribution of an additional predictor to $\mathbf{Q}_h$

Source	df	SSP$(q \times q)$	λ_k
x_0^*: (x_0 ignoring $x_1, \ldots, x_m$)	1	$\mathbf{u}_0 \mathbf{u}_0^t$	$\mathbf{u}_0^t [\mathbf{Q}_e^{(0)}]^{-1} \mathbf{u}_0$
x_1^*: (x_1 eliminating x_0 and ignoring $x_2, \ldots, x_m$)	1	$\mathbf{u}_1 \mathbf{u}_1^t$	$\mathbf{u}_1^t [\mathbf{Q}_e^{(1)}]^{-1} \mathbf{u}_1$
x_2^*: (x_2 eliminating x_0, x_1 and ignoring $x_3, \ldots, x_m$)	1	$\mathbf{u}_2 \mathbf{u}_2^t$	$\mathbf{u}_2^t [\mathbf{Q}_e^{(2)}]^{-1} \mathbf{u}_2$
⋮	⋮	⋮	⋮
x_m^*: (x_m eliminating $x_0, x_1, \ldots, x_{m-1}$)	1	$\mathbf{u}_m \mathbf{u}_m^t$	$\mathbf{u}_m^t [\mathbf{Q}_e^{(m)}]^{-1} \mathbf{u}_m$
Total regression	m	$\mathbf{Q}_h = \mathbf{U}^t \mathbf{U}$	
Residual	$\mathrm{N} - m$	$\mathbf{Q}_e = \mathbf{Y}^t \mathbf{Y} - \mathbf{U}^t \mathbf{U}$	
Total	N	$\mathbf{Q}_t = \mathbf{Y}^t \mathbf{Y}$	

Multiple determination. The ratio of regression to total sum of squares for the kth response, y_k, expresses the extent to which y_k is predicted by the $x_1, \ldots, x_p$. This quantity is the *coefficient of determination* R_k^2, of the kth response. It is readily obtained from the diagonal elements of $\mathbf{Q}_h$ and $\mathbf{Q}_t$, respectively:

$$\mathrm{R}_k^2 = \frac{[\mathbf{Q}_h]_{kk}}{[\mathbf{Q}_t]_{kk}}.$$

The positive square root of R_k^2 is the *multiple correlation coefficient*, R_k. This is the sample correlation coefficient between the kth response and the linear combination of predictors with coefficients $\hat{\boldsymbol{\beta}}_k$, that is of those predictors whose coefficients are the kth column of $\hat{\mathbf{B}}$. R_k is a biased estimate of the corresponding population quantity. If the number of predictors is large relative to N, the sample value R_k is upwardly biased. The *adjusted* coefficient of determination

R_k^{*2}, however, is bias corrected and is given by

$$R_k^{*2} = 1 - (1 - R_k^2)\frac{N-1}{N-p}.$$

To determine how much the ith antecedent variable x_i contributes to the prediction of the kth response, y_k, eliminating earlier predictors, we require

$$p_{ik} = \frac{[\mathbf{Q}_h^{(i)}]_{kk}}{[\mathbf{Q}_t]_{kk}}.$$

The sum of the p_{ik} over all p predictor variables is identically the coefficient of determination for the kth response, $\Sigma_i p_{ik} = R_k^2$.

4. Concluding remarks

In this section a number of general observations concerning the estimation, computation and interpretation of regression coefficients in multivariate regression analysis are made.

With respect to *estimation* we have seen that multivariate regression involves a straightforward extension of the procedures of multiple regression. The direct application of the familiar univariate estimation theory to q responses in itself, however, is indicative that the multivariate nature of the response has not been adequately taken into account in the process - the procedure neglects the intercorrelations among the responses. Sclove (1971) has demonstrated how by exploiting the covariance structure of the response variables improved parameter estimates can be obtained. Sclove's work has been extended by Brown and Zidek (1980) in the framework of a multivariate version of the Hoerl-Kennard ridge regression rule. In addition, van der Merwe and Zidek (1980) have proposed a multivariate generalization of the James-Stein estimator for regression coefficients. Van der Merwe and Zidek have shown how by implicitly employing the canonical correlation coefficients between linear composites of **x** and **y** together with the canonical variates of **y**, improved estimates for the $\hat{\beta}_{ik}$ of multivariate regression result.

Special difficulties in estimation and interpretation often arise in connection with field surveys or observational studies as opposed to designed experiments. In the first place, the predictor variables in nonexperimental contexts are usually random variables rather than variables whose values are fixed or selected in advance. Secondly, the data themselves are frequently contaminated by errors in the variables. It follows that much of the extensive body of theory developed from the fixed model is largely irrelevant in the context of field or observational studies. We shall briefly address the question of errors in the variables here, deferring our remarks concerning stochastic **X** until later in the section. Although both predictor and response variables are susceptible of error in nonexperimental studies, only those in the x's give rise to difficulties; errors in the y's are absorbed in the error terms $\mathbf{e}_j^t$ and do not require special treatment. We therefore need to consider only errors in the predictors. Measurement error has several undesirable effects, causing serious problems in estimation as well as increasing the error variance and decreasing the strength of the relationship between $\mathbf{x}$ and $\mathbf{y}$. In the presence of errors in the variables, the ordinary least squares estimates for the β_{ik} are inconsistent; that is to say the estimates are biased even for large samples. The problem of bias has led to the development of alternatives to the usual least squares estimators. Thus, by the use of *instrumental variables* or *two-stage least squares* it may be possible to overcome the inconsistency. Or the regression coefficients may be estimated by means of the *factor analysis* model (Malinvaud, 1970, p. 410; Lawley & Maxwell, 1973). Recently, even more promising procedures for dealing with errors-in-variables have been developed. Multivariate regression and factor analysis are themselves special cases of a class of more general *linear structural equation* (LISREL) models which permit the explicit treatment of measurement error (Jöreskog, 1977; Bentler, 1980; Browne, 1982). Of perhaps greater significance for surveys and observational studies are the *partial least squares* (PLS) approach of Wold (1966a, b, 1980) and the *alternating least squares* methods with optimal scaling features (ALSOS) of Forrest Young and his co-workers (e.g., see Young, 1981). These procedures not only make provision for errors of measurement but require weaker assumptions than those of structural equation models. The appeal of the new generation of techniques represented by linear structural equation, partial least squares and alternating least squares methods extends beyond their ability to deal effectively with errors in the variables. For example, they employ nonlinear means of parameter estimation and are applicable to categorical as well as to continuous variables, and to combinations of the two.

Difficulties in *computation* often arise in attempting to solve the normal equations (A1.3) explicitly. Not only are errors in the data squared in computing $\mathbf{S}_{11}=c\mathbf{X}^t\mathbf{X}$ and $\mathbf{S}_{12}=c\mathbf{X}^t\mathbf{Y}$ ($c\leq 1$, $c\neq 0$), but the accurate computation of such scalar-products matrices is difficult under any circumstances. Moreover, where $\mathbf{S}_{11}$ approaches singularity, the computation of $\mathbf{S}_{11}^{-1}$ will lead to further inaccuracy; small changes in $\mathbf{X}$ will then induce large changes in $\hat{\boldsymbol{\beta}}$. A numerically more stable alternative is provided by direct orthogonal-triangular or biorthogonal decomposition of $\mathbf{X}$ (see Chambers, 1974, 1977 Chap. 5; Stewart, 1978). Such an approach does not depend on $\mathbf{S}_{11}$ and $\mathbf{S}_{12}$ and so avoids the numerical problems associated with these matrices entirely.

In the *interpretation* of results, precautions which need to be observed in connection with the weights of multiple regression apply in multivariate regression also. Not only do the raw weights $\hat{\beta}_{ik}$ reflect contributions to regression but in addition are direct functions of the units of measurement and of the particular set of antecedent variables employed. The *standardized* weights, though still interdependent, are scale-free and so are more readily compared. For this and other reasons it is usual in multivariate regression to replace the SSP matrices of (A1.4) and Tables A1.1 and A1.2 by *correlation* matrices. This can be accomplished without the need for fundamental adjustment to the procedures or results outlined above. Precautions are also necessary in the interpretation of the random model. In the random version of

(A1.1) we saw that **X** and **E** are required to be independent. This is equivalent to specifying that **X** be a matrix of fixed constants whose elements are measured without error. While for stochastic **X** this condition at first sight appears to be unattainable, the distributional requirement can be validated by treating **X** *as if* it were fixed and interpreting all expectations and likelihoods as conditional on **X**. Justification for so doing is strongest in applications whose primary goal is to extract the maximum information from the available data. Exploratory surveys and observational studies, which above all tend to yield a stochastic **X** matrix, are precisely those for which this goal has most relevance. Errors-in-variables may be regarded as generating an important class of stochastic **X** matrices.

As multivariate linear regression can be fitted without difficulty, what is required now is for the method to be tested in practice. It would be worthwhile to determine whether the model in one form or another is appropriate for the analysis of ecological data, and, if so, under what conditions. The restriction to linearity in the parameters is not very important as $x_1, \ldots, x_p$ may be arbitrary functions of any set of observed variables. All too often one finds multiple regression analysis being used in circumstances in which multivariate regression might be more appropriate. Instances in which attempts are made to examine the dependence of $q \geq 2$ responses on $p \geq 2$ regressor variables by treating the responses $y_1, \ldots, y_q$ as if they were mutually uncorrelated and proceeding to a series of q multiple regression analyses are by no means uncommon in the literature. The implication here that the correlation matrix of the response variables is an identity matrix is rarely well-founded in practice. We have observed that the theory pertaining to the fixed version of (A1.1) is especially rich. The conditions which permit this theory to be exploited to best effect are those associated with controlled experiments. Surveys and observational studies present more difficult challenges, primarily because of complications arising from errors in the variables. Yet, certain results which hold when **X** is a matrix of fixed constants continue to hold when this is no longer true. Moreover, the model represented by (A1.1) may prove useful for model building or description, even in the absence of normal theory justification.

Further reading. Several general accounts of multivariate regression are available. Among these, mention is made of those by Anderson (1958, p. 178), Kshirsagar (1972, p. 304), Press (1972, p. 209), Finn (1974, pp. 92 & 134), Bock (1975, pp. 208 & 374), Timm (1975, p. 307), Gnanadesikan (1977, pp. 80 & 121), and Mardia, Kent and Bibby (1979, p. 307). The question of *estimation* in multivariate regression has been addressed by Sclove (1971), Robinson (1973), Brown and Zidek (1980), van der Merwe and Zidek (1980) and by Reilly and Patino-Leal (1981). Press (1980) has considered estimation from a Bayesian viewpoint while *inference* for the multivariate regression model has been examined by Fraser and Ng (1977). Zellner (1976) has examined estimation and inference under the assumption that the error terms have a joint *multivariate-t* distribution. *Stochastic* or *error in variables* forms of the general linear model have been discussed by, among others, Cochran (1970), Malinvaud (1970, p. 375), Kmenta (1971, p. 297), Moran (1971), Hodges and Moore (1972), Johnston (1972, p. 281), Lawley and Maxwell (1973), Kendall (1975, p. 109), Dolby (1976), Schmidt (1976, p. 93), Sockloff (1976), Bibby (1977), Mitra and Alam (1980) and Reilly and Patino-Leal (1981). *Variable selection* in multivariate regression has been investigated by McKay (1977, 1979) and Sparks, Coutsourides and Troskie (1983), who have extended Mallows' C_p plot to the multivariate case. Accounts of multivariate stepwise regression by *orthogonalization* of the model matrix have been given by Kshirsagar (1972, p. 320), Finn (1974, p. 137), Bock (1975, p. 222) and Timm (1975, p. 324). Kerlinger and Pedhazur (1973, p. 92) have discussed the nature and interpretation of *semi-partial* regression coefficients. The *assessment of dimensionality* in multivariate regression has been considered by Izenman (1975, 1980), while Gnanadesikan (1977, p. 263) and Izenman (1980) have described the calculation and analysis of *residuals* from a fitted model. Khuri and Conlon (1981) have developed a procedure for the *simultaneous optimization* of multiple responses, while Press (1972, p. 219) has described a *generalized* version of multivariate multiple regression. The *hypothesis* that certain regression coefficients are equal to zero in multivariate regression has been discussed by Burdick (1982) and tests based on a determinantal ratio and on canonical roots described and compared. Robinson (1973) has demonstrated that multivariate regression with *constraints on the rank* of the matrix of regression coefficients is equivalent to canonical correlation analysis. Finally, we observe that Muller (1982) has shown how the multivariate regression model can be used to *derive* canonical correlation analysis and has employed the model to demonstrate differences in the structure of the residual covariance matrices of the respective methods.

Computational aspects of linear least squares problems have been investigated or described by Golub (1969), Lawson and Hanson (1974), Chambers (1977, p. 101), Forsythe, Malcolm and Moler (1977, p. 192), Stewart (1978) and by Kennedy and Gentle (1980, p. 313). *Updating methods* for linear models including multivariate regression have been considered by Chambers (1975).

A.2 Data sets used in worked applications

Data on which five of the analyses of Chaps. 6–11 are based are included in this appendix. The data used in the analysis of herbivore-environment relationships in the Rwenzori National Park (Chap. 12) appear in Table 12.2. Data used in the grassland structure investigation of Chap. 10 are too extensive for convenient inclusion here.

Table A-1. Limestone grassland, Anglesey. Representation (percentage frequency in 10 cm × 20 cm units) of three species x_1, x_2 and x_3 in N=45 randomly selected sites together with the spatial coordinates y_1, y_2 and y_3 of these sites with respect to an arbitrary coordinate system

Site	x_1	x_2	x_3	y_1	y_2	y_3
1	48	0	0	−1	1	−1
2	60	68	84	−6	−13	78
3	52	0	0	2	9	18
4	40	16	0	−5	−1	5
5	48	8	0	12	−7	−84
6	48	0	0	3	−1	−3
7	76	68	0	−13	1	−13
8	48	0	0	−8	2	−16
9	84	99	76	−6	−12	72
10	44	12	0	12	−8	−96
11	84	4	4	7	−9	−63
12	64	0	0	−3	1	−3
13	52	44	12	−14	3	−42
14	72	4	0	−7	2	−14
15	28	0	0	4	−4	−16
16	48	8	0	3	−2	−6
17	52	4	0	−7	3	−21
18	56	8	0	1	−9	−9
19	80	76	40	−6	−10	60
20	24	56	0	12	−9	−108
21	48	0	0	9	−6	−54
22	80	0	0	3	8	24
23	68	44	0	−8	−7	56
24	68	8	0	10	−9	−90
25	24	40	0	11	−10	−110
26	40	0	0	4	8	32
27	64	4	0	−2	−1	2
28	64	60	52	−7	−9	63
29	28	8	0	−5	−5	25
30	24	8	0	−2	−2	4
31	56	0	0	3	5	15
32	28	0	0	−4	5	−20
33	52	0	0	−7	1	−7
34	72	0	0	7	−8	−56
35	48	60	36	−10	−5	50
36	52	0	0	−2	6	−12
37	40	0	0	7	−5	−35
38	20	0	0	−5	5	−25
39	32	4	0	−6	1	−6
40	28	16	0	−4	−6	24
41	52	20	0	1	−10	−10
42	32	0	0	1	−1	−1
43	36	4	0	3	7	21
44	20	4	0	7	−10	−70
45	64	44	36	−8	−10	80

Species $x_1, \ldots, x_3$: *Galium verum* L., *Dactylis glomerata* L., *Phleum bertolonii* DC.
Spatial coordinates $y_1, \ldots, y_3$: east-west, north-south, $y_3 = y_1 \cdot y_2$.

Table A-2. Limestone grassland, Anglesey. Estimates of the abundance (percentage frequency in 10 cm × 20 cm units) of eight species $x_1, \ldots, x_8$ and of the intensity of six soil variables $y_1, \ldots, y_6$ in N = 45 randomly selected sites

Site	x_1	x_2	x_3	x_4	x_5	x_6	x_7	x_8	y_1	y_2	y_3	y_4	y_5	y_6
1	4	0	0	36	28	24	99	68	3.9	0.5	1.0	1.95	3.90	0.50
2	92	84	0	8	0	0	84	4	4.5	1.6	1.9	7.20	8.55	3.04
3	9	0	0	52	4	40	96	68	3.3	0.9	1.1	2.97	3.63	0.99
4	52	0	0	52	12	28	96	24	4.5	1.2	3.0	5.40	13.50	3.60
5	99	0	36	88	52	8	72	0	6.4	0.9	0.6	5.76	3.84	0.54
6	12	0	0	20	40	40	88	68	3.4	0.7	0.9	2.38	3.06	0.63
7	72	0	20	72	24	20	72	8	8.3	0.6	1.3	4.98	10.79	0.78
8	80	0	0	48	16	28	92	8	5.2	0.7	1.5	3.64	7.80	1.05
9	80	76	4	8	12	0	84	0	6.4	1.6	2.1	10.24	13.44	3.36
10	92	0	40	72	36	12	84	0	6.4	0.9	1.3	5.76	8.32	1.17
11	28	4	0	16	56	28	96	56	4.4	0.9	1.3	3.96	5.72	1.17
12	8	0	0	36	68	8	99	28	3.9	0.7	1.1	2.73	4.29	0.77
13	99	12	4	84	36	12	88	8	7.2	0.8	0.7	5.76	5.04	0.56
14	40	0	0	68	12	8	88	24	3.2	0.9	1.6	2.88	5.12	1.44
15	28	0	0	36	64	28	99	56	4.9	0.3	0.9	1.47	4.41	0.27
16	28	0	0	28	44	20	88	32	4.2	0.6	1.6	2.52	6.72	0.96
17	80	0	0	52	20	32	96	20	5.0	0.9	1.3	4.50	6.50	0.78
18	84	0	0	76	44	16	96	0	5.4	1.5	2.7	8.10	14.58	4.05
19	88	40	12	8	24	8	92	0	6.9	1.2	1.2	8.28	8.28	1.44
20	99	0	60	88	28	0	80	0	5.7	1.2	1.3	6.84	7.41	1.56
21	12	0	0	36	16	12	88	76	5.1	0.4	0.9	2.04	4.59	0.36
22	0	0	0	20	8	0	99	60	3.7	0.7	1.5	2.59	5.55	1.05
23	88	0	12	72	32	16	88	0	7.5	1.0	1.8	7.50	13.50	1.08
24	56	0	4	32	56	4	96	16	4.9	0.6	1.0	2.94	4.90	0.60
25	99	0	40	60	20	4	56	4	5.5	0.9	1.2	4.95	6.60	1.08
26	12	0	0	28	4	4	99	72	4.2	0.7	1.3	2.94	5.46	0.91
27	28	0	0	48	64	4	99	28	3.3	0.5	1.3	1.65	4.29	0.65
28	92	52	0	40	64	8	96	4	5.1	1.4	1.6	7.14	8.16	2.24
29	80	0	0	68	40	12	80	8	6.6	0.7	1.0	4.62	6.60	0.70
30	32	0	0	56	28	36	84	24	6.4	0.7	1.2	4.48	7.68	0.84
31	40	0	0	60	8	36	96	56	4.6	0.8	1.4	3.68	6.44	0.89
32	44	0	0	44	8	20	96	24	4.3	0.8	1.2	3.44	5.16	0.96
33	48	0	0	72	20	12	99	32	4.2	1.0	1.3	4.20	5.46	1.30
34	48	0	0	8	44	8	92	56	4.3	1.0	1.3	4.30	5.59	1.30
35	99	36	20	56	8	4	24	0	7.2	0.8	0.9	5.76	6.48	0.72
36	15	0	4	36	4	28	99	44	4.0	0.9	1.5	3.60	6.00	1.35
37	8	0	0	20	16	12	99	56	3.9	0.5	1.2	1.95	4.68	0.60
38	28	0	0	24	16	12	99	36	3.3	0.7	1.8	2.31	5.94	1.26
39	52	0	0	48	12	28	99	32	3.8	1.0	1.5	3.80	5.70	1.50
40	92	0	4	56	12	16	70	4	8.1	0.7	1.3	5.67	10.53	0.91
41	92	0	8	52	56	8	99	4	5.0	1.6	1.5	8.00	7.50	2.40
42	4	0	0	44	24	4	99	60	3.4	0.5	1.3	1.70	4.42	0.65
43	16	0	0	36	0	24	99	76	3.9	0.4	1.3	1.56	5.07	0.52
44	76	0	0	48	12	36	96	32	6.2	0.7	1.0	4.34	6.20	0.70
45	96	36	4	28	28	8	88	4	5.3	1.2	1.8	6.36	9.54	2.16

Species $x_1, \ldots, x_8$: *Helictotrichon pubescens* (Huds.) Pilger, *Phleum bertolonii* DC., *Trifolium pratense* L., *Poterium sanguisorba* L., *Rhytidiadelphus squarrosus* (Hedw.) Warnst., *Hieracium pilosella* L., *Briza media* L., *Thymus drucei* Ronn.

Soil variables $y_1, \ldots, y_6$: soil depth (cm) d, extractable phosphate (ppm P) P, exchangeable potassium (m-equiv. K/100 g dry soil) K, d × P, d × K, P × K.

Table A-3. Lowland tropical rain forest, Guyana. Scores of $N=25$ sites on $p=6$ generalized vegetation variables $x_1, \ldots, x_6$ and estimates of the intensity of $q=9$ soil properties $y_1, \ldots, y_9$ in these sites

Site	x_1	x_2	x_3	x_4	x_5	x_6	y_1	y_2	y_3	y_4	y_5	y_6	y_7	y_8	y_9
1	.96	.12	.21	.03	.14	−.04	4.63	43.1	7.58	75.1	6.7	1.10	.13	8.49	1.031
2	.16	.02	.09	.97	.10	.07	4.37	45.8	21.20	85.5	7.0	1.03	.11	7.35	1.007
3	.86	.09	.15	.39	.22	.09	4.59	47.8	6.90	88.0	2.2	0.80	.10	5.41	1.000
4	.58	.06	.10	.07	.79	−.02	4.22	29.8	12.10	87.1	3.6	1.16	.11	6.29	1.007
5	.47	.04	.12	.08	.85	.13	3.89	33.8	16.00	87.1	2.7	1.19	.10	6.31	1.007
6	.76	.15	.62	.04	.11	−.02	4.20	25.7	7.80	83.6	3.4	1.26	.11	6.54	1.007
7	.67	.05	.20	.24	.21	.64	4.34	26.7	17.40	84.2	2.7	1.06	.10	5.90	1.101
8	.46	.84	.08	.16	.07	.02	4.32	45.8	7.22	95.3	0.7	1.52	.10	4.10	1.003
9	.04	.96	.03	.00	.04	.07	4.09	37.9	12.40	97.0	2.2	0.41	.06	2.68	1.000
10	.19	.03	.68	.39	.55	.01	4.46	47.8	14.50	87.9	3.5	0.67	.10	5.32	1.006
11	.94	.15	.13	.05	.20	.09	4.54	23.8	5.80	87.8	2.5	0.83	.10	5.52	1.005
12	.92	.10	.15	.07	.15	.21	4.49	40.9	6.60	83.0	3.5	0.97	.11	7.89	1.011
13	.11	.96	.03	−.01	.02	−.02	3.58	33.1	5.60	93.6	2.0	0.55	.08	4.42	1.001
14	.06	.99	.03	−.01	.00	.00	3.83	53.0	5.30	93.7	2.1	0.22	.09	2.36	1.000
15	.96	.08	.21	.01	.14	−.05	3.78	43.9	6.40	85.8	3.1	0.56	.10	6.38	1.005
16	.95	.10	.13	.07	.19	.09	4.08	28.4	6.80	85.3	2.2	0.75	.10	5.06	1.005
17	.95	.23	.09	.02	.12	−.10	4.12	31.1	7.33	86.9	1.9	0.65	.09	6.45	1.006
18	−.05	.98	.02	.00	.01	.02	3.85	48.5	5.20	90.8	2.0	0.28	.10	3.08	1.000
19	.63	.07	.72	.16	.15	.12	3.85	38.4	4.60	87.5	1.3	0.62	.10	4.54	1.004
20	.24	.01	.97	−.02	.02	.01	3.67	42.5	5.89	82.1	0.0	0.91	.10	6.21	1.008
21	.38	.85	.05	.01	.03	−.04	3.48	67.8	6.33	95.0	0.0	0.39	.11	4.00	1.000
22	.96	.15	.13	.04	.16	.04	3.73	50.6	4.44	88.9	3.1	0.63	.10	5.20	1.006
23	.96	.11	.12	.04	.14	.07	3.70	34.3	6.30	91.9	1.4	0.68	.10	5.82	1.003
24	.16	.06	.99	.01	.03	.04	3.77	47.0	6.28	81.7	9.1	0.60	.13	5.39	1.003
25	.78	.13	.46	.21	.17	.28	3.73	54.7	5.00	79.5	7.8	0.67	.10	5.53	1.009

Generalized vegetation variables (non-centered principal components of site-normalized basal area) $x_1, \ldots, x_6$: Greenheart (*Ocotea rodiaei* (Schomb.) Mez.), Wallaba (*Eperua falcata* Aubl.), Morabukea (*Mora gonggrijpii* (Kleinh.) Sandw.), *Mora* (*Mora excelsa* Bth.), *Pentaclethra* (*Pentaclethra macroloba* (Willd.) Kze.), *Eschweilera* (*Eschweilera sagotiana* Miers.).

Soil properties $y_1, \ldots, y_9$: pH, base saturation (%), phosphorus (ppm P), sand (% particles .05–.20 mm), silt (% particles .002–.05 mm), active acidity (KCl acid), potassium (m-equiv. K/100 g air dry soil), potential cation exchange capacity (CeCg), moisture retaining capacity (%H_2O in air dry soil).

Table A-4. Lowland tropical rain forest, Guyana. Scores of N = 25 sites on $p=6$ generalized vegetation variables – trees (> 5 cm dbh) $x_1, \ldots, x_6$ and $q=6$ generalized vegetation variables – seedlings (< 30 cm in height) $y_1, \ldots, y_6$

Site	x_1	x_2	x_3	x_4	x_5	x_6	y_1	y_2	y_3	y_4	y_5	y_6
1	.43	.10	.22	.45	.42	.59	.84	.04	.20	.18	.40	.08
2	.09	.05	.49	.11	.74	.23	.23	.10	.04	.23	.17	.88
3	.43	.12	.37	.17	.63	.46	.78	.06	.10	.21	.30	.48
4	.24	.04	.86	.07	.28	.25	.36	.10	.05	.77	.31	.15
5	.29	.02	.71	.09	.32	.48	.18	.11	.07	.68	.16	.33
6	.53	.12	.28	.34	.29	.55	.51	.07	.48	.15	.61	.20
7	.55	.01	.06	.23	.72	.06	.94	.02	.13	.09	.21	.07
8	.62	.58	.14	.22	.34	.01	.08	.34	.07	.07	.84	.28
9	.10	.95	.07	.04	.09	.06	.15	.67	.03	.21	.43	.23
10	.09	.08	.91	.24	.16	−.11	.06	.06	.02	.87	.01	−.03
11	.67	.08	.36	.15	.42	.30	.67	.06	.11	.06	.63	−.04
12	.62	.06	.20	.23	.69	.00	.40	.15	.08	.14	.69	.02
13	.10	.94	.04	.02	.01	.02	.06	.93	.03	−.04	.20	−.07
14	.02	.97	−.01	.01	.02	.02	.03	.99	.02	.04	−.04	.04
15	.64	.06	.20	.37	.50	.30	.40	.11	.46	.05	.73	.01
16	.67	.11	.30	.20	.57	−.02	.62	.18	.09	.10	.51	−.04
17	.90	.21	.08	.18	.11	−.01	.16	.17	.07	.10	.94	.05
18	.02	.98	.01	.01	.01	.03	.05	.95	.03	.01	.14	−.01
19	.33	.08	.20	.81	.39	.14	.14	.07	.96	.06	.18	.05
20	.25	.02	.08	.96	.09	.02	.08	.00	.99	.01	.09	−.01
21	.28	.87	.05	.06	.03	.02	.06	.82	.04	.22	.27	.16
22	.79	.14	.24	.17	.18	.38	.46	.12	.09	.13	.84	.07
23	.91	.15	.04	.17	.17	.16	.44	.21	.05	.16	.74	.12
24	.18	.08	.28	.68	.49	.38	.24	.02	.96	.04	.06	.03
25	.27	.05	.32	.45	.72	.21	.84	.07	.31	.23	.26	.19

Generalized vegetation variables – trees (non-centered principal components of site-normalized tree density) $x_1, \ldots, x_6$: Greenheart (*Ocotea rodiaei* (Schomb.) Mez.), Wallaba (*Eperua falcata* Aubl.), *Pentaclethra* (*Pentaclethra macroloba* (Willd.) Kze.), Morabukea (*Mora gonggrijpii* (Kleinh.) Sandw.), *Mora* (*Mora excelsa* Bth.)/*Eschweilera* (*Eschweilera sagotiana* Miers.), *Jessenia* (*Jessenia bataua* (Mart.) Burret.).

Generalized vegetation variables – seedlings (non-centered principal components of site-normalized seedling density) $y_1, \ldots, y_6$: Greenheart (*Ocotea rodiaei* (Schomb.) Mez.), Wallaba (*Eperua falcata* Aubl.), *Pentaclethra* (*Pentaclethra macroloba* (Willd.) Kze.), Morabukea (*Mora gonggrijpii* (Kleinh.) Sandw.), *Mora* (*Mora excelsa* Bth.), *Eschweilera* (*Eschweilera sagotiana* Miers.).

Table A-5. Nitrogen nutrition of eight grass species. Yield ($\log_{10}$ dry-weight (g) × 10) of eight grass species in five replicates (blocks) grown in sand culture at five levels of nitrogen, $x_1, \ldots, x_5$

Species	Block	z_1	z_2	z_3	z_4	z_5	z_6	z_7	x_1	x_2	x_3	x_4	x_5
Lolium	1	1	0	0	0	0	0	0	1.0128	1.7143	1.6440	2.0781	1.9647
perenne L.	2	1	0	0	0	0	0	0	.9445	1.5843	1.5276	2.0723	2.1158
	3	1	0	0	0	0	0	0	1.0453	1.4757	1.6170	1.7316	2.0948
	4	1	0	0	0	0	0	0	.9868	1.4564	1.5502	2.0697	2.2076
	5	1	0	0	0	0	0	0	.8261	1.3444	1.4886	1.8899	1.9484
Dactylis	1	0	1	0	0	0	0	0	.9542	1.5263	1.5888	1.5944	1.7016
glomerata L.	2	0	1	0	0	0	0	0	.9085	1.3444	1.5159	1.7738	2.0103
	3	0	1	0	0	0	0	0	.9638	1.5888	1.8756	1.9872	1.5955
	4	0	1	0	0	0	0	0	.5563	1.4843	1.8216	1.7427	1.9284
	5	0	1	0	0	0	0	0	1.0212	1.3201	1.4572	2.0165	2.0894
Phleum	1	0	0	1	0	0	0	0	.9777	1.3263	1.5159	1.8785	1.6721
bertolonii DC.	2	0	0	1	0	0	0	0	.8865	1.4757	1.5092	1.2945	1.4821
	3	0	0	1	0	0	0	0	.8633	1.5988	1.5132	1.6010	1.8751
	4	0	0	1	0	0	0	0	.6902	1.3962	1.3784	1.7903	1.7396
	5	0	0	1	0	0	0	0	.6232	1.2601	1.5999	1.6375	1.6563
Briza	1	0	0	0	1	0	0	0	.7083	1.2945	1.1206	1.4133	1.1818
media L.	2	0	0	0	1	0	0	0	.8325	1.2430	1.3927	1.5056	1.2648
	3	0	0	0	1	0	0	0	.8261	1.2900	1.0864	1.3997	1.0334
	4	0	0	0	1	0	0	0	.7404	.6721	1.0212	1.2765	1.0607
	5	0	0	0	1	0	0	0	.8451	.8513	.9085	.8388	1.0682
Koeleria	1	0	0	0	0	1	0	0	.5441	1.0607	.4843	1.2041	.7243
cristata	2	0	0	0	0	1	0	0	.7924	1.2380	1.2504	1.4065	1.2253
(L.) Pers.	3	0	0	0	0	1	0	0	.8513	1.3692	1.0645	1.4533	1.4548
	4	0	0	0	0	1	0	0	.5911	1.1106	.9445	1.2304	1.3444
	5	0	0	0	0	1	0	0	.6902	1.1271	1.1492	1.0531	1.1430
Festuca	1	0	0	0	0	0	1	0	.7324	1.1431	1.1004	1.1987	1.1303
ovina L.	2	0	0	0	0	0	1	0	.8325	1.0755	1.1271	1.2625	1.3118
	3	0	0	0	0	0	1	0	.8808	1.0682	1.2625	1.2648	1.3892
	4	0	0	0	0	0	1	0	.6128	.9823	1.0531	1.2788	1.2041
	5	0	0	0	0	0	1	0	.6812	1.0253	1.1732	1.1584	1.0453
Festuca	1	0	0	0	0	0	0	1	.3617	.7853	.2553	1.3522	1.5775
rubra L.	2	0	0	0	0	0	0	1	.4314	1.0492	1.0294	1.1644	1.3776
	3	0	0	0	0	0	0	1	.3617	1.0294	.9345	.6232	1.3711
	4	0	0	0	0	0	0	1	.0792	1.1875	1.1399	1.2430	1.0607
	5	0	0	0	0	0	0	1	.4624	1.1614	.7782	.8261	1.4346
Helictotrichon	1	0	0	0	0	0	0	0	.6628	1.1303	1.0043	1.3054	1.2625
pubescens	2	0	0	0	0	0	0	0	.8062	1.1875	1.2878	1.3404	1.3820
(Huds.) Pilger.	3	0	0	0	0	0	0	0	.5911	1.3522	1.2068	1.5011	1.5490
	4	0	0	0	0	0	0	0	.9638	1.4330	1.3010	1.6263	1.4346
	5	0	0	0	0	0	0	0	.7324	1.2742	1.1703	1.2553	1.2405

$x_1, x_2, \ldots, x_5$: species' response at 1, 9, 27, 81, 243 ppm nitrogen.
$z_1, z_2, \ldots, z_7$: indicator (dummy) variables for species generated by the scheme described on p. 198.

Table A-6. Some intermediate results in the canonical analysis of the data of Table A-5

Mean-corrected total SSP matrix $\mathbf{S}_{22} \equiv \mathbf{Q}_t^*$ not adjusted for blocks:

$$\begin{bmatrix} 1.7544 & & & & \\ .9960 & 2.0473 & & & \\ 1.6438 & 2.3093 & 4.2959 & & \\ 1.7627 & 2.2767 & 3.2664 & 4.8753 & \\ 1.4104 & 2.1512 & 3.0744 & 3.8597 & 5.1764 \end{bmatrix} = \mathbf{Q}_t^*$$

Mean-corrected total SSP matrix $\mathbf{S}_{22} \equiv \mathbf{Q}_t$ adjusted for blocks:

$$\begin{bmatrix} 1.6349 & & & & \\ .9116 & 1.9076 & & & \\ 1.5921 & 2.2217 & 3.9840 & & \\ 1.8043 & 2.2256 & 3.2690 & 4.6885 & \\ 1.3738 & 2.0790 & 2.8992 & 3.8513 & 5.0722 \end{bmatrix} = \mathbf{Q}_t$$

Mean-corrected hypothesis SSP matrix $\mathbf{S}_{21}\mathbf{S}_{11}^{-1}\mathbf{S}_{12} \equiv \mathbf{Q}_h$ for species:

$$\begin{bmatrix} 1.1966 & & & & \\ .8732 & 1.2735 & & & \\ 1.5884 & 1.7873 & 3.0667 & & \\ 1.7106 & 2.0525 & 3.1644 & 3.5920 & \\ 1.1885 & 2.1172 & 3.0431 & 3.5419 & 4.2375 \end{bmatrix} = \mathbf{Q}_h$$

Roots r_k^2 of $|\mathbf{Q}_h - r^2\mathbf{Q}_t^*| = 0$ and of $|\mathbf{Q}_h - r^2\mathbf{Q}_t| = 0$:

k	r_k^2 (unadjusted)	r_k^2 (adjusted)
1	.882	.910
2	.706	.724
3	.228	.298
4	.090	.093
5	.010	.013

A.3 Species composition of a limestone grassland community

Species used in the canonical variate analysis of Chap. 10. Nomenclature follows Clapham, Tutin and Warburg (1962) and Richards and Wallace (1950).

Agrostis tenuis Sibth.
Anthoxanthum odoratum L.
Anthyllis vulneraria L.
Briza media L.
Calluna vulgaris (L.) Hull
Carex caryophyllea Latour.
Carex flacca Schreb.
Centaurea nigra L.
Dactylis glomerata L.
Galium verum L.
Helianthemum chamaecistus Mill.
Helictotrichon pubescens (Huds.) Pilger
Hieracium pilosella L.
Holcus lanatus L.
Hypochaeris radicata L.
Koeleria cristata (L.) Pers.
Lotus corniculatus L.
Luzula campestris (L.) DC.
Phleum bertolonii DC.
Plantago lanceolata L.
Poterium sanguisorba L.
Sieglingia decumbens (L.) Bernh.
Taraxacum laevigatum (Willd.) DC.
Thymus drucei Ronn.
Trifolium pratense L.
Trifolium repens L.
Viola riviniana Rchb.
Cladonia impexa Harm.
Dicranum scoparium Hedw.
Psuedoscleropodium purum (Hedw.) Fleisch.
Rhytidiadelphus squarrosus (Hedw.) Warnst.

References

Afriat, S.N. (1957). Orthogonal and oblique projectors and the characteristics of pairs of vector spaces. *Proceedings of the Cambridge Philosophical Society,* **53**, 800–816

Afriat, S.N. (1960). *Statistical projection: an investigation of the role of orthogonal projectors in regression theory.* Research Memorandum 16, Economic Research Program, Princeton University. Princeton, New Jersey

Afriat, S.N., Sastry, M.V.R. and Tinter, G. (1975). *Studies in correlation: multivariate analysis and econometrics.* Vandenhoeck and Ruprecht, Göttingen

Akaike, H. (1976). Canonical correlation analysis of time series and the use of an information criterion. In R.K. Mehra and D.G. Lainiotis (eds.), *System identification–advances and case studies.* Academic Press, New York, pp. 27–95

Albrecht, G.H. (1980). Multivariate analysis and the study of form, with special reference to canonical variate analysis. *American Zoologist,* **20**, 670–693

Alldredge, J.R., and Gilb, N.S. (1976). Ridge regression: an annotated bibliography. *International Statistical Review,* **44**, 355–360

Alpert, M.I. and Peterson, R.A. (1972). On the interpretation of canonical analysis. *Journal of Marketing Research,* **9**, 187–192

Altman, D.G. (1978). Plotting probability ellipses. *Applied Statistics,* **27**, 347–349

Anderberg, M.R. (1973). *Cluster analysis for applications.* Academic Press, New York

Anderson, T.W. (1958). *An introduction to multivariate statistical analysis.* Wiley, New York

Andrews, D.F. (1972). Plots of high-dimensional data. *Biometrics,* **28**, 125–136

Andrews, D.F., Gnanadesikan, R. and Warner, J.L. (1971). Transformations of multivariate data. *Biometrics,* **27**, 825–840

Ashton, E.H., Flinn, R.M. and Oxnard, C.E. (1975). The taxonomic and functional significance of overall body proportions in primates. *Journal of Zoology, London,* **175**, 73–105

Ashton, E.H., Flinn, R.M., Oxnard, C.E. and Spence, T.F. (1976). The adaptive and classificatory significance of certain quantitative features of the forelimb in primates. *Journal of Zoology, London,* **179**, 515–556

Ashton, E.H., Healy, M.J.R. and Lipton, S. (1957). The descriptive use of discriminant functions in physical anthropology. *Proceedings of the Royal Society of London B,* **146**, 552–572

Atkinson, A.C. (1983). Diagnostic regression analysis and shifted power transformations. *Technometrics,* **25**, 23–33.

Austin, M.P. (1968). An ordination study of a chalk-grassland community. *Journal of Ecology,* **89**, 408–425

Austin, M.P. (1972). Models and analysis of descriptive vegetation data. In J.N.R. Jeffers (ed.), *Mathematical models in ecology.* Blackwell, Oxford, pp. 61–86

Austin, M.P. (1977). Use of ordination and other multivariate descriptive methods to study succession. *Vegetatio,* **35**, 165–176

Bachacou, J., Chessel, D. and Croze, J.P. (1979). Étude des structure spatiale en forêt alluviale rhénane. III. Dispersion interspecifique et analyse des correspondances. *Oecologia Plantarum,* **14**, 371–388

Baggaley, A.R. (1981). Multivariate analysis: an introduction for consumers of behavioral research. *Evaluation Review,* **5**, 123–131

Bagozzi, R.P., Fornell, C. and Larker, D.F. (1981). Canonical correlation analysis as a special case of a structural relations model. *Multivariate Behavioral Research,* **16**, 437–454

Barcikowski, R.S. and Stevens, J.P. (1975). A Monte Carlo study of the stability of canonical correlations, canonical weights and canonical variate-variable correlations. *Multivariate Behavioral Research,* **10**, 353–364

Bargmann, R.E. (1962). *Representative ordering and selection of variables.* Part A. Virginia Polytechnic Institute. Cooperative Research Project No. 1132. U.S. Office of Education

Bargmann, R.E. (1979). Structural analysis of singular matrices using union-intersection statistics. In L. Orlóci, C.R. Rao and W.M Stiteler (eds.), *Multivariate methods in ecological work.* International Cooperative Publishing House, Fairland, Maryland, pp. 1–9

Barkham, J.P. and Norris, J.M. (1970). Multivariate procedures in an investigation of vegetation and soil relations of two beech woodlands, Cotswold Hills, England. *Ecology,* **51**, 630–639

Barnett, V. (1978a). The study of outliers: purpose and model. *Applied Statistics,* **27**, 242–250

Barnett, V. (1978b). Multivariate outliers: Wilks' test and distance measures. *Bulletin de l-Institut International de Statistique,* **47**, 37–40

Barnett, V. (1979). Some outlier tests for multivariate samples. *South African Statistical Journal,* **13**, 29–52

Barnett, V. (1983a). Reduced distance measures and transformations in processing multivariate outliers. *Australian Journal of Statistics,* **25**, 1–12

Barnett, V. (1983b). Principles and methods for handling outliers in data sets. In T. Wright (ed.), *Statistical methods and the improvement of data quality.* Academic Press, Orlando, Florida, pp. 131–166.

Barnett, V, and Lewis, T. (1963). A study of the relation between G.C.E. and degree results (with discussion). *Journal of the Royal Statistical Society A,* **126**, 187–226

Barnett, V. and Lewis, T. (1978). *Outliers in statistical data.* Wiley, New York

Bartlett, M.S. (1937). Some examples of statistical methods in agriculture and applied biology. *Journal of the Royal Statistical Society (Supplement),* **4**, 137–183

Bartlett, M.S. (1938). Further aspects of the theory of multiple regression. *Proceedings of the Cambridge Philosophical Society,* **34**, 33–40.

Bartlett, M.S. (1947). Multivariate analysis. *Journal of the Royal Statistical Society (Supplement),* **9**, 176–197

Bartlett, M.S. (1948). Internal and external factor analysis. *British Journal of Psychology (Statistical Section),* **1**, 73–81

Bartlett, M.S. (1951). The goodness of fit of a single hypothetical discriminant function in the case of several groups. *Annals of Eugenics, London,* **16**, 199–214

Bartlett, M.S. (1965). Multivariate statistics. In T.H. Waterman and H.J. Morowitz (eds.), *Theoretical and mathematical biology.* Blaisdell, New York, pp. 201–224

Bartlett, M.S. (1975). *The statistical analysis of spatial pattern.* Chapman and Hall, London

Beaton, A.E. (1978). Salvaging experiments: interpreting least squares in non-random samples. In D. Hogben and D.W. Fife (eds.), *Computer science and statistics: proceedings of the 10th symposium on the interface.* National Bureau of Standards, Washington, D.C., pp. 137–145

Beauchamp, J.J., Begovich, C.L., Kane, V.E. and Wolf, D.A. (1980). Application of discriminant analysis and generalized distance measures to uranium exploration. *Mathematical Geology,* **12**, 539–558

Bebbington, A.C. (1978). A method of bivariate trimming for robust estimation of the correlation coefficient. *Applied Statistics,* **27**, 221–226

Bebbington, A.C. and Smith, T.M.F. (1977). The effect of survey design on multivariate analysis. In C.A. O'Muircheartaigh and C. Payne (eds.), *The analysis of survey data. Volume 2, Model fitting.* Wiley, Chichester, pp. 175–192

Beckman, R.J. and Cook, R.D. (1983). Outlier ....s (with discussion). *Technometrics,* **25**, 119–163

Belsley, D.A., Kuh, E. and Welsch, R.E. (1980). *Regression diagnostics: identifying influential data and sources of collinearity.* Wiley, New York

Bentler, P.M. (1980). Multivariate analysis with latent variables: causal modeling. *Annual Review of Psychology,* **31**, 419–456

Bentler, P.M. and Huba, G.J. (1982). Symmetric and asymmetric rotations in canonical correlation analysis: new methods with drug variable examples. In N. Hirschberg and L.G. Humphreys (eds.), *Multivariate applications in the social sciences.* Erlbaum, New Jersey, pp. 21–46

Benzécri, J.-P. (1969). Statistical analysis as a tool to make patterns emerge from data. In S. Watanabe (ed.), *Methodologies of pattern recognition.* Academic Press, New York, pp. 35–60

Benzécri, J.-P. (1977). Histoire et préhistoire de l'analyse des données. Partie V, L'analyse des correspondances. *Les Cahiers de l'Analyse des Données,* **2**, 9–40

Benzécri, J.-P. et collaborateurs. (1973). *L'analyse des données. Tome 2, L'analyse des correspondances.* Dunod, Paris

Benzécri, J.-P. et collaborateurs. (1980). *L'analyse des données. Tome 1, La taxinomie 3e.* Dunod, Paris

Beran, R. (1979). Testing for ellipsoidal symmetry of a multivariate density. *The Annals of Statistics,* **7**, 150–162

Berge, J.M.F. Ten. (1977). Orthogonal Procrustes rotation for two or more matrices. *Psychometrika,* **42**, 267–276

Berry, K.J. and Mielke, P.W. (1983). Computation of finite population parameters and approximate probability values for multi-response permutation procedures (MRPP). *Communications in Statistics–Simulation and Computation,* **B12**, 83–107

Bibby, J. (1977). The general linear model–a cautionary tale. In C.A. O'Muircheartaigh and C. Payne (eds.), *The analysis of survey data. Vol 2, Model fitting.* Wiley, Chichester, pp. 35–79

Björck, Å. (1978). Comment on the iterative refinement of least-squares solutions. *Journal of the American Statistical Association,* **73**, 161–166

Björck, Å. and Golub, G.H. (1973). Numerical methods for computing angles between linear subspaces. *Mathematics of Computation,* **27**, 579–594

Blanc, F., Chardy, P., Laurec, A. et Reys, J.-P. (1976). Choix des métriques qualitatives en analyse d'inertie. Implications en écologie marine benthique. *Marine Biology,* **35**, 49–67

Bock, R.D. (1960). *Methods and applications of optimal scaling.* The University of North Carolina Psychometric Laboratory Research Memorandum No. 25

Bock, R.D. (1975). *Multivariate statistical methods in behavioral research.* McGraw-Hill, New York

Bonin, G. et Roux, M. (1978). Utilisation de l'analyse factorielle des correspondances dans l'étude phyto-écologique des quelques pelouses de l'Apennin lucano-calabrais. *Oecologia Plantarum,* **13**, 121–138

Bourlière, F. (1965). Densities and biomass of some ungulate populations in Eastern Congo and Rwanda, with notes on population structure and lion/ungulate ratios. *Zoologica Africana,* **1**, 199–207

Box, G.E.P. and Cox, D.R. (1964). An analysis of transformations. *Journal of the Royal Statistical Society B,* **26**, 211–252

Box, G.E.P. and Tidwell, P.W. (1962). Transformation of the independent variable. *Technometrics,* **4**, 531–550

Bradu, D. and Gabriel, K.R. (1978). The biplot as a diagnostic tool for models of two-way tables. *Technometrics,* **20**, 47–68

Bradu, D. and Grine, F.E. (1979). Multivariate analysis of Diaemodontine crania from South Africa and Zambia. *South African Journal of Science,* **75**, 441–448

Breiman, L. and Friedman, J.H. (1982). *Estimating optimal transformations for multiple regression and correlation.* Statistics Department, Stanford University Technical Report ORION-10

Brillinger, D.R. (1967). Contribution on the discussion of E.J. Williams' paper, The analysis of association among many variates (p. 236). *Journal of the Royal Statistical Society B,* **29**, 199–242

Brillinger, D.R. (1969). The canonical analysis of stationary time series. In P.R. Krishnaiah (ed.), *Multivariate analysis II.* Academic Press, New York, pp. 331–350

Brillinger, D.R. (1975). *Time series: data analysis and theory.* Holt, Rinehart and Winston, New York

Brillinger, D.R. (1981). *Time series: data analysis and theory.* Expanded edition. Holden-Day, San Francisco

Brown, K.R. (1979). Multivariate assessment of phentic relationships within the tribe Luciliini (Diptera: Calliphoridae). *Australian Journal of Zoology,* **27**, 465–477

Brown, P.J. and Zidek, J.V. (1980). Adaptive multivariate ridge regression. *The Annals of Statistics* **8**, 64–74

Browne, M.W. (1979). The maximum-likelihood solution in inter-battery factor analysis. *British Journal of Mathematical and Statistical Psychology,* **32**, 75–86

Browne, M.W. (1980). Factor analysis of multiple batteries by maximum likelihood. *British Journal of Mathematical and Statistical Psychology,* 33, 184–199

Browne, M.W. (1982). Covariance structures. In D.M. Hawkins (ed.), *Topics in applied multivariate analysis.* Cambridge University Press, Cambridge, pp. 72–141

Bryson, R.A. and Kutzbach, J.E. (1974). On the analysis of pollen-climate canonical transfer functions. *Quaternary Research*, **4**, 162–174

Bunch, J.R. and Nielsen, C.P. (1978). Updating the singular value decomposition. *Numerische Mathematik*, **31**, 111–129

Bunch, J.R., Nielsen, C.P. and Sorensen D.A. (1978). Rank one modification of the symmetric eigenproblem. *Numerische Mathematik*, **31**, 31–48.

Burdick, R.K. (1982). A note on the multivariate general linear test. *The American Statistician*, **36**, 131–132

Burg, E. van der, and Leeuw, J. de (1983). Non-linear canonical correlation. *British Journal of Mathematical and Statistical Psychology*, **36**, 54–80

Burnaby, T.P. (1971). The skeletal forms of some African shrews. Quoted from R.E. Blackith and R.A. Reyment, *Multivariate Morphometrics*. Academic Press, London, pp. 105–107

Burt, R.L., Reid, R. and Williams, W.T. (1976). Exploration for, and utilization of, collections of tropical pasture legumes. I. The relationship between agronomic performance and climate of origin in introduced *Stylosanthes* spp. *Agro-Ecosystems*, **2**, 293–307

Burwash, R.A. and Culbert, R.R. (1976). Multivariate geochemical and mineral patterns in the Precambrian basement of western Canada. *Canadian Journal of Earth Sciences*, **13**, 1–18

Businger, P.A. and Golub, G.H. (1969). Algorithm 358: singular value decomposition of a complex matrix. *Communications of the Association for Computing Machinery*, **12**, 564–565

Buyse, M. (1983). Les differentes approches conduisant a l'analyse des correspondances. *Biometrie Praximetrie*, **23**, 1–26

Buzas, M.A. (1971). Analysis of species densities by the multivariate general linear model. *Limology and Oceanography*, **16**, 667–670

Buzas, M.A. (1972). Biofacies analysis of presence or absence data through canonical variate analysis. *Journal of Paleontology*, **46**, 55–77

Cailliez, F. and Pagès, J-P. (1976). *Introduction à l'analyse des données*. Societè de Mathématiques Appliquées et de Sciences Humaines, Paris

Campbell, N.A. (1976) A multivariate approach to variation in Microfilariae: examination of the species *Wucheria lewisi* and demes of the species *W. bancrofti*. *Australian Journal of Zoology*, **24**, 105–114

Campbell, N.A. (1978a). The influence function as an aid in outlier detection in discriminant analysis. *Applied Statistics*, **27**, 251–258

Campbell, N.A. (1978b). Multivariate analysis in biological anthropology: some further considerations. *Journal of Human Evolution*, **7**, 197–203

Campbell, N.A. (1979a). *Canonical variate analysis: some practical aspects*. Ph.D. thesis, University of London

Campbell, N.A. (1979b). Some practical aspects of canonical variate analysis. *BIAS*, **6**, 7–18

Campbell, N.A. (1980a). Shrunken estimators in discriminant and canonical variate analysis. *Applied Statistics*, **29**, 5–14

Campbell, N.A. (1980b). Robust procedures in multivariate analysis. I. Robust covariance estimation. *Applied Statistics*, **29**, 231–237

Campbell, N.A. (1980c). On the study of the Border Cave remains: statistical comments. *Current Anthropology*, **21**, 532–535

Campbell, N.A. (1981a). Graphical comparison of covariance matrices. *Australian Journal of Statistics*, **23**, 21–37

Campbell, N.A. (1981b). Personal communication

Campbell, N.A. (1982). Robust procedures in multivariate analysis. II. Robust canonical variate analysis. *Applied Statistics*, **31**, 1–8

Campbell, N.A. (1984). Canonical variate analysis with unequal covariance matrices: generalizations of the usual solution. *Mathematical Geology*, **16**, 109–124

Campbell, N.A. and Atchley, W.R. (1981). The geometry of canonical variate analysis. *Systematic Zoology*, **30**, 268–280

Campbell, N.A. and Dearn, J.M. (1980). Attitudinal variation in, and morphological divergence between, three related species of grasshopper, *Praxibulus* sp., *Kosciuscola cognatus* and *K. usitatus* (Orthoptera: Acrididae). *Australian Journal of Zoology*, **28**, 103–118

Campbell, N.A. and Kitchener, D.J. (1980). Morphological divergence in the genus *Eptesicus* (Microchiroptera: Vespertilionidae) in western Australia: a multivariate approach. *Australian Journal of Zoology*, **28**, 457–474

Campbell, N.A. and Mahon, R.J. (1974). A multivariate study of variation in two species of rock crab of the genus *Leptograpsus*. *Australian Journal of Zoology*, **22**, 417–425

Campbell, N.A. and Palmer, M.J. (1984). The role of shrunken estimators in canonical correlation analysis. *Submitted for publication*

Campbell, N.A. and Reyment, R.A. (1978). Discriminant analysis of a Cretaceous foraminifer using shrunken estimators. *Mathematical Geology*, **10**, 347–359

Campbell, N.A. and Reyment, R.A. (1980). Robust multivariate procedures applied to the interpretation of atypical individuals of a Cretaceous foraminifer. *Cretaceous Research*, **1**, 207–221

Campbell, N.A. and Saunders, D.A. (1976). Morphological variation in the white-tailed black cockatoo, *Calptorhynchus baudinii*, in western Australia: a multivariate approach. *Australian Journal of Zoology*, **24**, 589–595

Campbell, N.A. and Tomenson, J.A. (1983). Canonical variate analysis for several sets of data. *Biometrics*, **39**, 425–435

Campbell, N.A. and Woodings, T.L. (1981). Improved diagnostic output from statistical packages. *Bulletin of the International Statistical Institute*, **49** (Pt. 1), 279–293

Carney, E.J. (1975). Ridge estimates for canonical analysis. In J.W. Frane (ed.), *Computer science and statistics: proceedings of the 8th symposium on the interface*. University of California, Los Angeles, pp. 252–256

Carroll, J.D. (1968). A generalized of canonical correlation analysis to three or more sets of variables. *Proceedings of 76th annual convention of the American Psychological Association*, pp. 227–228.

Carroll, J.D. and Chang, J.J. (1970). Analysis of individual differences in multidimensional scaling via an N-Way generalization of 'Eckart-Young' decomposition *Psychometrika*, **35**, 283–319

Cassie, R.M. (1972). Fauna and sediments of an intertidal and mud-flat: an alternative multivariate analysis. *Journal of Experimental Marine Biology and Ecology*, **9**, 55–64

Cassie R.M. and Michael, A.D. (1968). Fauna and sediments of an intertidal mudflat: a multivariate analysis. *Journal of Experimental Marine Biology and Ecology*, **2**, 1–23

Chambers, J.M. (1974). Linear regression computations: some numerical and statistical aspects. *Bulletin of the International Statistical Institute*, **39**, 246–254

Chambers, J.M. (1975). Updating methods for linear models for the addition or deletion of observations. In J.N. Srivastava (ed.), *A survey of statistical design and linear models*. North-Holland, Amsterdam.

Chambers, J.M. (1977). *Computational methods for data analysis*. Wiley, New York

Chambers, J.M., Cleveland, W.S., Kleiner, B. and Tukey, P.A. (1983). *Graphical methods for data analysis*. Wadsworth, Belmont, California

Chen, H.J. and Kettenring, J.R. (1972). *CANON: a computer program package for the multi-set canonical correlation analysis*. Bell Laboratories Memorandum. Bell Laboratories, Murray Hill, New Jersey

Chernick, M.R. (1983). Influence functions, outlier detection and data editing. In T. Wright (ed.), *Statistical methods and the improvement of data quality*. Academic Press, Orlando, Florida, pp. 167–176

Chmielewski, M.A. (1981). Elliptically symmetric distributions: a review and bibliography. *International Statistical Review*, **49**, 67–74

Clapham, A.R., Tutin, T.G. and Warburg, E.F. (1962). *Flora of the British Isles*. 2nd edn. Cambridge University Press, Cambridge

Claringbold, P.J. (1958). Multivariate quantal analysis. *Journal of the Royal Statistical Society B*, **20**, 398–405

Clarkson, D.B. (1979). Estimating the standard errors of rotated factor loadings by jackknifing. *Psychometrika*, **44**, 297–314

Cliff, A.D. and Ord, J.K. (1981 a). The effects of spatial autocorrelation on geographical modelling. In R.G. Craig and M.L. Labovitz (eds.), *Future trends in geomathematics*. Pion, London, pp. 108–137

Cliff, A.D. and Ord, J.K. (1981 b). *Spatial processes: models and applications*. Pion, London

Cliff, N. and Krus, D.J. (1976). Interpretation of canonical analysis: rotated vs. unrotated solutions. *Psychometrika*, **41**, 35–42

Cochran, W.G. (1970). Some effects of errors of measurement on multiple correlation. *Journal of the American Statistical Association*, **65**, 22–34

Coe, M.J., Cumming, D.H. and Phillipson, J. (1976). Biomass and production of large African herbivores in relation to rainfall and primary production. *Oecologia* (Berlin), **22**, 341–354

Cohen, A., Gnanadesikan, R., Kettenring, J.R. and Landwehr, J.M. (1977). Methodological developments in some applications of clustering. In P.R. Krishnaiah (ed.), *Applications of statistics*. North-Holland, New York, pp. 141–162

Cohen, C. (1972). Eigenvalue problems in multivariate statistics: a unified approach. In M.O. Locks (ed.), *Computer science and statistics: proceedings of the 5th symposium on the interface*. Oklahoma State University, pp. 150–157

Cohen, C. (1977). The generalized symmetric eigenproblem in multivariate statistical models. *Communications in Statistics–Theory and Methods,* **A6,** 277–288

Cohen, C. and Ben-Israel, A. (1969). On the computation of canonical correlations. *Cahiers Centre d'Études de Reserche Opérationnelle,* **11**, 121–132

Cohen, J. (1982). Set correlation as a general multivariate data-analytic method. *Multivariate Behavioral Research,* **17**, 301–341

Coleman, D., Holland, P., Kaden, N., Klema, V. and Peters, S.C. (1980). A system of subroutines for iteratively reweighted least squares computations. *ACM Transactions on Mathematical Software,* **6**, 327–336

Cook, R.D. and Wang, P.C. (1983). Transformations and influential cases in regression. *Technometrics,* **25**, 337–343

Cook, R.D. and Weisberg, S. (1982). *Residuals and influence in regression.* Chapman and Hall, London

Cooley, W.W. and Lohnes, P.R. (1971). *Multivariate data analysis.* Wiley, New York

Corsten, L.C.A. (1976a). Matrix approximation, a key to the application of multivariate methods. *Proceedings of the 9th international biometrics conference.* Part I, pp. 61–77. Raleigh, North Carolina

Corsten, L.C.A. (1976b). Canonical correlation in incomplete blocks. In S. Ikeda and others (eds.), *Essays in probability and statistics.* Shinko Tsusho, Tokyo, pp. 125–154

Corsten, L.C.A. and Gabriel, K.R. (1976). Graphical exploration in comparing variance matrices. *Biometrics,* **32**, 851–863

Cox, C. and Gabriel, K.R. (1982). Some comparisons of biplot display and pencil-and-paper exploratory data analysis methods. In R.L. Launer and A.F. Siegel (eds.), *Modern data analysis.* Academic Press, New York, pp. 45–82

Cox, D.R. and Small, N.J.H. (1978). Testing multivariate normality. *Biometrika,* **65**, 263–272

Coxon, A.P.M. (1982). *The user's guide to multidimensional scaling with special reference to the MDS(X) library of computer programs.* Heinemann Educational Books, London

Cramer, E.M. (1973). A simple derivation of the canonical correlation equations. *Biometrics,* **29**, 379–380

Cronbach, L.J. (1971). Validity. In R.L. Thornike (ed.), *Educational Measurement.* American Council on Education, Washington. D.C.

Dale, M. (1975). On objectives of methods or ordination. *Vegetatio,* **30**, 15–32

Daniel, C. and Wood, F.S. (1980). *Fitting equations to data–computer analysis of multifactorial data.* 2nd ed. Wiley, New York

Daniel, J., Gragg, W.B., Kaufman, L. and Stewart, G.W. (1976). Reorthogonalization and stable algorithms for updating the Gram-Schmidt QR factorization. *Mathematics of Computation,* **30**, 772–795

Darlington, R.B., Weinberg, S.L. and Walberg, H.J. (1973). Canonical variate analysis and related techniques. *Review of Educational Research,* **43,** 433–454

David, M., Campiglio, C. and Darling, R. (1974). Progresses in *R*- and *Q*-mode analysis: correspondence analysis and its application to the study of geological processes. *Canadian Journal of Earth Sciences,* **11**, 131–146

David, M., Dagbert, M. and Beauchemin, Y. (1977). Statistical analysis in geology: correspondence analysis method. *Quarterly Journal of the Colorado School of Mines,* **72**, 1–60

Davis, A.W. (1980). On the effects of multivariate nonnormality on Wilks's likelihood ratio criterion. *Biometrika,* **67**, 419–427

Davis, A.W. (1982). On the effects of moderate multivariate nonnormality on Roy's largest root test. *Journal of the American Statistical Association,* **77**, 896–900

Davis, T.A.W. and Richards, P.W. (1933). The vegetation of Moraballi Creek, British Guiana: an ecological study of a limited area of tropical rain forest. Part I. *Journal of Ecology,* **21**, 350–384

Davis, T.A.W. and Richards, P.W. (1934). The vegetation of Moraballi Creek, British Guiana: an ecological study of a small area of tropical rain forest. Part II. *Journal of Ecology,* **22**, 106–155

Dawson, B. (1977). *The sampling distribution of the canonical redundancy statistic.* Ph.D. thesis, University of Illinois

Dawson-Saunders, B. (1982). Correcting for bias in the canonical redundancy statistic. *Educational and Psychological Measurement,* **42**, 131–143

Dawson-Saunders, B. and Tatsuoka, M.M. (1983). The effect of affine transformation on redundancy analysis. *Psychometrika,* **48**, 299–302

Delany, M.H. and Healy, M.J.R. (1966). Variation in the white-toothed shrews (*Crocidura* spp.). *Proceedings of the Royal Society of London B,* **164**, 63–74

Dempster, A.P. (1966). Estimation in multivariate analysis. In P.R. Krishnaiah (ed.), *Multivariate analysis.* Academic Press, New York, pp. 315–334

Dempster, A.P. (1969). *Elements of continuous multivariate analysis.* Addison-Wesley, Reading, Massachusetts

Dempster, A.P. and Gasko-Green, M. (1981). New tools for residual analysis. *The Annals of Statistics,* **9**, 945–959

Dempster, A.P., Laird, N.M. and Rubin, D.B. (1977). Maximum likelihood from incomplete data via the EM algorithm (with discussion). *Journal of the Royal Statistical Society B,* **39**, 1–38

DeSarbo, W.S. (1981). Canonical/redundancy factoring analysis. *Psychometrika,* **46**, 307–329

DeSarbo, W.S., Hausman, R.E., Lin, S. and Thompson, W. (1982). Constrained canonical analysis. *Psychometrika,* **47**, 489–516

Deville, J.-C. and Saporta, G. (1983). Correspondence analysis, with an extension towards nominal time series. *Journal of Econometrics,* **22**, 169–189

Devlin, S.J., Gnanadesikan, R. and Kettenring, J.R. (1975). Robust estimation and outlier detection with correlation coefficients. *Biometrika,* **62**, 531–545

Devlin, S.J., Gnanadesikan, R. and Kettenring, J.R. (1981). Robust estimation of dispersion matrices and principal components. *Journal of the American Statistical Association,* **76**, 354–362

Diaconis, P. and Efron, B. (1983). Computer-intensive methods in statistics. *Scientific American,* **284**, 116–130

Diaconis, P. and Friedman, J.H. (1983). M and N plots. In M.H. Rizui, J.S. Rustagi and D. Siegmund (eds.), *Recent advances in statistics–papers in honor of Herman Chernoff on his sixtieth birthday.* Academic Press, New York, pp. 425–447

Digby, P.G.N. and Gower, J.C. (1981). Ordination between- and within-groups applied to soil classification. In D.F. Merriam (ed.), *Down-to-earth statistics: solutions looking for geological problems.* Syracuse University Geology Contribution, Syracuse, New York, pp. 63–75

Dijkstra, T. (1983). Some comments on maximum likelihood and partial least squares methods. *Journal of Econometrics,* **22**, 67–90

DiPillo, P.J. (1979). Biased discriminant analysis: evaluation of the optimum probability of misclassification. *Communications in Statistics–Theory and Methods,* **A8**, 1447–1457

Dolby, G.R. (1976). The ultrastructural relationship: a synthesis of the functional and structural relations. *Biometrika,* **63**, 39–50

Dongarra, J.J., Bunch, J.R., Moler, C.B. and Stewart, G.W. (1979). *LINPACK users' guide.* Society for Industrial and Applied Mathematics, Philadelphia

Donoho, D.L. and Huber, P.J. (1983). The notion of breakdown point. In P.J. Bickel, K.A. Doksum and J.L. Hodges, Jr. (eds.), *A festschrift for Erich L.Lehmann.* Wadsworth, Belmont, California, pp. 157–184

Draper, N.R. and John, J.A. (1981). Influential observations and outliers in regression. *Technometrics,* **23**, 21–26

Draper, N.R. and van Nostrand, R.C. (1979). Ridge regression and James-Stein estimation: review and comments. *Technometrics,* **21**, 451–466

Drehmer, D.E. and Morris, G.W. (1981). Cross-validation with small samples: an algorithm for computing Gollob's estimator. *Educational and Psychological Measurement,* **41**, 195–200

Dunn, J.E. (1981). Data-based transformations in multivariate analysis. In D.E. Capen (ed.), *The use of multivariate statistics in studies of wildlife habitat.* USDA Forest Service General Technical Report RM-87, Rocky Mountain Forest Range Experiment Station, Fort Collins, Colorado, pp. 93–102

Dunn, J.E. and Tubbs, J.D. (1980). VARSTAB: a procedure for determining homoscedastic transformations of multivariate normal populations. *Communications in Statistics–Simulation and Computation,* **B9**, 589–598

Eastment, H.T. and Krzanowski, W.J. (1982). Cross-validatory choice of the number of components from a principal component analysis. *Technometrics,* **24**, 73–77

Eckart, C. and Young, G. (1936). The approximation of one matrix by another of lower rank. *Psychometrika,* **1**, 211–218

Efron, B. (1979a). Bootstrap methods: another look at the jackknife. *The Annals of Statistics,* **7**, 1–26

Efron, B. (1979b). Computers and the theory of statistics: thinking the unthinkable. SIAM *Review,* **21**, 460–480

Efron, B. (1981). Nonparametric standard errors and confidence intervals (with discussion). *The Canadian Journal of Statistics,* **9**, 139–172

Efron, B. (1982). *The jackknife, the bootstrap and other resampling plans.* Society for Industrial and Applied Mathematics, Philadelphia

Efron, B. and Gong, G. (1981). Statistical theory and the computer. In W.F. Eddy (ed.), *Computer science and statistics: proceedings of the 13th symposium on the interface.* Springer-Verlag, New York, pp. 3–7

Efron, B. and Gong, G. (1983). A leisurely look at the bootstrap, the jackknife, and cross-validation. *The American Statistician,* **37**, 36–49

Escoufier, Y. (1973). Le traitement des variables vectorielles. *Biometrics,* **29**, 751–760

Falkenhagen, E.R. and Nash, S.W. (1978). Multivariate classification in provenance research. *Silvae Genetica,* **27**, 14–23

Fanshawe, D.B. (1952). *The vegetation of British Guiana: a preliminary review.* Imperial Forestry Institute Paper 29. Oxford

Fatti, L.P., Hawkins, D.M. and Raath, E.L. (1982). Discriminant analysis. In D.M. Hawkins (ed.), *Topics in applied multivariate analysis.* Cambridge University Press, Cambridge, pp. 1–71

Fearn, T. (1983). A misuse of ridge regression in the calibration of a near infrared reflectance instrument. *Applied Statistics,* **32**, 73–79

Field, C.R. (1968). A comparative study of the food habits of some wild ungulates in the Queen Elizabeth National Park, Uganda. *Symposium of the Zoological Society of London,* **21**, 135–151

Field, C.R. (1972). The food habits of wild ungulates in Uganda by analyses of stomach contents. *East African Wildlife Journal,* **10**, 17–42

Field, C.R. and Laws, R.M. (1970). The distribution of the larger herbivores in the Queen Elizabeth National Park, Uganda. *Journal of Applied Ecology,* **7**, 273–294

Fienberg, S.E. (1979). The use of chi-squared statistics for categorical data problems. *Journal of the Royal Statistical Society B,* **41**, 54–64

Filliben, J.J. (1975). The probability plot correlation coefficient test for normality. *Technometrics,* **17**, 111–117

Finn, J.D. (1974). *A general model for multivariate analysis.* Holt, Rinehart and Winston, New York

Fisher, R.A. (1936). The use of multiple measurements in taxonomic problems. *Annals of Eugenics,* **7**, 179–188

Fisher, R.A. (1938). *Statistical methods for research workers.* 7th edn. Oliver and Boyd, Edinburgh, pp. 285–298

Fisher, R.A. (1940). The precision of discriminant functions. *Annals of Eugenics, London,* **10**, 422–429

Fornell, C. (1979). External single-set components analysis of multiple criterion/multiple predictor variables. *Multivariate Behavioral Research,* **14**, 323–338

Fornell, C. (1982). A second generation of multivariate analysis–an overview. In C. Fornell (ed.), *A second generation of multivariate analysis. Vol. 1, Methods.* Praeger, New York, pp. 1–21

Forsythe, G.E., Malcolm, M.A. and Moler, C.B. (1977). *Computer methods for mathematical computations.* Prentice-Hall, Englewood Cliffs, New Jersey

Francis, J.G.F. (1961). The QR transformation. Part I. *Computer Journal,* **4**, 265–271

Francis, J.G.F. (1962). The QR transformation. A unitary analog to the LR transformation. Part II. *Computer Journal,* **4**, 332–345

Fraser, D.A.S. and Ng, K.W. (1977). Interference for the multivariate regression model. In P.R. Krishnaiah (ed.), *Multivariate analysis, IV.* North-Holland, Amsterdam, pp. 35–53

Friedman, J.H. and Rafsky, L.C. (1981). Graphics for the multivariate two-sample problem. *Journal of the American Statistical Association,* **76**, 275–295

Friedman, J.H. and Rafsky, L.C. (1983). Graph-theoretic measures of multivariate association and prediction. *The Annals of Statistics,* **11**, 377–391

Friedman, J.H. and Stuetzle, W. (1981). Projection pursuit regression. *Journal of the American Statistical Association,* **76**, 817–823

Friedman, J.H. and Stuetzle, W. (1982). Projection pursuit methods for data analysis. In R.L. Launer and A.F. Siegel (eds.), *Modern data analysis*. Academic Press, New York, pp. 123–147

Friedman, J.H. and Tukey, J.W. (1974). A projection pursuit algorithm for exploratory data analysis. *IEEE Transactions on Computers*, **C-23**, 881–890

Fujikoshi, Y. and Veitch, L.G. (1979). Estimation of dimensionality in canonical correlation analysis. *Biometrika*, **66**, 345–351

Furnival, G. and Wilson, R. (1974). Regression by leaps and bounds. *Technometrics*, **16**, 499–511

Gabriel, K.R. (1971). The biplot graphic display of matrices with application to principal component analysis. Biometrika, **58**, 453–467

Gabriel, K.R. (1972). Analysis of meteorological data by means of canonical decomposition and biplots. *Journal of Applied Meteorology*, **11**, 1071–1077

Gabriel, K.R. (1978a). Least squares approximation of matrices by additive and multiplicative models. *Journal of the Royal Statistical Society B*, **40**, 186–196

Gabriel K.R. (1978b). The complex correlation biplot. In S. Shye (ed.), *Theory construction and data analysis in the social sciences*. Jossey-Bass, San Francisco, pp. 350–370

Gabriel, K.R. (1979). Some statistical issues in weather experimentation. *Communications in Statistics–Theory and Methods*, **A8**, 975–1015

Gabriel, K.R. (1981). Biplot display of multivariate matrices for inspection of data and diagnosis. In V. Barnett (ed.), *Interpreting multivariate data*. Wiley, Chichester, pp. 147–173

Gabriel, K.R. and Zamir, S. (1979). Lower rank approximation of matrices by least squares with any choice of weights. *Technometrics*, **21**, 489–498

Garbow, B.S., Boyle, J.M., Dongarra, J.J. and Moler, C.B. (1977). *Matrix eigensystem routines: EISPACK guide extension*. Springer-Verlag, Berlin

Garrett, R.G. (1983). Opportunities for the 80's. *Mathematical Geology*, **15**, 385–398

Gauch, H.G. and Stone, E.L. (1979). Vegetation and soil pattern in a mesophytic forest at Ithaca, New York. *The American Midland Naturalist*, **102**, 332–345

Gauch, H.G. and Wentworth, T.R. (1976). Canonical analysis as an ordination technique. *Vegetatio*, **33**, 17–22

Geisser, S. (1965). Bayesian estimation in multivariate analysis. *Annals of Mathematical Statistics*, **36**, 150–159

Geisser, S. (1975). The predictive sample reuse method with applications. *Journal of the American Statistical Association*, **70**, 320–328

Geisser, S. (1977). Discrimination, allocatory and separatory, linear aspects. In J. van Ryzin (ed.), *Classification and clustering*. Academic Press, New York, pp. 301–330

Gelvin, B.R. (1980). Morphometric affinities of *Gigantopithecus*. *American Journal of Physical Anthropology*, **53**, 541–568

Gerbrands, J.J. (1981). On the relationships between SVD, KLT and PCA. *Pattern Recognition*, **14**, 375–381

Gerson, M. (1975). The techniques and uses of probability plotting. *The Statistician*, **24**, 235–257

Gifi, A. (1981). *Nonlinear multivariate analysis* (prepublication edition). Department of Data Theory, University of Leiden, 2312 TW Leiden, The Netherlands

Gittins, R. (1965). Multivariate approaches to a limestone grassland community. I. A stand ordination. *Journal of Ecology*, **53**, 385–401

Gittins, R. (1969). The application of ordination techniques. In I.H. Rorison (ed.), *Ecological aspects of the mineral nutrition of plants*. Blackwell, Oxford, pp. 37–66

Gittins, R. and Ogden, J. (1977). *A reconnaissance survey of lowland tropical rain forest in Guyana*. Unpublished manuscript.

Glahn, H. (1968). Canonical analysis and its relation to discriminant analysis and multiple regression. *Journal of the Atmospheric Sciences*, **25**, 23–31

Glahn, H.R. (1969). Some relationships derived from canonical correlation theory. *Econometrika*, **37**, 252–256

Gleason, T.C. (1976). On redundancy in canonical analysis. *Psychological Bulletin*, **83**, 1004–1006

Gleason, T.C. and Staelin, R. (1973). Improving the metric quality of questionnaire data. *Psychometrika*, **38**, 393–410

Glynn, W.J. and Muirhead, R.J. (1978). Inference in canonical correlation analysis. *Journal of Multivariate Analysis*, **8**, 468–478

Gnanadesikan, R. (1977). *Methods for statistical data analysis of multivariate observations*. Wiley, New York

Gnanadesikan, R. (1980). Graphical methods for internal comparisons in ANOVA and MANOVA. In P.R. Krishnaiah (ed.), *Handbook of statistics. Vol. 1, Analysis of variance*. North-Holland, Amsterdam, pp.133–177

Gnanadesikan, R. and Kettenring, J.R. (1972). Robust estimates, residuals and outlier detection with multiresponse data. *Biometrics,* **28**, 81–124

Goldstein, R.A. and Grigal D.F. (1972). Definition of vegetation structure by canonical analysis. *Journal of Ecology,* **60**, 277–284

Gollob, H.F. (1967). *Cross-validation using samples of size one*. Paper presented at the American Psychological Association meeting in Washington, D.C., September 1967

Golub, G.H. (1969). Matrix decompositions and statistical computations. In R.C. Milton and J.A. Nelder (eds.), *Statistical computation*. Academic Press, New York, pp. 365–397

Golub, G.H. and Kahan, W. (1965). Calculating the singular values and pseudo-inverse of a matrix. *SIAM Journal of Numerical Analysis,* **2**, 205–224

Golub, G.H., Klema, V. and Peters, S.C. (1980). Rules and software for detecting rank degeneracy. *Journal of Econometrics,* **12**, 41–48

Golub, G.H. and Reinsch, C. (1970). Singular value decomposition and least squares solutions. *Numerische Mathematik,* **14**, 403–420

Gong, G. (1982). Some ideas on using the bootstrap in assessing model variability. In K.W. Heiner, R.S. Sacher and J.W. Wilkinson (eds.), *Computer science and statistics: proceedings of the 14th symposium on the interface*. Springer-Verlag, New York, pp. 169–173

Good, I.J. (1965). *The estimation of probabilities: an essay on modern Bayesian methods*. MIT Press, Cambridges, Massachusetts

Good, I.J. (1969). Some applications of the singular decomposition of a matrix. *Technometrics,* **11**, 823–831; erratum, **12**, 721

Goodall, D.W. (1970). Statistical plant ecology. *Annual Review of Ecology and Systematics,* **1**, 99–124

Goodman, L.A. (1981). Association models and canonical correlation in the analysis of cross-classifications having ordered categories. *Journal of the American Statistical Association,* **76**, 320–334

Gordon, A.D. (1981). *Classification: methods for the exploratory analysis of multivariate data*. Chapman and Hall, London.

Gordon, A.D. (1982). Numerical methods in Quaternary palaeoecology. V. Simultaneous graphical representation of the levels and taxa in a pollen diagram. *Review of Palaeobotany and Palynology,* **37**, 155–183

Gower, J.C. (1966). Some distance properties of latent root and vector methods in multivariate analysis. *Biometrika,* **53**, 315–328

Gower, J.C. (1968). Adding a point to vector diagrams in multivariate analysis. *Biometrika,* **55**, 582–585

Gower, J.C. (1972). Measures of taxonomic distance and their analysis. In J.S. Weiner and J. Huizinga (eds.), *The assessment of population affinities in man*. Clarendon Press, Oxford, pp. 1–24

Gower, J.C. (1973). Contribution to discussion of R. Gnanadesikan's paper 'Graphical methods for informal inference in multivariate data analysis'. *Bulletin of the International Statistical Institute, Proceedings,* **39**, 195–205 and 233–235

Gower, J.C. (1975). Generalized Procrustes analysis. *Psychometrika,* **40**, 33–51

Gower, J.C. (1976). Growth-free canonical variates and generalized variances. *Bulletin of the Geological Institutions of the University of Uppsala, N.S.* **7**, 1–10

Gower, J.C. (1977). The analysis of three-way grids. In P. Slater (ed.), *The measurement of interpersonal space by grid technique. Vol. 2, Dimensions of intrapersonal space*. Wiley, London, pp. 163–173

Gower, J.C. (1984). Multivariate analysis: ordination, multidimensional scaling and allied topics. In W. Ledermann (ed.) *Handbook of applicable mathematics, Vol. 6*. Wiley, Chichester

Gower, J.C. and Digby, P.G.N. (1981). Expressing complex relationships in two dimensions. In V. Barnett (ed.), *Interpreting multivariate data*. Wiley, Chichester, pp. 83–118

Green, P.E. (1976). *Mathematical tools for applied multivariate analysis*. Academic Press, New York

Green, P.E. (1978). *Analyzing multivariate data*. The Dryden Press, Illinois

Green, P.J. (1981). Peeling bivariate data. In V. Barnett (ed.), *Interpreting multivariate data*. Wiley, Chichester, pp. 3–19

Greenacre, M.J. (1981). Practical correspondence analysis. In V. Barnett (ed.), *Interpreting multivariate data*. Wiley, Chichester, pp. 119–146

Greenacre, M.J. (1984). *Theory and applications of correspondence analysis*. Academic Press, New York
Greenacre, M.J. and Degos, L. (1977). Correspondence analysis of HLA gene frequency data from 124 population samples. *American Journal of Human Genetics*, **29**, 60–75
Greenacre, M.J. and Underhill, L.G. (1982). Scaling a data matrix in a low-dimensional Euclidean space. In D.M. Hawkins (ed.), *Topics in applied multivariate analysis*. Cambridge University Press, Cambridge, pp. 183–268
Gunst, R.F. (1983). Regression analysis with multicollinear predictor variables: definition, detection and effects. *Communications in Statistics–Theory and Methods*, **12**, 2217–2260
Gunst, R.F. and Mason, R.L. (1980). *Regression analysis and its applications: a data-oriented approach*. Marcel Dekker, New York.
Gupta, R.D. and Kabe, D.G. (1979). Goodness-of-fit tests of hypothesized canonical variables. *The Canadian Journal of Statistics*, **7**, 91–96
Guttman, L. (1941). The quantification of a class of attributes: a theory and method of scale construction. In P. Horst et al. (eds.), *The prediction of personal adjustment*. The Social Science Research Council, New York. Bulletin No. 48, pp. 319–348
Guttman, L. (1946). An approach for quantifying paired comparison and rank order. *Annals of Mathematical Statistics*, **17**, 144–163
Guttman, L. (1968). A general non-metric technique for finding the smallest coordinate space for a configuration of points. *Psychometrika*, **33**, 469–506
Haberman, S.J. (1981). Tests for independence in two-way contingency tables based on canonical correlation and on linear-by-linear interaction. *The Annals of Statistics*, **9**, 1178–1186
Hall, C.E. (1969). Rotation of canonical variates in multivariate analysis of variance. *Journal of Experimental Education*, **38**, 31–38
Hall, C.E. (1977). Some elements of the design of multivariate experiments. *Journal of Experimental Education*, **45**, 26–37
Hampel, F.R. (1971). A general qualitative definition of robustness. *Annals of Mathematical Statistics*, **42**, 1887–1896
Hampel, F.R. (1974). The influence curve and its role in robust estimation. *Journal of the American Statistical Association*, **62**, 1179–1186
Hampel, F.R. (1977). *Modern trends in the theory of robustness*. Research Report No. 13, Swiss Federal Institute of Technology, Zurich
Hanumara, R.C. and Thompson, W.A. (1968). Percentage points of the extreme roots of a Wishart matrix. *Biometrika*, **55**, 505–512
Harman, H.H. (1967). *Modern factor analysis*. 2nd edn. The University of Chicago Press, Chicago
Harner, E.J. and Whitmore, R.C. (1981). Robust principal component and discriminant analysis of two grassland bird species' habitat. In D.E. Capen (ed.), *The use of multivariate statistics in studies of wildlife habitat*. USDA Forest Service General Technical Report RM-87, Rocky Mountain Forest and Range Experiment Station, Fort Collins, Colorado, pp. 209–221
Harris, R.J. (1975). *A primer of multivariate statistics*. Academic Press, New York
Hatheway, W.H. (1971). Contingency-table analysis of rain forest vegetation. In G.P. Patil, E.C. Pielou and W.E. Waters (eds.), *Statistical ecology, Vol. 3*. Pennsylvania State University Press, University Park, Pennsylvania, pp. 217–313
Hawkins, D.M. (1980). *Identification of outliers*. Chapman and Hall, London
Hawkins, D.M. (1981). A new test for multivariate normality and homoscedasticity. *Technometrics*, **23**, 105–110
Hawkins, D.M. and Merriam, D.F. (1974). Zonation of multivariate sequences of digitized geological data. *Mathematical Geology*, **6**, 263–269
Hawkins, D.M., Muller, M.W. and Ten Krooden, J.A. (1982). Cluster analysis. In D.M. Hawkins (ed.), *Topics in applied multivariate analysis*. Cambridge University Press, Cambridge, pp. 303–356
Hawkins, D.M. and Ten Krooden, J.A. (1979). Zonation of sequences of heteroscedastic multivariate data. *Computers and Geosciences*, **5**, 189–194
Healy, M.J.R. (1965). Computing a discriminant function from within-sample dispersions *Biometrics*, **21**, 1011–1012
Heiser, W.J. and Meulman, J. (1983a). Analyzing rectangular tables by joint and constrained multidimensional scaling. *Journal of Econometrics*, **22**, 139–167
Heiser, W.J. and Meulman, J. (1983b). Constrained multidimensional scaling, including confirmation. *Applied Psychological Measurement*, **7**, 381–404

Henderson, H.V. and Velleman P.F. (1981). Building multiple regression models interactively. *Biometrics,* **37**, 391–411

Herman, J.B., Dunham, R.B. and Hulin, C.L. (1975). Organizational structure, demographic characteristics, and employee responses. *Organizational Behavior and Human Performance,* **13**, 206–232

Hill, M.A. and Dixon, W.J. (1982). Robustness in real life: a study of clinical laboratory data. *Biometrics,* **38**, 377–396

Hill, M.O. (1973). Reciprocal averaging: an eigenvector method of ordination. *Journal of Ecology,* **61**, 237–251

Hill, M.O. (1974). Correspondence analysis: a neglected multivariate method. *Applied Statistics,* **23**, 340–354

Hill, M.O. and Smith, A.J.E. (1976). Principal component analysis of taxonomic data with multi-state discrete characters. *Taxon,* **25**, 249–255

Hinkley, D. (1977). On quick choice of power transformation. *Applied Statistics,* **26**, 67–69

Hirschfeld, H.O. (1935). A connection between correlation and contingency. *Proceedings of the Cambridge Philosophical Society,* **31**, 520–524

Hoaglin, D.C. and Welsch, R.E. (1978). The hat matrix in regression and ANOVA. *The American Statistician,* **32**, 17–22

Hocking, R.R. (1983). Developments in linear regression methodology: 1959–1982 (with discussion). *Technometrics,* **25**, 219–249

Hocking, R.R. and Marx, D.L. (1979). Estimation with incomplete data: an improved computational method and the analysis of nested data. *Communications in Statistics–Theory and Methods,* **A8**, 1155–1181

Hocking, R.R. and Pendleton, O.J. (1983). The regression dilemma. *Communications in Statistics–Theory and Methods,* **A12**, 497–527

Hocking, R.R. and Smith, W.B. (1968). Estimation of the parameters in the multivariate normal distribution with missing observations. *Journal of the American Statistical Association,* **63**, 159–173

Hodges, S.D. and Moore, P.G. (1972). Data uncertainties and least squares regression. *Journal of Applied Statistics,* **21**, 185–195

Hoerl, A.E. and Kennard, R.W. (1970). Ridge regression: biased estimation for nonorthogonal problems. *Technometrics,* **12**, 55–67

Hoerl, A.E. and Kennard, R.W. (1981). Ridge regression 1980: advances, algorithms and applications. *American Journal of Mathematical and Management Science,* **1**, 5–83

Hogg, R.V. (1979). Statistical robustness: one view of its use in applications today. *The American Statistician,* **33**, 108–115

Hohwald, J. and Heiberger, R.M. (1978). Two conceptualizations of discriminant analysis and their implementation in computer programs. In D. Hogben and D.W. Fife (eds.), *Computer science and statistics: proceedings of the 10th symposium on the interface.* National Bureau of Standards, Washington, D.C., pp. 389–394

Hope, K. (1968). *Methods of multivariate analysis.* University of London Press

Hopper, S.D. and Campbell, N.A. (1977). A multivariate morphometric study of species relationships in kangaroo paws (*Anigozanthos* Labill. and *Macropidia* Drumm. ex Harv.: Haemodoraceae). *Australian Journal of Botany,* **25**, 523–544

Hopper, S.D., Campbell, N.A. and Caputi, N. (1984). Geographical variation in fruits, leaves and buds of *Eucalyptus caesia. Submitted for publication*

Horst, P. (1935). Measuring complex attitudes. *Journal of Social Psychology,* **6**, 369–374

Horst, P. (1961 a). Relations among *m* sets of measures. *Psychometrika,* **26,** 129–150

Horst, P. (1961 b). Generalized canonical correlations and their applications to experimental data. *Journal of Clinical Psychology* (Monograph Supplement), **14**, 331–347

Hotelling, H. (1935). The most predictable criterion. *Journal of Educational Psychology,* **26**, 139–142

Hotelling, H. (1936). Relations between two sets of variates. *Biometrika,* **28**, 321–377

Howarth, R.J. and Earle, S.A.M. (1979). Application of a generalized power transformation to geochemical data. *Mathematical Geology,* **11**, 45–62

Huba, G.J., Newcomb, M.D. and Bentler, P.M. (1981). Comparison of canonical correlation and interbattery factor analysis on sensation seeking and drug use domains. *Applied Psychological Measurement,* **5**, 291–306

Huba, G.J., Palisoc, A.L. and Bentler, P.M. (1982). ORSIM2: a Fortran program for symmetric and asymmetric orthogonal rotation of canonical variates and interbattery factors. *The American Statistician,* **36**, 62

Huba, G.J., Wingard, J.A. and Bentler, P.M. (1980). Longitudinal analysis of the role of peer support, adult models, and peer subcultures in begining adolescent substance use: an application of setwise canonical correlation methods. *Multivariate Behavioral Research*, **15**, 259–279

Hubbard, C.E. (1954). *Grasses: a guide to their structure, identification, uses and distribution in the British Isles*. Penguin Books, Harmondsworth

Huber, P.J. (1977). *Robust statistical procedures*. Society for Industrial and Applied Mathematics, Philadelphia

Huber, P.J. (1981). *Robust statistics*. Wiley, New York

Huberty, C.J. (1975). Discriminant analysis. *Review of Educational Research*, **45**, 543–598

Huberty, C.J. (1983). Some univariate-multivariate generalizations. *Educational and Psychological Measurement*, **43**, 705–721

Huseby, J.R., Schwertman, N.C. and Allen, D.M. (1980). Computation of the mean vector and covariance matrix for incomplete multivariate data. *Communications in Statistics–Simulation and Computation*, **B9**, 301–309

Huynh, H. (1975). On canonical analysis with fallible data. *Psychometrika*, **40**, 575–577

International Mathematical and Statistical Libraries. (1979). *Library reference manual*. ISML, Houston

Izenman, A.J. (1975). Reduced rank regression for the multivariate linear model. *Journal of Multivariate Analysis*, **5**, 248–264

Izenman, A.J. (1980). Assessing dimensionality in multivariate regression. In P.R. Krishnaiah (ed.), *Handbook of statistics. Vol. 1, Analysis of variance*. North-Holland, Amsterdam, pp. 571–591

James, M. (1979). The generalized inverse form of canonical correlation. *Communications in Statistics–Theory and Methods*, **A8**, 561–568

Jeffers, J.N.R. (1978). *An introduction to systems analysis: with ecological applications*. Arnold, London

Jewell, N.P. (1982). Some multiple *Q–Q* plotting procedures. In R.L. Launer and A.F. Siegel (eds.), *Modern data analysis*. Academic Press, New York, pp. 13–35.

Jewell, N.P. and Bloomfield, P. (1983). Canonical correlations of past and future for time series: definitions and theory. *The Annals of Statistics*, **11**, 837–847

Jewell, N.P., Bloomfield, P. and Bartmann, F.C. (1983). Canonical correlations of past and future for time series: bounds and computation. *The Annals of Statistics*, **11**, 848–855

Johansson, J.K. (1981). An extension of Wollenberg's redundancy analysis. *Psychometrika*, **46**, 93–103

Johnson, R.M. (1963). On a theorem stated by Eckart and Young. *Psychometrika*, **28**, 259–263

Johnson, R.W. (1980). Studies of a vegetation transect through brigalow (*Acacia harpophylla*) forest in central Queensland. *Australian Journal of Ecology*, **5**, 287–307

Johnston, J. (1972). *Econometric methods*. 2nd edn. McGraw-Hill, New York

Joiner, B.L. and Hall, D.L. (1983). The ubiquitous role of f'/f in efficient estimation of location. *The American Statistician*, **37**, 128–133

Jolicoeur, P. (1959). Multivariate geographical variation in the wolf *Canus lupus* L. *Evolution*, **13**, 283–299

Jones, L.V. and Bock, R.D. (1960). Multiple discriminant analysis applied to 'ways to live' ratings from six cultural groups. *Sociometry*, **23**, 162–176

Jöreskog, K.G. (1977). Structural equation models in the social sciences: specification, estimation and testing. In P.R. Krishnaiah (ed.), *Applications of statistics*. North-Holland, Amsterdam, pp. 265–287

Kafadar, K., Rice, J. and Spiegelman, C. (1983). One-sided trimming in small samples with asymmetric contamination. *Communications in Statistics–Theory and Methods*, **12**, 477–496

Kaiser, H.F. and Cerny, B.A. (1980). On the canonical analysis of contingency tables. *Educational and Psychological Measurement*, **40**, 95–99

Kakusho, O. and Mizoguchi, R. (1983). A new algorithm for non-linear mapping with applications to dimension and cluster analyses. *Pattern Recognition*, **16**, 109–117

Kariya, T. (1981). Robustness of multivariate tests. *The Annals of Statistics*, **9**, 1267–1275

Keller, W.J. and Wansbeek, T. (1983). Multivariate methods for quantitative and qualitative data. *Journal of Econometrics*, **22**, 91–111

Kendall, M.G. (1975). *Multivariate analysis*. Griffin, London

Kendall, M.G. and Stuart, A. (1973). *The advanced theory of statistics. Vol. 2, Inference and relationship*. 3rd edn. Hafner, New York

Kendall, M.G. and Stuart, A. (1976). *The advanced theory of statistics. Vol. 3, Design and analysis, and time-series*. 3rd edn. Hafner, New York

Kennedy, W.J., Jr. and Gentle, J.E. (1980). *Statistical computing*. Marcel Dekker, New York

Kerlinger, F.N. and Pedhazur, E.J. (1973). *Multiple regression in behavioral research*. Holt, Rinehart and Winston, New York

Kettenring, J.R. (1971). Canonical analysis of several sets of variables. *Biometrika,* **58**, 433–451

Khatri, C.G. (1976). A note on multiple and canonical correlation for a singular covariance matrix. *Psychometrika,* **41**, 465–470

Khatri, C.G. (1978). Some optimization problems with applications to canonical correlations and sphericity tests. *Journal of Multivariate Analysis,* **8**, 453–467

Khuri, A.I. and Conlon, M. (1981). Simultaneous optimization of multiple responses represented by polynomial regression functions. *Technometrics,* **23**, 363–375

Kimber, A.C. (1983). Trimming in gamma samples. *Applied Statistics,* **32**, 7–14

Kmenta, J. (1971). *Elements of econometrics*. The Macmillan Company, New York

Knapp, T.R. (1978). Canonical correlation analysis: a general parametric significance-testing system. *Psychological Bulletin,* **85**, 410–416

Koenker, R. and Bassett, G. (1978). Regression quantiles. *Econometrica,* **46**, 33–50.

Kowal, R.R., Lechowicz, M.J. and Adams, M.S. (1976). The use of canonical analysis to compare response curves in physiological ecology. *Flora,* **165**, 29–46

Kowalski, C.J. (1972). On the effects of non-normality on the distribution of the sample product-moment correlation coefficient. *Applied Statistics,* **21**, 1–12

Koziol, J.A. (1982). A class of invariant procedures for assessing multivariate normality. *Biometrika,* **69**, 423–427

Koziol, J.A. (1983). On assessing multivariate normality. *Journal of the Royal Statistical Society B,* **45**, 358–361

Krus, D.J., Reynolds, T.J. and Krus, P.H. (1976). Rotation in canonical variate analysis. *Educational and Psychological Measurement,* **36**, 725–730

Kruskal, J.B. (1969). Toward a practical method which helps uncover the structure of a set of multivariate observations by finding the linear transformation which optimizes a new 'index of condensation'. In R.C. Milton and J.A. Nelder (eds.), *Statistical computation*. Academic Press, New York, pp. 427–440

Kruskal, J.B. and Wish, M. (1978). *Multidimensional scaling*. Sage Publications, Beverly Hills

Kruskal, W. and Mosteller, F. (1979a). Representative sampling. I. Non-scientific literature. *International Statistical Review,* **47**, 13–24

Kruskal, W. and Mosteller, F. (1979b). Representative sampling. II. Scientific literature, excluding statistics. *International Statistical Review,* **47**, 111–127

Kruskal, W. and Mosteller, F. (1979c). Representative sampling. III. The current statistical literature. *International Statistical Review,* **47**, 245–265

Kruskal, W. and Mosteller, F. (1980). Representative sampling. IV. The history of the concept in statistics, 1895–1939. *International Statistical Review,* **48**, 169–195

Krzanowski, W.J. (1971). The algebraic basis of classical multivariate methods. *The Statistician,* **20**, 51–61

Krzyśko, M. (1979). Discriminant variables. *Biometrical Journal,* **21**, 227–241

Kshirsagar, A.M. (1972). *Multivariate analysis*. Marcel Dekker, New York

Lachenbruch, P.A. (1975). *Discriminant analysis*. Hafner Press, New York

Lachenbruch, P.A. (1982). Robustness of discriminant functions. *Proceedings of the 7th annual SUGI conference*. SAS Institute, Cary, North Carolina, pp. 626–632

Lachenbruch, P.A. and Goldstein, M. (1979). Discriminant analysis. *Biometrics,* **35**, 69–85

Lachenbruch, P.A. and Mickey, M.R. (1968). Estimation of error rates in discriminant analysis. *Technometrics,* **10**, 1–11

Lacoste, A. et Roux, M. (1971). L'analyse multidimensionelle en phytosociologie et en écologie. Application à des données de l'étage subalpin des Alpes maritimes. I. L'analyse des données floristique. *Oecologia Plantarum,* **6**, 353–369

Lacoste, A. et Roux, M. (1972). L'analyse multidimensionelle en phytosociologie et en écologie. Application à des données de l'étage subalpin des Alpes maritimes. II. L'analyse des données ecologiques et l'analyse globale. *Oecologia Plantarum,* **7**, 125–146

Lambert, D.L. (1977). *P-values: asymptotics and robustness*. Ph.D. thesis. Department of Statistics, University of Rochester, New York

Lancaster, H.O. (1957). Some properties of the bivariate normal distribution considered in the form of a contingency table. *Biometrika,* **44**, 289–292

Lancaster, H.O. (1958). The structure of bivariate distributions. *Annals of Mathematical Statistics,* **29**, 719–736
Lancaster, H.O. (1963). Canonical correlations and partitions of χ^2. *Quarterly Journal of Mathematics,* **14**, 220–224
Lancaster, H.O. (1965). The Helmert matrices. *American Mathematical Monthly,* **72**, 4–12
Lancaster, H.O. (1966). Kolmogorov's remark on the Hotelling canonical correlations. *Biometrika,* **53**, 585–588
Lancaster, H.O. (1969). *The chi-squared distribution.* Wiley, New York
Landwehr, J.M. (1983). Using partial residual plots to detect nonlinearity in multiple regression. Unpublished manuscript. Bell Laboratories, Murray Hill, New Jersey
Larsen, W.A. and McCleary, S.J. (1972). The use of partial residual plots in regression analysis. *Technometrics,* **14**, 781–790
Laurec, A., Chardy, P., de la Salle, P. and Richaert, M. (1979). Use of dual structures in inertia analysis ecological implications. In L. Orlóci, C.R. Rao and W.M. Stiteler (eds.), *Multivariate methods in ecological work.* International Co-operative Publishing House, Fairland, Maryland, pp. 127–174
Lawley, D.N. (1959). Tests of significance in canonical analysis. *Biometrika,* **46**, 59–66
Lawley, D.N. and Maxwell, A.E. (1973). Regression and factor analysis. *Biometrika,* **60**, 331–338
Lawson, C.L. and Hanson, R.J. (1974). *Solving least squares problems.* Prentice-Hall, Englewood Cliffs, New Jersey
Lebart, L. (1976). The significance of eigenvalues issued from correspondence analysis. *Proceedings in computational statistics (COMPSTAT).* Physica Verlag, Vienna, pp. 38–45
Lebart, L. and Morineau, A. (1982). *SPAD système portable pour l'analyse des données.* CESIA, Paris
Lebart, L., Morineau, A. and Tabard, N. (1977). *Techniques de la description statistique.* Dunod, Paris
Lebart, L., Morineau, A. and Warwick, K.M. (1984). *Multivariate descriptive analysis.* Wiley, New York
Lee, P.J. (1969). The theory and application of canonical trend surfaces. *Journal of Geology,* **77**, 303–318
Lee, P.J. (1981). The most predictable surface (MPS) mapping method in petroleum exploration. *Bulletin of Canadian Petroleum Geology,* **29**, 224–240
Lee, R., McCabe, D.J. and Graham, W.K. (1983). Multivariate relationships between job characteristics and job satisfaction in the public sector: a triple cross-validation study. *Multivariate Behavioral Research,* **18**, 47–62
Lee, S.-Y. (1978). Generalizations of the partial, part and bipartial canonical correlation analysis. *Psychometrika,* **43**, 427–431
Lee, S.-Y. (1979). Errata to generalizations of the partial, part and bipartial canonical correlation analysis. *Psychometrika,* **44**, 131
Lehmann, E.L. (1975). *Nonparametrics: statistical methods based on ranks.* Holden-Day, San Francisco
Leong, K., Nishisato, S. and Wolf, R.G. (1982). *OPSCALE, a Fortran program for analysis of categorial data by dual scaling.* International Education Services, Chicago
Leeuw, J. de (1973). *Canonical analysis of categorial data.* Ph.D. thesis, University of Leiden, The Netherlands
Leeuw, J. de (1982). A note on partitioned determinants. *Psychometrika,* **47**, 531–534
Lewis, E.L. and Mouw, J.T. (1978). *The use of contrast coefficients.* Supplement to McNeil, Kelley and McNeil, *Testing research hypotheses using multiple linear regression.* Southern Illinois University Press, Carbondale
Li, J.C.R. (1964). *Statistical inference, Vol. 1.* Edwards Bros., Ann Arbor, Michigan
Lisowski, F.P., Albrecht, G.H. and Oxnard, C.E. (1974). The form of the talus in some higher primates: a multivariate study. *American Journal of Physical Anthropology,* **41**, 191–216
Lisowski, F.P., Albrecht, G.H. and Oxnard, C.E. (1976). African fossil tali: further multivariate morphometric studies. *American Journal of Physical Anthropology,* **45**, 5–18
Little, R.J.A. and Rubin, D.B. (1983). Missing data in large data sets. In T. Wright (ed.), *Statistical methods and the improvement of data quality.* Academic Press, Orlando, Florida, pp. 215–243
Lock, J.M. (1977). The vegetation of Rwenzori National Park, Uganda. *Botanische Jahrbücher,* **98**, 372–448

Lubischew, A.A. (1962). On the use of discriminant functions in taxonomy. *Biometrics,* **18**, 455–477
Lyttkens, E. (1972). Regression aspects of canonical correlation. *Journal of Multivariate Analysis,* **2**, 418–439
McBratney, A.B. and Webster, R. (1981). Detection of ridge and furrow pattern by spectral analysis of crop yield. *International Statistical Review,* **49**, 45–52
McCabe, G.P. (1975). Computations for variable selection in discriminant analysis. *Technometrics,* **17**, 103–109
McDonald, R.P. (1968). A unified treatment of the weighting problem. *Psychometrika,* **33**, 351–381
McHenry, C.E. (1978). Computation of a best subset in multivariate analysis. *Applied Statistics,* **27**, 291–296
McKay, R.J. (1977). Variable selection in multivariate regression: an application of simultaneous test procedures. *Journal of the Royal Statistical Society B,* **39**, 371–380
McKay, R.J. (1979). The adequacy of variable subsets in multivariate regression. *Technometrics,* **21**, 475–479
McKay, R.J. and Campbell, N.A. (1982a). Variable selection techniques in discriminant analysis. I. Description. *British Journal of Mathematical and Statistical Psychology,* **35**, 1–29
McKay, R.J. and Campbell, N.A. (1982b). Variable selection techniques in discriminant analysis. II. Allocation. *British Journal of Mathematical and Statistical Psychology,* **35**, 30–41
McKeon, J.J. (1965). *Canonical analysis: some relations between canonical correlation, factor analysis, discriminant function analysis and scaling theory.* Psychometric Monograph 13. University of Chicago Press, Chicago
McNeill, K. (1974). The multiple linear regression approach to 'chi-square' hypotheses. *The Journal of Experimental Education,* **43**, 53–55
Madsen, K.S. (1977). A growth curve model for studies in morphometrics. *Biometrics,* **33**, 659–669
Mahalanobis, P.C., Majumdar, D.N. and Rao, C.R. (1949). Anthropometric survey of the United Provinces, 1941: a statistical study. *Sankhyā,* **9**, 90–324
Malinvaud, E. (1970). *Statistical methods of econometrics.* North-Holland, Amsterdam
Malkovich, J.F. and Afifi, A.A. (1973). On tests for multivariate normality. *Journal of the American Statistical Association,* **68**, 176–179
Mallows, C.L. (1973). Some comments on C_p. *Technometrics,* **15**, 661–675
Mallows, C.L. (1979). Robust methods–some examples of their use. *The American Statistician,* **33**, 179–184
Malmgren, B.A. (1979). Multivariate normality tests of planktonic foraminiferal data. *Mathematical Geology,* **11**, 285–297
Mandel, J. (1982). Use of the singular value decomposition in regression analysis. *The American Statistician,* **36**, 15–24
Marasinghe, M.G. and Kennedy, W.J., Jr. (1982). Direct methods for generating extreme characteristic roots of certain random matrices. *Communications in Statistics–Simulation and Computation,* **B11**, 527–542
Mardia, K.V. (1970). Measures of multivariate skewness and kurtosis with applications. *Biometrika,* **57**, 519–530
Mardia, K.V. (1974). Applications of some measures of multivariate skewness and kurtosis to testing normality and to robustness studies. *Sankhyā B,* **36**, 115–128
Mardia, K.V. (1975). Assessment of multinormality and the robustness of Hotelling's T^2 test. *Applied Statistics,* **24**, 163–171
Mardia, K.V. (1977). Mahalanobis distance and angles. In P.R. Krishnaiah (ed.), *Multivariate analysis IV.* North-Holland, Amsterdam, pp. 495–511
Mardia, K.V. (1980). Tests of univariate and multivariate normality. In P.R. Krishnaiah (ed.), *Handbook of statistics. Vol. 1, Analysis of variance.* North-Holland, Amsterdam, pp. 279–320
Mardia, K.V., Kent, J.T. and Bibby, J.M. (1979). *Multivariate analysis.* Academic Press, London
Mardia, K.V. and Zemroch, P.J. (1975). Measures of multivariate skewness and kurtosis. Algorithm AS 84. *Applied Statistics,* **24**, 262–265
Maronna, R.A. (1976). Robust *M*-estimators of multivariate location and scatter. *The Annals of Statistics,* **4**, 51–67
Marquardt, D.W. and Snee, R.E. (1975). Ridge regression in practice. *The American Statistician,* **12**, 3–19

Marriott, F.H.C. (1974). *The interpretation of multiple observations.* Academic Press, London

Matloff, N.S. (1980). The jackknife. Algorithm AS 148. *Applied Statistics,* **29**, 115–117

Maung, K. (1941). Measurement of association in a contingency table with special reference to the pigmentation of hair and eye colour of Scottish school children. *Annals of Eugenics, London,* **11**, 189–205

Maxwell, A.E. (1961). Canonical variate analysis when the variables are dichotomous. *Educational and Psychological Measurement,* **21**, 259–271

Maxwell, A.E. (1973). Tests of association in terms of matrix algebra. *British Journal of Mathematical and Statistical Psychology,* **26**, 155–166

Maxwell, A.E. (1977). Multiple regression and poorly conditioned matrices. *British Journal of Mathematical and Statistical Psychology,* **30**, 210–213

Mendoza, J.L., Markos, V.H. and Gonter, R. (1978). A new perspective on sequential testing procedures in canonical analysis: a Monte Carlo evaluation. *Multivariate Behavioral Research,* **13**, 371–382

Meredith, W. (1964). Canonical correlation with fallible data. *Psychometrika,* **29**, 55–65

Meredith, W. and Tisak, J. (1982). Canonical analysis of longitudinal and repeated measures data with stationary weights. *Psychometrika,* **47**, 47–67

Merembeck, B.F. and Turner, B.J. (1980). Directed canonical analysis and the performance of classifiers under its associated linear transformation. *IEEE Transactions on Geoscience and Remote Sensing,* **GE-18**, 190–196

Mielke, P.W. (1979). Some parametric, nonparametric and permutation inference procedures resulting from weather modification experiments. *Communications in Statistics–Theory and Methods,* **A8**, 1083–1096

Mielke, P.W., Berry, K.J. and Johnson, E.S. (1976). Multi-response permutation procedures for *a priori* classifications. *Communications in Statistics–Theory and Methods,* **A5**, 1409–1424

Mielke, P.W., Berry, K.J. and Brier, G.W. (1981). Application of multi-response permutation procedures for examining seasonal changes in monthly sea-level pressure patterns. *Monthly Weather Review,* **109**, 120–126

Mielke, P.W., Berry, K.J., Brockwell, P.J. and Williams, J.S. (1981). A class of nonparametric tests based on multiresponse permutation procedures. *Biometrika,* **68**, 720–724

Mielke, P.W. and Iyer, H.K. (1982). Permutation techniques for analyzing multi-response data from randomized block experiments. *Communications in Statistics–Theory and Methods,* **A11**, 1427–1437

Miller, J.K. (1969). *The development and application of bi-multivariate correlations: a measure of statistical association between multivariate measurement sets.* Ed.D. thesis. Faculty of Education Studies, State University of New York at Buffalo

Miller, J.K. (1975). The sampling distribution and a test for the significance of the bimultivariate redundancy statistic: a Monte Carlo study. *Multivariate Behavioral Research,* **10**, 233–244

Miller, Robert G. (1962). Statistical prediction by discriminant analysis. *Meteorological Monographs,* **4**, No. 25

Miller, Rupert G. (1974). The jackknife–a review. *Biometrika,* **61**, 1–15

Miller, Rupert G. (1978). The jackknife: survey and applications. *Proceedings of the 23rd conference on the design of experiments in army research development and testing,* pp. 371–391

Misra, R.K. (1980). Discriminant functions for linear comparisons of means when covariance matrices are unequal. *Biometrical Journal,* **22**, 755–758

Mitra, A. and Alam, K. (1980). Measurement error in regression analysis. *Communications in Statistics–Theory and Methods,* **A9**, 717–723

Miyata, M. (1970). Complex generalization of canonical correlation and its application to a sea-level study. *Journal of Marine Research,* **28**, 202–214

Moler, C.B. (1978). Three research problems in numerical linear algebra. In G.H. Golub and J. Oliger (eds.), *Numerical analysis. Proceedings of symposia in applied mathematics,* **22**. American Mathematical Society, Providence, Rhode Island, pp. 1–18

Moran, P.A.P. (1971). Estimating structural and functional relationships. *Journal of Multivariate Analysis,* **1**, 232–255

Morrison, D.F. (1976). *Multivariate statistical methods.* 2nd edn. McGraw-Hill, New York

Mosier, C.I. (1951). Problems and designs of cross-validation. *Educational and Psychological Measurement,* **11**, 5–11

Mosteller, F. and Tukey, J.W. (1977). *Data analysis and regression.* Addison-Wesley, Reading, Massachusetts

Muirhead, R.J. (1982). *Aspects of multivariate statistical theory.* Wiley, New York

Muirhead, R.J. and Waternaux, C.M. (1980). Asymptotic distributions in canonical correlation analysis and other multivariate procedures for nonnormal populations. *Biometrika,* **67**, 31–43

Mulaik, S.A. (1972). *The foundations of factor analysis.* McGraw-Hill, New York

Muller, K.E. (1980a). *The general linear model for canonical correlation.* Mimeo. Department of Biostatistics, University of North Carolina, Chapel Hill

Muller, K.E. (1980b). *Maximizing between set covariance: an alternative to canonical correlation.* Mimeo. Department of Biostatistics, University of North Carolina, Chapel Hill

Muller, K.E. (1981). Relationships between redundancy analysis, canonical correlation and multivariate regression. *Psychometrika,* **46**, 139–142

Muller, K.E. (1982). Understanding canonical correlation through the general linear model and principal components. *The American Statistician,* **36**, 342–354

Mutombo, F.K. (1973). *Traitement des données manquantes et rationalisation d'un réseau de stations de mesures.* Doctoral thesis. Université Pierre et Marie Curie, Paris

Nahikian, H.M. (1964). *A modern algebra for biologists.* University of Chicago Press, Chicago

Nair, K.R. and Mukerji, H.K. (1960). Classification of natural and plantation teak (*Tectona grandis*) grown at different localities of India and Burma with respect to its physical and mechanical properties. *Sankhyā,* **22**, 1–20

Nash, S.W. (1983). Personal communication

Neter, J., Wasserman, W. and Kutner, M.H. (1983). *Applied linear regression models.* Irwin, Illinois

Nishisato, S. (1978). Optimal scaling of paired comparison and rank order data: an alternative to Guttman's formulation. *Psychometrika,* **43**, 263–271

Nishisato, S. (1980). *Analysis of categorical data: dual scaling and its applications.* University of Toronto Press, Toronto

Nishisato, S. and Sheu, Wen-Jenn. (1980). Piecewise method of reciprocal averages for dual scaling of multiple-choice data. *Psychometrika,* **45**, 467–478

Nishisato, S. (1984). Forced classification: a simple application of a quantification method. *Psychometrika,* **49**, 25–36

Nora-Chouteau, Ch. (1974). *Une méthode de reconstitution et d'analyse des données incomplètes.* Doctoral thesis. Université Pierre et Marie Curie, Paris

Noy-Meir, I. (1973). Data transformations in ecological ordination. I. Some advantages of non-centering. *Journal of Ecology,* **61**, 329–341

Noy-Meir, I., Walker, D. and Williams, W.T. (1975). Data transformations in ecological ordination. II. On the meaning of data standardization. *Journal of Ecology,* **63**, 779–800

Numerical Algorithms Group. (1977). *Library reference manual.* NAG, Oxford

Odgen, J. (1966). *Ordination studies on a small area of tropical rain forest.* M.Sc. thesis, University of Wales

Odgen, J. (1977a). Personal communication

Odgen, J. (1977b). *Tropical rain forest in the Guyana lowlands: description and dynamics.* Unpublished manuscript. Australian National University, Canberra

O'Grady, K.E. (1982). Measures of explained variance: cautions and limitations. *Psychological Bulletin,* **92**, 766–777

Okiria, R. (1980). Habitat exploitation by the bushbuck in Rwenzori National Park. *African Journal of Ecology,* **18**, 11–17

Olkin, I. and Tomsky, J.L. (1981). A new class of multivariate tests based on the union-intersection principle. *The Annals of Statistics,* **9**, 792–802

O'Neill, M.E. (1978a). Asymptotic distributions of the canonical correlations from contingency tables. *Australian Journal of Statistics,* **20**, 75–82

O'Neill, M.E. (1978b). Distributional expansions for canonical correlations from contingency tables. *Journal of the Royal Statistical Society B,* **40**, 303–312

O'Reilly, F.J. and Mielke, P.W. (1980). Asymptotic normality of MRPP statistics from invariance principles of *U*-statistics. *Communications in Statistics–Theory and Methods,* **A9**, 629–637

Ottestad, P. (1975). Component analysis: an alternative system. *International Statistical Review,* **43**, 81–108

Oxnard, C.E. (1973). Some locomotor adaptations among lower primates. *Symposium of the Zoological Society of London,* **33**, 255–299

Oxnard, C.E. (1979). The morphological behavioral interface in the extant primates: some implications for systematics and evolution. In M.E. Morbeck, H. Preuschoft and N. Gomberg (eds.), *Environment, behavior and morphology: dynamic interactions in primates.* Fischer, New York, pp. 209–227

Oxnard, C.E. (1983). The structure of anatomical fragments and their combination: a problem underlying the assessment of the *Australopithecines.* In K.J. Reichs (ed.), *Hominid origins: inquiries past and present.* University Press of America, pp. 141–205

Parr, W.C. and Schucany, W.R. (1980). The jackknife: a bibliography. *International Statistical Review,* **48**, 73–78

Patterson, D.A., Pirkle, F.L., Johnson, M.E., Bement, T.R., Stablein, N.K. and Jackson, C.K. (1981). Discriminant analysis applied to aerial radiometric data and its application to uranium favorability in south Texas. *Mathematical Geology,* **13**, 535–568

Pearce, S.C. (1969). Multivariate techniques of use in biological research. *Experimental Agriculture,* **5**, 67–77

Pearson, E.S. and Hartley, H.O. (1972). *Biometrika tables for statisticians. Vol. 2.* Cambridge University Press, Cambridge

Perreault, W.D., Jr. and Young, F.W. (1980). Alternating least squares optimal scaling: analysis of nonmetric data in marketing research. *Journal of Marketing Research,* **17**, 1–13

Phillips, B.F., Campbell, N.A. and Wilson, B.R. (1973). A multivariate study of geographic variation in the whelk *Dicathais. Journal of Experimental Marine Biology and Ecology,* **11**, 27–69

Pielou, E.C. (1977). *Mathematical ecology.* Wiley-Interscience, New York

Pillai, K.C.S. and Hsu, Y.-S. (1979). Exact robustness studies of the test of independence based on four multivariate criteria and their distribution problems under violations. *Annals of the Institute of Mathematical Statistics A,* **31**, 85–101

Porebsky, O.R. (1966). Discriminatory and canonical analysis of technical college data. *British Journal of Mathematical and Statistical Psychology,* **19**, 215–236

Portnoy, S. (1982). Review of *Robust statistics* by P.J. Huber (1981). *Technometrics,* **24**, 163

Press, S.J. (1972). *Applied multivariate analysis.* Holt, Rinehart and Winston, New York

Press, S.J. (1980). Bayesian inference in MANOVA. In P.R. Krishnaiah (ed.), *Handbook of statistics. Vol. 1, Analysis of variance.* North-Holland, Amsterdam, pp. 117–132

Pruzek, R.M. (1971). Methods and problems in the analysis of multivariate data. *Review of Educational Research,* **41**, 163–190

Quenouille, M. (1949). Approximate tests of correlation in time series. *Journal of the Royal Statistical Society B,* **11**, 18–84

Ramsay, J.O. (1982). When the data are functions. *Psychometrika,* **47**, 379–396

Ramsay, J.O. and de Leeuw, J. (1983). Review of *Introduction à l'analyse des données* by F. Cailliez and J.-P. Pagès. *Psychometrika,* **48**, 147–151

Rao, B.R. (1969). Partial canonical correlations. *Trabajso de Estadistica y de Investigacion Operative,* **20**, 211–219

Rao, C.R. (1948). The utilization of multiple measurements in problems of biological classification. *Journal of the Royal Statistical Society B,* **10**, 159–203

Rao, C.R. (1952). *Advanced statistical methods in biometric research.* Wiley, New York

Rao, C.R. (1960). Multivariate analysis: an indispensable statistical aid in applied research. *Sankhyā,* **22**, 317–338

Rao, C.R. (1973). *Linear statistical inference and its applications.* Wiley, New York

Rao, C.R. (1979). Separation theorems for singular values of matrices and their applications in multivariate analysis. *Journal of Multivariate Analysis,* **9**, 362–377

Rao, C.R. (1980). Matrix approximation and reduction of dimensionality in multivariate statistical analysis. In P.R. Krishnaiah (ed.), *Multivariate analysis, V.* North-Holland, Amsterdam, pp. 3–22

Rao, C.R. (1981). A lemma on *g*-inverse of a matrix and computation of correlation coefficients in the singular case. *Communications in Statistics–Theory and Methods,* **A10**, 1–10

Rao, C.R. (1983). Multivariate analysis: some reminiscences on its origin and development. *Sankhyā B,* **45**, 284–299

Reilly, P.M. and Patino-Leal, H. (1981). A Bayesian study of the error-in-variables model. *Technometrics,* **23**, 221–231

Rempe, U. and Weber, E.E. (1972). An illustration of the principal ideas of MANOVA. *Biometrics*, **28**, 235–238
Renshaw, E. and Ford, E.D. (1983). The interpretation of process from pattern using two-dimensional spectral analysis: methods and problems of interpretation. *Applied Statistics*, **32**, 51–63
Reyment, R.A. (1961). Quadrivariate principal component analysis of *Globergerina yegudensis*. *Stockholm Contributions in Geology*, **8**, 17–26
Reyment, R.A. (1976). Chemical components of the environment and Late Campanian microfossil frequencies. *Geologiska Föreningens i Stockholm Förhandlingar*, **98**, 322–328
Reyment, R.A. (1979). Multivariate methods in statistical paleoecology. In L. Orlóci, C.R. Rao and W.M. Stiteler (eds.), *Multivariate methods in ecological work*. International Co-operative Publishing House, Fairland, Maryland, pp. 211–235
Reyment, R.A. (1980). *Morphometric methods in biostratigraphy*. Academic Press, London
Reyment, R.A. (1982). Size and shape variation in some Japanese Upper Turonian (Cretaceous) ammonites. *Stockholm Contributions in Geology*, **37**, 201–214
Reyment, R.A. and Banfield, C.F. (1976). Growth-free canonical variates applied to fossil foraminifers. *Bulletin of the Geological Institutions of the University of Uppsala, N.S.* **7**, 11–21
Reynolds, T.J. and Jackofsky, E.F. (1981). Interpreting canonical analysis: the use of orthogonal transformations. *Educational and Psychological Measurement*, **41**, 661–671
Richards, P.W. and Wallace, E.C. (1950). An annotated list of British mosses. *Transactions of the British Bryological Society*. **1**, i–xxxi
Rincon-Gallardo, S., Quesenberry, C.P. and O'Reilly, F.J. (1979). Conditional probability integral transformations and goodness-of-fit tests for multivariate normal distributions. *The Annals of Statistics*, **7**, 1052–1057
Robert, P. and Escoufier, Y. (1976). A unifying tool for linear multivariate statistical methods: the *RV*-coefficient. *Applied Statistics*, **25**, 257–265
Robinson, P.M. (1973). Generalized canonical analysis for time series. *Journal of Multivariate Analysis*, **3**, 141–160
Robson, D.S. (1959). A simple method for constructing orthogonal polynomials when the independent variable is unequally spaced. *Biometrics*, **15**, 187–191
Rosenbaum, P.R. (1984). From association to causation in observational studies: the role of tests of strongly ignorable treatment assignment. *Journal of the American Statistical Association*, **79**, 41–48
Roy, J. (1958). Step-down procedures in multivariate analysis. *Annals of Mathematical Statistics*, **29**, 1177–1187
Roy, S.N. (1957). *Some aspects of multivariate analysis*. Wiley, New York
Royston, J.P. (1983). Some techniques for assessing multivariate normality based on the Shapiro-Wilk *W*. *Applied Statistics*, **32**, 121–133
Rozeboom, W.W. (1965). Linear correlations between sets of variables. *Psychometrika*, **30**, 57–71
Rozeboom, W.W. (1979). Ridge regression: bonanza or beguilement. *Psychological Bulletin*, **86**, 242–249
Sabo, S.R. (1980). Niche and habitat relations in subalpine bird communities of the White Mountains of New Hampshire. *Ecological Monographs*, **50**, 241–259
Sachs, H.M., Webb, T., III and Clark, D.R. (1977). Paleoecological transfer functions. *Annual Review of Earth and Planetary Sciences*, **5**, 159–178
Saxena, A.K. (1978). Complex multivariate statistical analysis: an annotated bibliography. *International Statistical Review*, **46**, 209–214
Schmidt, P. (1976). *Econometrics*. Marcel Dekker, New York
Schuenemeyer, J.H. and Bargmann, R.E. (1978). Maximum eccentricity as a union-intersection test statistic in multivariate analysis. *Journal of Multivariate Analysis*, **8**, 268–273
Schultz, J.P. (1960). *Ecological studies on rain forest in northern Surinam*. Amsterdam
Schwager, S.J. and Margolin, B.H. (1982). Detection of multivariate normal outliers. *The Annals of Statistics*, **10**, 943–954
Sclove, S.L. (1971). Improved estimation of parameters in multivariate regression. *Sankhyā A*, **33**, 61–66
Scobey, P. and Kabe, D.G. (1980). Some correlation optimization problems of econometrics, statistics and psychology. *Biometrical Journal*, **22**, 651–670
Scott, D.W. and Koopman, R.F. (1977). *Rotation of canonical variates using nonidentical orthogonal*

transformations. Unpublished manuscript. Psychology Department, Simon Fraser University, Burnaby

Seal H.L. (1964). *Multivariate statistical analysis for biologists*. Methuen, London

Seif, E., Evans, J.C. and Balaam, L.N. (1979). A multivariate procedure for classifying environments according to their interaction with genotypes. *Australian Journal of Agricultural Research*, **30**, 1021–1026

Shepard, R.N. and Carroll, J.D. (1966). Parametric representation of nonlinear data structures. In P.R. Krishnaiah (ed.), *Multivariate analysis*. Academic Press, New York, pp. 561–592

Sibson, R. (1978). Studies in the robustness of multidimensional scaling: procrustes statistics. *Journal of the Royal Statistical Society B*, **40**, 234–238

SIROMATH (1982). ASSAYPACK. *Computer programs for the robust statistical analysis of multi-element assay data*. SIROMATH Pty. Ltd., 71 York Street, Sydney, N.S.W., Australia 2000

Skinner, H.A. (1977). Exploring relationships among multiple data sets. *Multivariate Behavioral Research*, **12**, 199–222

Skinner, H.A. (1978). The art of exploring predictor-criterion relationships. *Psychological Bulletin*, **85**, 327–337

Skinner, H.A. and Sheu, Wen-Jenn. (1982). Dimensional analysis of rank-order and categorial data. *Applied Psychological Measurement*, **6**, 41–45

Small, N.J.H. (1978). Plotting squared radii. *Biometrika*, **65**, 657–658

Small, N.J.H. (1980). Marginal skewness and kurtosis in testing multivariate normality. *Applied Statistics*, **29**, 85–87

Smith, B.T., Boyle, J.M., Dongarra, J.J., Garbow, B.S., Ikebe, Y., Klema, V.C. and Moler, C.B. (1976). *Matrix eigensystem routines: EISPACK guide*. 2nd edn. Springer-Verlag, Berlin

Smith, G. (1980). An example of ridge regression difficulties. *Canadian Journal of Statistics*, **8**, 217–225

Smith, P.L. (1979). Splines as a useful and convenient statistical tool. *The American Statistician*, **33**, 57–62

Smith, T.M.F. (1983). On the validity of inferences from non-random samples. *Journal of the Royal Statistical Society A*, **146**, 394–403

Smith, W.B. and Riggs, M.W. (1982). Approaches to estimating covariance matrices with partial multinomial data. *Proceedings of the 7th annual SUGI conference*. SAS Institute, Cary, North Carolina, pp. 616–625

Sockloff, A.L. (1976). The analysis of nonlinearity via linear regression with polynomial and product variables: an examination. *Review of Educational Research*, **46**, 267–291

Sparks, R.S., Coutsourides, D. and Troskie, L. (1983). The multivariate C_p. *Communications in Statistics–Theory and Methods*, **A12**, 1775–1793

Srikantan, K.S. (1970). Canonical analysis between nominal measurements. *Journal of the American Statistical Association*, **65**, 284–292

Statistical Task Force to the Weather Modification Advisory Board, (1978). *The management of weather resources. Vol. II, The role of statistics*. Washington, D.C., U.S. Government Printing Office

Stavig, G.R. (1977). The semistandardized regression coefficient. *Multivariate Behavioral Research*, **12**, 255–258

Stavig, G.R. and Acock, A.C. (1981). Applying the semistandarized regression coefficient to factor, canonical and path analysis. *Multivariate Behavioral Research*, **16**, 207–213

Stenson, H.H. (1982). Multidimensional psychophysics. In N. Hirschberg and L.G. Humphreys (eds.), *Multivariate applications in the social sciences*. Lawrence Erlbaum, New Jersey, pp. 47–59

Stephenson, R.B., Beatty, J.C. and Gentleman, J.F. (1981). Interactive graphical analysis for multivariate data. In W.F. Eddy (ed.), *Computer science and statistics: proceedings of the 13th symposium on the interface*. Springer-Verlag, New York, pp. 316–319

Sterling, T.D. (1979). Statistical problems in multi-variate (etiological) surveys. *The Canadian Journal of Statistics*, **7**, 205–215

Stewart, D.K. and Love, W.A. (1968). A general canonical correlation index. *Psychological Bulletin*, **70**, 160–163

Stewart, G.W. (1973). *Introduction to matrix computations*. Academic Press, New York

Stewart, G.W. (1978). Orthogonal transformations in regression calculations. In D. Hogben and D.W. Fife (eds.), *Computer science and statistics: proceedings of the 10th symposium on the interface*. National Bureau of Standards, Washington, D.C., pp. 189–190

Stigler, S.M. (1977). Do robust estimates work with real data. *The Annals of Statistics*, **5**, 1055–1098

Stone, M. (1974). Cross-validity choice and assessment of statistical predictions (with discussion). *Journal of the Royal Statistical Society B,* **36**, 111–147

Stone, M. (1977a). An asymptotic equivalence of choice of model by cross-validation and Akaike's criterion. *Journal of the Royal Statistical Society B,* **39**, 44–47

Stone, M. (1977b). Asymptotics for and against cross-validation. *Biometrika,* **64**, 29–35

Strang, G. (1980). *Linear algebra and its applications.* 2nd edn. Academic Press, New York

Switzer, P. (1980). Extensions of linear discriminant analysis for statistical classification of remotely sensed satellite imagery. *Mathematical Geology,* **12**, 367–376

Tansley, A.G. (1939). *The British islands and their vegetation.* Cambridge University Press, Cambridge

Tatsuoka, M.M. (1971). *Multivariate analysis: techniques for educational and psychological research.* Wiley, New York

Teil, H. (1975). Correspondence analysis: an outline of its method. *Mathematical Geology,* **7**, 3–12

Teil, H. and Cheminee, J.L. (1975). Application of correspondence factor analysis to the study of major and trace elements in the Erta Ale Chain (Afar, Ethiopia). *Mathematical Geology,* **7**, 13–30

Thissen, D., Baker, L. and Wainer, H. (1981). Influenced-enhanced scatterplots. *Psychological Bulletin,* **90**, 179–184

Thorndike, R.M. (1976). Studying canonical analysis: comments on Barcikowski and Stevens. *Multivariate Behavioral research,* **11**, 249–253

Thorndike, R.M. (1977). Canonical analysis and predictor selection. *Multivariate Behavioral Research,* **12**, 75–87

Thorndike, R.M. (1978). *Correlational procedures for research.* Gardner Press, New York

Thorndike, R.M. and Weiss, D.J. (1973). A study of the stability of canonical correlations and canonical components. *Educational and Psychological Measurement,* **33**, 123–134

Thorndike, R.M. and Weiss, D.J. (1983). An empirical investigation of step-down canonical correlation with cross-validation. *Multivariate Behavioral Research,* **18**, 183–196

Thornton, D.D. (1971). The effect of complete removal of hippopotamus on grassland in the Queen Elizabeth National Park, Uganda. *East African Wildlife Journal,* **9**, 47–55

Timm, N.H. (1975). *Multivariate analysis with applications in education and psychology.* Wadsworth, Belmont, California

Timm, N.H. and Carlson, J.E. (1976). Part and bipartial canonical correlation analysis. *Psychometrika,* **41**, 159–176

Titterington, D.M. (1975). Optimal design: some geometric aspects of *D*-optimality. *Biometrika,* **62**, 313–320

Titterington, D.M. (1978). Estimation of correlation coefficients by ellipsoidal trimming. *Applied Statistics,* **27**, 227–234

Tsianco, M.C., Gabriel, K.R., Odoroff, C.L. and Plumb, S. (1981). BGRAPH: a program for biplot multivariate graphics. In W.F. Eddy (ed.), *Computer science and statistics: proceedings of the 13th symposium on the interface.* Springer-Verlag, New York, pp. 344–347

Tucker, L.R. (1958). An inter-battery method of factor analysis. *Psychometrika,* **23**, 111–136

Tukey, J.W. (1958). Bias and confidence in not quite large samples (abstract). *Annals of Mathematical Statistics,* **29**, 614

Tukey, J.W. (1969). Analyzing data: sanctification or detective work? *American Psychologist,* **24**, 83–91

Tukey, P.A. and Tukey, J.W. (1981). Data-driven view selection; agglomeration and sharpening. In V. Barnett (ed.), *Interpreting multivariate data.* Wiley, Chichester, pp. 215–243

Tyler, D.E. (1982). On the optimality of the simultaneous redundancy transformations. *Psychometrika,* **47**, 77–86

Tyler, D.E. (1983). Robustness and efficiency properties of scatter matrices. *Biometrika,* **70**, 411–412

Urbakh, V.Yu. (1971). Linear discriminant analysis: loss of discriminating power when a variate is omitted. *Biometrics,* **27**, 531–534

Van de Geer, J.P. (1971). *Introduction to multivariate analysis for the social sciences.* Freeman, San Francisco

Van de Geer, J.P. (1981). *Varieties of the canonical solution.* Unpublished manuscript. Department of Data Theory, University of Leiden

Van de Geer, J.P. (1984). Linear relations among *k* sets of variables. *Psychometrika,* **49**, 79–94

van den Wollenberg, A.L. (1977). Redundancy analysis: an alternative for canonical analysis. *Psychometrika,* **42**, 207–219 van der Merwe, A. and Zidek, J.V. (1980). Multivariate regression analysis and canonical variates. *The Canadian Journal of Statistics,* **8**, 27–39

van Hecke, P., Impens, I., Goossens, R. and Hebrant, F. (1980). Multivariate analysis of multispectral remote sensing data on grasslands from different soil types. *Vegetatio,* **42**, 165–170

van Rijckevorsel, J. (1982). Canonical analysis with B-splines. In H. Caussinus, P. Ettinger and R. Tomassone (eds.), *COMPSTAT 1982.* Part I. Physica Verlag, Wien

Velleman, P.F. and Welsch, R.E. (1981). Efficient computing of regression diagnostics. *The American Statistician,* **35**, 234–242

Vinod, H.D. (1976). Canonical ridge and econometrics of joint production. *Journal of Econometrics,* **4**, 147–166

Vinod, H.D. (1978). A survey of ridge regression and related techniques for improvements over ordinary least squares. *Review of Econometrics and Statistics,* **60**, 121–131

Vinod, H.D. and Ullah, A. (1981). *Recent advances in regression methods.* Marcel Dekker, New York

Walker, E.P., Warnick, F., Lange, K.I., Uible, H.E., Hamlet, S.E., Davis, M.A. and Wright, P.F. (1968). *Mammals of the world.* 2nd edn. revised by J.L. Paradiso. The Johns Hopkins Press, Baltimore

Wall, F.J. (1968). *The generalized variance ratio of the U-statistic.* The Dikewood Corporation. Albuquerque, New Mexico

Warren, W.G. (1971). Correlation or regression: bias or precision. *Applied Statistics,* **20**, 148–164

Webb, D.A. (1954). Is the classification of plant communities either possible or desirable? *Botanisk Tidsskrift,* **51**, 362–370

Webb, T., III and Bryson, R.A. (1972). Late- and postglacial climatic change in the northern midwest, USA: quantitative estimates derived from fossil pollen spectra by multivariate statistical analysis. *Quaternary Research,* **2**, 70–115

Webb, T., III and Clark, D.R. (1977). Calibrating micropaleontological data in climatic terms: a critical review. *Annals of the New York Academy of Sciences,* **288**, 93–118

Webster, R. (1977a). *Quantitative and numerical methods in soil classification and survey.* Clarendon Press, Oxford

Webster, R. (1977b). Canonical correlation in pedology: how useful. *Journal of Soil Science,* **28**, 196–221

Webster, R. (1979). Exploratory and descriptive uses of multivariate analysis in soil survey. In N. Wrigley (ed.), *Statistical applications in the spatial sciences.* Pion, London, pp. 286–306

Webster, R. and McBratney, A.B. (1981). Soil segment overlap in character space and its implications for soil classification. *Journal of Soil Science,* **32**, 133–147

Weinberg, S.L. and Darlington, R.B. (1976). Canonical analysis when the number of variables is large relative to sample size. *Journal of Educational Statistics,* **1**, 313–332

Weisberg, S. (1980). *Applied linear regression.* Wiley, New York

Weiss, D.J. (1972). Canonical correlation analysis in counseling psychology research. *Journal of Counseling Psychology,* **19**, 241–252

Welch, B.L. (1939). Note on discriminant functions. *Biometrika* **31**, 218–220

Wilk, M.B. and Gnanadesikan, R. (1968). Probability plotting methods for the analysis of data. *Biometrika,* **55**, 1–17

Wilk, M.B., Gnanadesikan, R. and Huyett, M.J. (1962). Probability plots for the gamma distribution. *Technometrics,* **4**, 1–15

Wilkinson, L. (1977). Confirmatory rotation of MANOVA canonical variates. *Multivariate Behavioral Research,* **12**, 487–494

Willan, A.R. and Watts, D.G. (1978). Meaningful multicollinearity measures. *Technometrics,* **20**, 407–412

Williams, B.K. (1981). Discriminant analysis in wildlife research: theory and applications. In D.E. Capen (ed.), *The use of multivariate statistics in studies of wildlife management.* USDA Forest Service General Technical Report RM-87, Rocky Mountain Forest and Range Experiment Station, Fort Collins, Colorado, pp. 59–71

Williams, B.K. (1983). Some observations on the use of discriminant analysis in ecology. *Ecology,* **64**, 1283–1291

Williams, E.J. (1952). Use of scores for the analysis of association in contingency tables. *Biometrika,* **39**, 274–280

Williams, E.J. (1967). The analysis of association among many variates (with discussion). *Journal of the Royal Statistical Society B,* **29**, 199–242
Williams, W.T. (1971). Strategy and tactics in the acquisition of ecological data. *Proceedings of the Ecological Society of Australia,* **6**, 57–62
Williams, W.T. (1976). Other ordination procedures. In W.T. Williams (ed.), *Pattern analysis in agricultural research.* CSIRO, Melbourne and Elsevier, Amsterdam, pp. 59–69
Williams, W.T., Edye, L.A., Burt, R.L. and Grof, B. (1973). The use of ordination techniques in the preliminary evaluation of *Stylosanthes* accessions. *Australian Journal of Agricultural Research,* **24**, 715–731
Winsberg S. and Ramsay, J.O. (1980). Monotonic transformations to additivity using splines. *Biometrika,* **67**, 669–674
Winsberg, S. and Ramsay, J.O. (1983). Monotone spline transformation for dimension reduction. *Psychometrika,* **48**, 575–595
Wold, H. (1966a). Nonlinear estimation by iterative least squares procedures. In F.N. David (ed.), *Research papers in statistics.* Wiley, London, pp. 411–444
Wold, H. (1966b). Estimation of principal components and related models by iterative least squares. In P.R. Krishnaiah (ed.), *Multivariate analysis.* Academic Press, New York, pp. 391–420
Wold, H. (1980). Model construction by partial least squares. In J. Kmenta and J.B. Ramsey (eds.), *Evaluation of econometric models.* Academic Press, New York, pp. 47–74
Wold, H. (1982). Systems under indirect observation using PLS. In C. Fornell (ed.), *A second generation multivariate analysis. Vol. 1, Methods.* Praeger, New York, pp. 325–347
Wold, S. (1974). Spline functions in data analysis. *Technometrics,* **16**, 1–11
Wood, D.A. and Erskine, J.A. (1976). Strategies in canonical correlation with application to behavioral data. *Educational and Psychological Measurement,* **36**, 861–878
Yoaciel, S.M. and Orsdol, K. van (1981). The influence of environmental changes on an isolated topi (*Damaliscus lunatus jimela* Matschie) population in the Ishasha sector of the Rwenzori National Park, Uganda. *African Journal of Ecology,* **19**, 169–174
Young, F.W. (1981). Quantitative analysis of qualitative data. *Psychometrika,* **46**, 357–388
Young, F.W., Leeuw, J. de and Takane, Y. (1976). Regression with qualitative and quantitative variables: an alternating least squares method with optimal scaling features. *Psychometrika,* **41**, 505–529
Zellner, A. (1976). Bayesian and non-Bayesian analysis of the regression model with multivariate Student-*t* error terms. *Journal of the American Statistical Association,* **71**, 400–405

Species' index

Author index

Subject index